Proceedings of the IJSSD Symposium 2012 on Progress in Structural Stability and Dynamics

Edited by

C.M. Wang, National University of Singapore
Y.B. Yang, Taiwan University
J.N. Reddy, Texas A&M University

Southeast University Press
Nanjing, 2012

Disclaimer

No responsibility is assumed for any injury and/or damage to persons or property as a matter of products liability, negligence or otherwise, or from any use or operation of any methods, products, instructions or ideas contained in the proceedings materials herein.

Printed by Southeast University Press, Nanjing, China

IJSSD Symposium 2012

on Progress in Structural Stability and Dynamics

14–16 April 2012, Nanjing, China

Symposium Organising Committee

C.M. Wang, National University of Singapore, Singapore

Y.B. Yang, Taiwan University, Taiwan, China

J.N. Reddy, Texas A&M University, USA

S.L. Chan, Hong Kong Polytechnic University, Hong Kong, China

G.P. Shu, Southeast University, China

J.Y. Lu, Southeast University, China

S.G. Fan, Southeast University, China

R.H. Lu, Southeast University, China

Preface

International Journal of Structural Stability (IJSSD) has been in existence since 2001. The aim of the journal is to provide a unique forum for the publication and rapid dissemination of original research on stability and dynamics of structures. In support of the journal, conferences and symposia have been organised regularly. The first International Conference on Structural Stability and Dynamics (ICSSD) was held in Taipei in 2000, and it was followed by conferences in Singapore (2002), Orlando, USA (2005) and Jaipur, India (2012). A smaller IJSSD Symposium, that piggy back on the International Conference on Advances in Steel Structures (ICASS), was organised in Hong Kong in 2009. The IJSSD Chief Editors are very grateful to Prof. G.P. Shu (of Southeast University) and Prof. S.L. Chan (of Hong Kong Polytechnic University) for inviting us to continue in this very successful synergetic embedding of the IJSSD symposium in ICASS. This time round, we have 24 papers for presentation at the IJSSD Symposium 2012 to be held in conjunction with the 7th ICASS in Nanjing, China.

This IJSSD Symposium Proceedings consists of 24 papers which cover a broad range of topics such as structural stability and dynamics of thin-walled structural members, elasticas, functionally graded beams and plates, composite structures, spherical shells, bridges, floating structures, carbon nanotubes, graphene sheets and numerical techniques for dynamic analyses.

We hope that the research findings described in this volume of proceedings will inspire researchers, engineers and designers to conceptualize and build even more awesome structures for the betterment of mankind.

C.M. Wang, Y.B. Yang and J.N. Reddy

Editors of IJSSD Symposium Proceedings

Contents

MODIFIED COUPLE STRESS THEORIES OF FUNCTIONALLY GRADED BEAMS AND PLATES

* J. N. Reddy and J. Kim

Department of Mechanical Engineering, Texas A & M University
College Station, Texas 77843-3123 USA
* Email: jnreddy@tamu.edu

KEYWORDS

Functionally graded materials, modified couple stress theory, shear deformable beams and plates, the Von Karman nonlinearity.

ABSTRACT

In this paper an overview of general third-order beam and plate theories that account for (a) geometric nonlinearity, (b) microstructure-dependent size effects, and (c) two-constituent material variation through the thickness (i. e., functionally graded material beams and plates) is presented. A detailed derivation of the equations of motion, using Hamilton's principle, is presented, and it is based on a modified couple stress theory, power-law variation of the material through the thickness, and the von Karman nonlinear strains. The modified couple stress theory includes a material length scale parameter that can capture the size effect in a functionally graded material. The governing equations of motion derived herein for a general third-order theory with geometric nonlinearity, microstructure dependent size effect, and material gradation through the thickness are specialized to classical and shear deformation beam and plate theories available in the literature. The theory presented herein also can be used to develop finite element models and determine the effect of the geometric nonlinearity, microstructure-dependent size effects, and material grading through the thickness on bending and post-buckling response of elastic beams and plates.

INTRODUCTION

The next generation of material systems used in space and other structures as well as in MEMS and NEMS feature *thermo-mechanical coupling*, *functionality*, *intelligence*, and *miniaturization*. These systems may operate under varying conditions. When functionally graded material systems are used in nano- and micro-devices, it is necessary to account for the microstructure-dependent size effect and the geometric nonlinearity. Since beam and plate structural elements are commonly used in these devices and structures, it is useful to develop refined theories of plates that account for size effects, material gradation through thickness, and geometric nonlinearity.

In the context of plate theories, no plate theory exists that accounts for shear deformation while not requiring shear correction factors, material variation through plate thickness, includes microstructure-dependent size effects, and geometric nonlinearity. This very fact motivated the present study. The objective of the current paper is to develop a general third-order plate theory that accounts for through-thickness power-law variation of a two-constituent material with temperature-dependent material properties, modified couple stress theory, and the von Karman nonlinear strains. In particular, we extend the modified couple stress theory of Yang et al. [1] (also see [2-6]) to the case of functionally graded plates using the third-order plate kinematics of Reddy [7-11], and Bose and Reddy [12]. Since most nanoscale devices involve plate-like elements that may be functionally graded and undergo moderately large rotations, the newly developed plate theory can be used to capture the size effects in functionally graded microplates. Moreover, the bending-extensional coupling is captured through the von Karman nonlinear strains.

MODIFIED COUPLE STRESS MODEL

The couple stress theory proposed by Yang et al. [1] is a modification of the classical couple stress theory. They established that the couple stress tensor is symmetric and the symmetric curvature tensor is the only proper conjugate strain measure to have a contribution to the total strain energy of the body. The two main advantages of the modified couple stress theory over the classical couple stress theory are the inclusion of a symmetric couple stress tensor and the involvement of only one length scale parameter, which is a direct consequence of the fact that the strain energy density function depends only on the strain and the symmetric part of the curvature tensor (see Ma, Gao, and Reddy [3-5] and Reddy [6]). According to the modified couple stress theory, the virtual strain energy δU can be written as

$$\delta U = \int_V \delta\varepsilon : \sigma + \delta\chi : m \mathrm{d}V = \int_V \delta\varepsilon_{ij} : \sigma_{ij} + \delta\chi_{ij} : m_{ij} \, \mathrm{d}V \tag{1}$$

where summation on repeated indices is implied; here σ_{ij} denotes the cartesian components of (the symmetric part of) the stress tensor, ε_{ij} are the strain components, m_{ij} are the components of the deviatoric part of the symmetric couple stress tensor, and χ_{ij} are the components of the symmetric curvature tensor

$$\chi_{ij} = \frac{1}{2}\left(\frac{\partial\omega_i}{\partial x_j} + \frac{\partial\omega_j}{\partial x_i}\right) = -\frac{1}{2} e_{ijk} \frac{\partial u_j}{\partial x_k} \tag{2}$$

FUNCTIONALLY GRADED MATERIALS

Consider a plate of total thickness h. The x and y coordinates are taken in the midplane, denoted with Ω, and the z-axis is taken normal to the plate, as shown in Figure 1. We assume that the material of the plate is isotropic but varies from one kind of material on one side, $z = -h/2$, to another material on the other side, $z = h/2$, as indicated in Figure 2. A typical material property P of the FGM through the plate thickness is assumed to be represented by a power-law (see Praveen and Reddy [13])

$$P(z, T) = [P_c(T) - P_m(T)]f(z) + P_m(T), \quad f(z) = \left(\frac{1}{2} + \frac{z}{h}\right)^n \tag{3}$$

where $P_c(T)$ and $P_m(T)$ are the values of a typical material property P, such as the modulus, density, and conductivity, of the ceramic material and metal, respectively; n denotes the volume fraction exponent, called power-law index. When $n = 0$, we obtain the single-material plate [with the property $P_c(T)$].

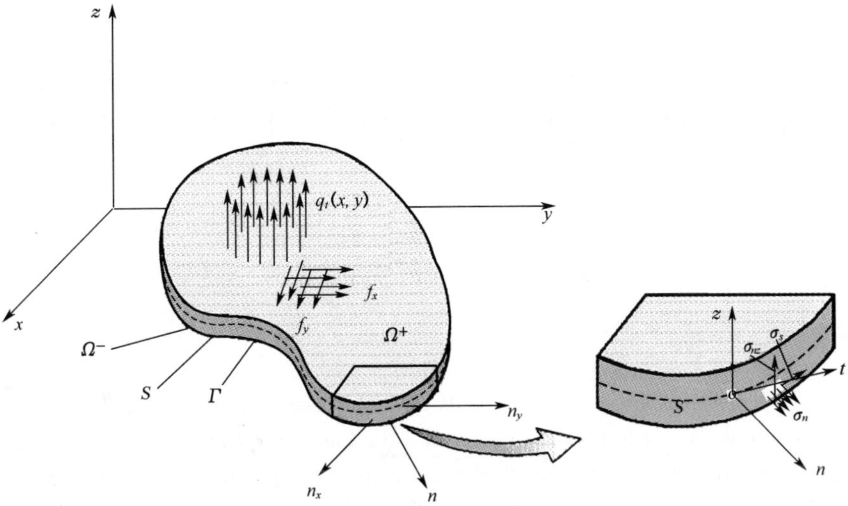

Figure 1 Geometry of a plate loaded with forces.

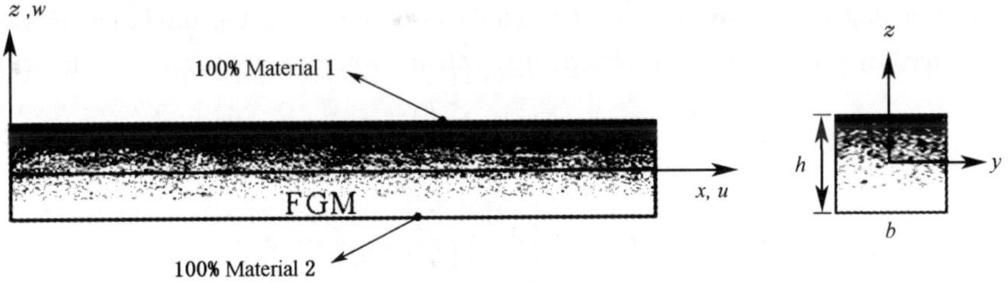

Figure 2 Through-thickness functionally graded plate.

When FGMs are used in high-temperature environment, the material properties are functions of temperature, and they can be expressed as

$$P_\alpha(T) = c_0(c_{-1}T^{-1} + 1 + c_1T + c_2T^2 + c_3T^3),\ \alpha = c \text{ or } m \tag{4}$$

where c_0 is a constant appearing in the cubic fit of the material property with temperature; and c_{-1}, c_1, c_2, and c_3 coefficients obtained after factoring out c_0 from the cubic curve fit of the property. For the analysis with constant properties, the material properties were all evaluated at 25.15℃.

A GENERAL THIRD-ORDER THEORY

Here develop a general third-order theory for the deformation of the plate first and then specialize to the well-known plate theories. We restrict the formulation to linear elastic material behavior, small strains, and moderate rotations and displacements, so that there is no geometric update of the domain, that is, the integrals posed on the deformed configuration are evaluated using the undeformed domain and there is no difference between the Cauchy stress tensor and the second Piola—Kirchhoff stress tensor.

The equations of motion are obtained using Hamilton's principle. The three-dimensional problem is reduced to two-dimensional one by assuming a displacement field that is explicit in the thickness coordinate z. We begin with the following displacement field

$$
\begin{aligned}
u_1(x, y, z, t) &= u(x, y, t) + z\theta_x + z^2\phi_x + z^3\psi_x \\
u_2(x, y, z, t) &= v(x, y, t) + z\theta_y + z^2\phi_y + z^3\psi_y \\
u_3(x, y, z, t) &= w(x, y, t) + z\theta_z + z^2\phi_z
\end{aligned} \tag{5}
$$

where (u, v, w) are the displacements along the coordinate lines of a material point on the xy-plane, i. e., $u(x, y, t) = u_1(x, y, 0, t)$, $v(x, y, t) = u_2(x, y, 0, t)$, $w(x, y, t) = u_3(x, y, 0, t)$ and

$$
\begin{aligned}
&\theta_x = \left(\frac{\partial u_1}{\partial z}\right)_{z=0}, \theta_y = \left(\frac{\partial u_2}{\partial z}\right)_{z=0}, \theta_z = \left(\frac{\partial u_3}{\partial z}\right)_{z=0} \\
&2\phi_x = \left(\frac{\partial^2 u_1}{\partial z^2}\right)_{z=0}, \quad 2\phi_y = \left(\frac{\partial^2 u_2}{\partial z^2}\right)_{z=0}, \quad 2\phi_z = \frac{\partial^2 u_3}{\partial z^2}, \\
&6\psi_x = \frac{\partial^3 u_1}{\partial z^3}, 6\psi_y = \frac{\partial^3 u_2}{\partial z^3}
\end{aligned} \tag{6}
$$

The reason for expanding the inplane displacements up to the cubic term and the transverse displacement up to the quadratic term in z is to obtain a quadratic variation of the transverse shear strains $\gamma_{xz} = 2\varepsilon_{xz}$ and $\gamma_{yz} = 2\varepsilon_{yz}$ through the plate thickness. Note that all three displacements contribute to the quadratic variation. In the most general case represented by the displacement field in Eqn. (5), there are 11 generalized displacements $(u, v, w, \theta_x, \theta_y, \theta_z, \phi_x, \phi_y, \phi_z, \psi_x, \psi_y)$ and, therefore, 11 differential equations will be required to determine them.

The von Karman nonlinear strain-displacement relations associated with the displacement field in Eqn. (17) can be obtained by assuming that the strains are small and rotations are moderately large; that is, we assume

$$
\begin{aligned}
&\left(\frac{\partial u_\alpha}{\partial x}\right)^2 \approx 0, \left(\frac{\partial u_\alpha}{\partial y}\right)^2 \approx 0, \left(\frac{\partial u_3}{\partial x}\right)^2 \approx \left(\frac{\partial w}{\partial x}\right)^2 \\
&\left(\frac{\partial u_3}{\partial y}\right)^2 \approx \left(\frac{\partial w}{\partial y}\right)^2, \left(\frac{\partial u_3}{\partial x}\right)\left(\frac{\partial u_3}{\partial y}\right) \approx \frac{\partial w}{\partial x}\frac{\partial w}{\partial y}
\end{aligned} \tag{7}
$$

for $\alpha = 1, 2$. Thus the nonzero strains of the general third-order theory with the von Karman nonlinearity are

$$
\begin{Bmatrix} \varepsilon_{xx} \\ \varepsilon_{yy} \\ \varepsilon_{xy} \end{Bmatrix} = \begin{Bmatrix} \varepsilon_{xx}^{(0)} \\ \varepsilon_{yy}^{(0)} \\ \varepsilon_{xy}^{(0)} \end{Bmatrix} + z\begin{Bmatrix} \varepsilon_{xx}^{(1)} \\ \varepsilon_{yy}^{(1)} \\ \varepsilon_{xy}^{(1)} \end{Bmatrix} + z^2\begin{Bmatrix} \varepsilon_{xx}^{(2)} \\ \varepsilon_{yy}^{(2)} \\ \varepsilon_{xy}^{(2)} \end{Bmatrix} + z^3\begin{Bmatrix} \varepsilon_{xx}^{(3)} \\ \varepsilon_{yy}^{(3)} \\ \varepsilon_{xy}^{(3)} \end{Bmatrix} \tag{8}
$$

$$
\begin{Bmatrix} \varepsilon_{zz} \\ \gamma_{xz} \\ \gamma_{yz} \end{Bmatrix} = \begin{Bmatrix} \varepsilon_{zz}^{(0)} \\ \gamma_{xz}^{(0)} \\ \gamma_{yz}^{(0)} \end{Bmatrix} + z\begin{Bmatrix} \varepsilon_{zz}^{(1)} \\ \gamma_{xz}^{(1)} \\ \gamma_{yz}^{(1)} \end{Bmatrix} + z^2\begin{Bmatrix} \varepsilon_{zz}^{(2)} \\ \gamma_{xz}^{(2)} \\ \gamma_{yz}^{(2)} \end{Bmatrix} \tag{9}
$$

with

$$
\begin{Bmatrix} \varepsilon_{xx}^{(0)} \\ \varepsilon_{yy}^{(0)} \\ \gamma_{xy}^{(0)} \end{Bmatrix} = \begin{Bmatrix} \frac{\partial u}{\partial x} + \frac{1}{2}\left(\frac{\partial w}{\partial x}\right)^2 \\ \frac{\partial v}{\partial y} + \frac{1}{2}\left(\frac{\partial w}{\partial y}\right)^2 \\ \frac{\partial u}{\partial y} + \frac{\partial v}{\partial x} + \frac{\partial w}{\partial x}\frac{\partial w}{\partial y} \end{Bmatrix}, \begin{Bmatrix} \varepsilon_{xx}^{(1)} \\ \varepsilon_{yy}^{(1)} \\ \gamma_{xy}^{(1)} \end{Bmatrix} = \begin{Bmatrix} \frac{\partial \theta_x}{\partial x} \\ \frac{\partial \theta_y}{\partial y} \\ \frac{\partial \theta_x}{\partial y} + \frac{\partial \theta_y}{\partial x} \end{Bmatrix} \tag{10}
$$

$$
\begin{Bmatrix} \varepsilon_{xx}^{(2)} \\ \varepsilon_{yy}^{(2)} \\ \gamma_{xy}^{(2)} \end{Bmatrix} = \begin{Bmatrix} \frac{\partial \phi_x}{\partial x} \\ \frac{\partial \phi_y}{\partial y} \\ \frac{\partial \phi_x}{\partial y} + \frac{\partial \phi_y}{\partial x} \end{Bmatrix}, \begin{Bmatrix} \varepsilon_{xx}^{(3)} \\ \varepsilon_{yy}^{(3)} \\ \gamma_{xy}^{(3)} \end{Bmatrix} = \begin{Bmatrix} \frac{\partial \psi_x}{\partial x} \\ \frac{\partial \psi_y}{\partial y} \\ \frac{\partial \psi_x}{\partial y} + \frac{\partial \psi_y}{\partial x} \end{Bmatrix} \tag{11}
$$

$$\begin{Bmatrix}\varepsilon_{zz}^{(0)}\\ \gamma_{xz}^{(0)}\\ \gamma_{yz}^{(0)}\end{Bmatrix}=\begin{Bmatrix}\theta_z\\ \theta_x+\dfrac{\partial w}{\partial x}\\ \theta_y+\dfrac{\partial w}{\partial y}\end{Bmatrix},\ \begin{Bmatrix}\varepsilon_{zz}^{(1)}\\ \gamma_{xz}^{(1)}\\ \gamma_{yz}^{(1)}\end{Bmatrix}=\begin{Bmatrix}\phi_z\\ 2\phi_x+\dfrac{\partial \theta_z}{\partial x}\\ 2\phi_y+\dfrac{\partial \theta_z}{\partial y}\end{Bmatrix},\ \begin{Bmatrix}\varepsilon_{zz}^{(2)}\\ \gamma_{xz}^{(2)}\\ \gamma_{yz}^{(2)}\end{Bmatrix}=\begin{Bmatrix}0\\ 3\psi_x+\dfrac{\partial \phi_z}{\partial x}\\ 3\psi_y+\dfrac{\partial \phi_z}{\partial y}\end{Bmatrix} \tag{12}$$

In view of the displacement field in Eq. (5), components of the rotation vector and curvature tensor take the form (with $\omega_1=\omega_x$, $\omega_2=\omega_y$, $\omega_3=\omega_z$, $\chi_{11}=\chi_{xx}$, $\chi_{22}=\chi_{yy}$, and so on)

$$\begin{aligned}
\omega_x &= \frac{1}{2}\left[\left(\frac{\partial w}{\partial y}+z\frac{\partial \theta_z}{\partial y}+z^2\frac{\partial \phi_z}{\partial y}\right)-(\theta_y+2z\phi_y+3z^2\psi_y)\right]=\omega_x^{(0)}+z\omega_x^{(1)}+z^2\omega_x^{(2)}\\
\omega_y &= \frac{1}{2}\left[(\theta_x+2z\phi_x+3z^2\psi_x)-\left(\frac{\partial w}{\partial x}+z\frac{\partial \theta_z}{\partial x}+z^2\frac{\partial \phi_z}{\partial x}\right)\right]=\omega_y^{(0)}+z\omega_y^{(1)}+z^2\omega_y^{(2)}\\
\omega_z &= \frac{1}{2}\left[\left(\frac{\partial v}{\partial x}+z\frac{\partial \theta_y}{\partial x}+z^2\frac{\partial \phi_y}{\partial x}+z^3\frac{\partial \psi_y}{\partial x}\right)-\left(\frac{\partial u}{\partial y}+z\frac{\partial \theta_x}{\partial y}+z^2\frac{\partial \phi_x}{\partial y}+z^3\frac{\partial \psi_x}{\partial y}\right)\right]\\
&= \omega_z^{(0)}+z\omega_z^{(1)}+z^2\omega_z^{(2)}+z^3\omega_z^{(3)}
\end{aligned} \tag{13}$$

where

$$\begin{aligned}
&\omega_x^{(0)}=\frac{1}{2}\left(\frac{\partial w}{\partial y}-\theta_y\right),\ \omega_x^{(1)}=\frac{1}{2}\left(\frac{\partial \theta_z}{\partial y}-2\phi_y\right),\ \omega_x^{(2)}=\frac{1}{2}\left(\frac{\partial \phi_z}{\partial y}-3\psi_y\right),\ \omega_y^{(0)}=\frac{1}{2}\left(\theta_x-\frac{\partial w}{\partial x}\right)\\
&\omega_y^{(1)}=\frac{1}{2}\left(2\phi_x-\frac{\partial \theta_z}{\partial x}\right),\ \omega_y^{(2)}=\frac{1}{2}\left(3\psi_x-\frac{\partial \phi_z}{\partial x}\right),\ \omega_z^{(0)}=\frac{1}{2}\left(\frac{\partial v}{\partial x}-\frac{\partial u}{\partial y}\right),\ \omega_z^{(1)}=\frac{1}{2}\left(\frac{\partial \theta_y}{\partial x}-\frac{\partial \theta_x}{\partial y}\right)\\
&\omega_z^{(2)}=\frac{1}{2}\left(\frac{\partial \phi_y}{\partial x}-\frac{\partial \phi_x}{\partial y}\right),\ \omega_z^{(3)}=\frac{1}{2}\left(\frac{\partial \psi_y}{\partial x}-\frac{\partial \psi_x}{\partial y}\right)
\end{aligned} \tag{14}$$

and

$$\begin{aligned}
\chi_{xx} &= \chi_{xx}^{(0)}+z\chi_{xx}^{(1)}+z^2\chi_{xx}^{(2)}\\
\chi_{yy} &= \chi_{yy}^{(0)}+z\chi_{yy}^{(1)}+z^2\chi_{yy}^{(2)}\\
\chi_{zz} &= \chi_{zz}^{(0)}+z\chi_{zz}^{(1)}+z^2\chi_{zz}^{(2)}\\
\chi_{xy} &= \chi_{xy}^{(0)}+z\chi_{xy}^{(1)}+z^2\chi_{xy}^{(2)}\\
\chi_{xz} &= \chi_{xz}^{(0)}+z\chi_{xz}^{(1)}+z^2\chi_{xz}^{(2)}+z^3\chi_{xz}^{(3)}\\
\chi_{yz} &= \chi_{yz}^{(0)}+z\chi_{yz}^{(1)}+z^2\chi_{yz}^{(2)}+z^3\chi_{yz}^{(3)}
\end{aligned} \tag{15}$$

with

$$\begin{aligned}
\chi_{xx}^{(0)} &= \frac{1}{2}\frac{\partial}{\partial x}\left(\frac{\partial w}{\partial y}-\theta_y\right),\ \chi_{xx}^{(1)}=\frac{1}{2}\frac{\partial}{\partial x}\left(\frac{\partial \theta_z}{\partial y}-2\phi_y\right),\ \chi_{xx}^{(2)}=\frac{1}{2}\frac{\partial}{\partial x}\left(\frac{\partial \phi_z}{\partial y}-3\psi_y\right)\\
\chi_{yy}^{(0)} &= \frac{1}{2}\frac{\partial}{\partial y}\left(\theta_x-\frac{\partial w}{\partial x}\right),\ \chi_{yy}^{(1)}=\frac{1}{2}\frac{\partial}{\partial y}\left(2\phi_x-\frac{\partial \theta_z}{\partial x}\right),\ \chi_{yy}^{(2)}=\frac{1}{2}\frac{\partial}{\partial y}\left(3\psi_x-\frac{\partial \phi_z}{\partial x}\right)\\
\chi_{zz}^{(0)} &= \frac{1}{2}\left(\frac{\partial \theta_y}{\partial x}-\frac{\partial \theta_x}{\partial y}\right),\ \chi_{zz}^{(1)}=\left(\frac{\partial \phi_y}{\partial x}-\frac{\partial \phi_x}{\partial y}\right),\ \chi_{zz}^{(2)}=\frac{3}{2}\left(\frac{\partial \psi_y}{\partial x}-\frac{\partial \psi_x}{\partial y}\right)\\
\chi_{xy}^{(0)} &= \frac{1}{2}\left[\frac{\partial}{\partial y}\left(\frac{\partial w}{\partial y}-\theta_y\right)+\frac{\partial}{\partial x}\left(\theta_x-\frac{\partial w}{\partial x}\right)\right],\ \chi_{xy}^{(1)}=\frac{1}{2}\left[\frac{\partial}{\partial y}\left(\frac{\partial \theta_z}{\partial y}-2\phi_y\right)+\frac{\partial}{\partial x}\left(2\phi_x-\frac{\partial \theta_z}{\partial x}\right)\right]\\
\chi_{xy}^{(2)} &= \frac{1}{2}\left[\frac{\partial}{\partial y}\left(\frac{\partial \phi_z}{\partial y}-3\psi_y\right)+\frac{\partial}{\partial x}\left(3\psi_x-\frac{\partial \phi_z}{\partial x}\right)\right],\ \chi_{xz}^{(0)}=\frac{1}{2}\left[\left(\frac{\partial \theta_z}{\partial y}-2\phi_y\right)+\frac{\partial}{\partial x}\left(\frac{\partial v}{\partial x}-\frac{\partial u}{\partial y}\right)\right]\\
\chi_{xz}^{(1)} &= \left(\frac{\partial \phi_z}{\partial y}-3\psi_y\right)+\frac{1}{2}\frac{\partial}{\partial x}\left(\frac{\partial \theta_y}{\partial x}-\frac{\partial \theta_x}{\partial y}\right),\ \chi_{xz}^{(2)}=\frac{1}{2}\frac{\partial}{\partial x}\left(\frac{\partial \phi_y}{\partial x}-\frac{\partial \phi_x}{\partial y}\right),\ \chi_{xz}^{(3)}=\frac{1}{2}\frac{\partial}{\partial x}\left(\frac{\partial \psi_y}{\partial x}-\frac{\partial \psi_x}{\partial y}\right)\\
\chi_{yz}^{(0)} &= \frac{1}{2}\left[\left(2\phi_x-\frac{\partial \theta_z}{\partial x}\right)+\frac{\partial}{\partial y}\left(\frac{\partial v}{\partial x}-\frac{\partial u}{\partial y}\right)\right],\ \chi_{yz}^{(1)}=\left(3\psi_x-\frac{\partial \phi_z}{\partial x}\right)+\frac{1}{2}\frac{\partial}{\partial y}\left(\frac{\partial \theta_y}{\partial x}-\frac{\partial \theta_x}{\partial y}\right)\\
\chi_{yz}^{(2)} &= \frac{1}{2}\frac{\partial}{\partial y}\left(\frac{\partial \phi_y}{\partial x}-\frac{\partial \phi_x}{\partial y}\right),\ \chi_{yz}^{(3)}=\frac{1}{2}\frac{\partial}{\partial y}\left(\frac{\partial \psi_y}{\partial x}-\frac{\partial \psi_x}{\partial y}\right)
\end{aligned} \tag{16}$$

(17)

The equations of motion are obtained by using the principle of virtual displacements or Hamilton's principle (see Reddy [14])

$$\int_0^T (\delta K - \delta U - \delta V)\,dt = 0 \tag{18}$$

where δK is the virtual kinetic energy, δU is the virtual strain energy, and δV is the virtual work done by external forces.

The details of deriving the expressions for the virtual energies are not given here due to the space restrictions (see Reddy and Kim [15]). The equations of motion are

$$\delta u:\ \frac{\partial M_{xx}^{(0)}}{\partial x}+\frac{\partial M_{xy}^{(0)}}{\partial y}+\frac{1}{2}\frac{\partial}{\partial y}\left(\frac{\partial M_{xz}^{(0)}}{\partial x}+\frac{\partial M_{yz}^{(0)}}{\partial y}\right)+F_x^{(0)}+\frac{1}{2}\frac{\partial c_z^{(0)}}{\partial y}=m_0\ddot{u}+m_1\ddot{\theta}_x+m_2\ddot{\phi}_x+m_3\ddot{\psi}_x$$

$$\delta v:\ \frac{\partial M_{xy}^{(0)}}{\partial x}+\frac{\partial M_{yy}^{(0)}}{\partial y}-\frac{1}{2}\frac{\partial}{\partial x}\left(\frac{\partial M_{xz}^{(0)}}{\partial x}+\frac{\partial M_{yz}^{(0)}}{\partial y}\right)+F_y^{(0)}-\frac{1}{2}\frac{\partial c_z^{(0)}}{\partial x}=m_0\ddot{v}+m_1\ddot{\theta}_y+m_2\ddot{\phi}_y+m_3\ddot{\psi}_y$$

$$\delta w:\ \frac{\partial}{\partial x}\left(\frac{\partial w}{\partial x}M_{xx}^{(0)}+\frac{\partial w}{\partial y}M_{xy}^{(0)}\right)+\frac{\partial}{\partial y}\left(\frac{\partial w}{\partial x}M_{xy}^{(0)}+\frac{\partial w}{\partial y}M_{yy}^{(0)}\right)+\frac{\partial M_{xz}^{(0)}}{\partial x}+\frac{\partial M_{yz}^{(0)}}{\partial y}$$
$$-\frac{1}{2}\frac{\partial}{\partial y}\left(\frac{\partial M_{xx}^{(0)}}{\partial x}+\frac{\partial M_{xy}^{(0)}}{\partial y}\right)+\frac{1}{2}\frac{\partial}{\partial x}\left(\frac{\partial M_{yy}^{(0)}}{\partial y}+\frac{\partial M_{xy}^{(0)}}{\partial x}\right)+F_z^{(0)}+\frac{1}{2}\left(\frac{\partial c_y^{(0)}}{\partial x}-\frac{\partial c_x^{(0)}}{\partial y}\right)$$
$$=m_0\ddot{w}+m_1\ddot{\theta}_z+m_2\ddot{\phi}_z$$

$$\delta\theta_x:\ \frac{\partial M_{xx}^{(1)}}{\partial x}+\frac{\partial M_{xy}^{(1)}}{\partial y}-M_{xz}^{(0)}+\frac{1}{2}\left(\frac{\partial M_{xy}^{(0)}}{\partial x}+\frac{\partial M_{yy}^{(0)}}{\partial y}-\frac{\partial M_{zz}^{(0)}}{\partial y}\right)+\frac{1}{2}\frac{\partial}{\partial y}\left(\frac{\partial M_{xz}^{(1)}}{\partial x}+\frac{\partial M_{yz}^{(1)}}{\partial y}\right)$$
$$+F_x^{(1)}+\frac{1}{2}c_y^{(0)}+\frac{1}{2}\frac{\partial c_z^{(1)}}{\partial y}=m_1\ddot{u}+m_2\ddot{\theta}_x+m_3\ddot{\phi}_x+m_4\ddot{\psi}_x$$

$$\delta\theta_y:\ \frac{\partial M_{xy}^{(1)}}{\partial x}+\frac{\partial M_{yy}^{(1)}}{\partial y}-M_{yz}^{(0)}-\frac{1}{2}\left(\frac{\partial M_{xx}^{(0)}}{\partial x}+\frac{\partial M_{xy}^{(0)}}{\partial y}-\frac{\partial M_{zz}^{(0)}}{\partial x}\right)-\frac{1}{2}\frac{\partial}{\partial x}\left(\frac{\partial M_{xz}^{(1)}}{\partial x}+\frac{\partial M_{yz}^{(1)}}{\partial y}\right)$$
$$+F_y^{(1)}-\frac{1}{2}c_x^{(0)}-\frac{1}{2}\frac{\partial c_z^{(1)}}{\partial x}=m_1\ddot{v}+m_2\ddot{\theta}_y+m_3\ddot{\phi}_y+m_4\ddot{\psi}_y$$

$$\delta\phi_x:\ \frac{\partial M_{xx}^{(2)}}{\partial x}+\frac{\partial M_{xy}^{(2)}}{\partial y}-2M_{xz}^{(1)}+\left(\frac{\partial M_{xy}^{(1)}}{\partial x}+\frac{\partial M_{yy}^{(1)}}{\partial y}-\frac{\partial M_{zz}^{(1)}}{\partial y}-M_{yz}^{(0)}\right)+\frac{1}{2}\frac{\partial}{\partial y}\left(\frac{\partial M_{xz}^{(2)}}{\partial x}+\frac{\partial M_{yz}^{(2)}}{\partial y}\right)$$
$$+F_x^{(2)}+c_y^{(1)}+\frac{1}{2}\frac{\partial c_z^{(2)}}{\partial y}=m_2\ddot{u}+m_3\ddot{\theta}_x+m_4\ddot{\phi}_x+m_5\ddot{\psi}_x$$

$$\delta\phi_y:\ \frac{\partial M_{xy}^{(2)}}{\partial x}+\frac{\partial M_{yy}^{(2)}}{\partial y}-2M_{yz}^{(1)}-\left(\frac{\partial M_{xx}^{(1)}}{\partial x}+\frac{\partial M_{xy}^{(1)}}{\partial y}-\frac{\partial M_{zz}^{(1)}}{\partial x}-M_{xz}^{(0)}\right)-\frac{1}{2}\frac{\partial}{\partial x}\left(\frac{\partial M_{xz}^{(2)}}{\partial x}+\frac{\partial M_{yz}^{(2)}}{\partial y}\right)$$
$$+F_y^{(2)}-c_x^{(1)}-\frac{1}{2}\frac{\partial c_z^{(2)}}{\partial x}=m_2\ddot{v}+m_3\ddot{\theta}_y+m_4\ddot{\phi}_y+m_5\ddot{\psi}_y$$

$$\delta\psi_x:\ \frac{\partial M_{xx}^{(3)}}{\partial x}+\frac{\partial M_{xy}^{(3)}}{\partial y}-3M_{xz}^{(2)}+\frac{3}{2}\left(\frac{\partial M_{xy}^{(2)}}{\partial x}+\frac{\partial M_{yy}^{(2)}}{\partial y}-\frac{\partial M_{zz}^{(2)}}{\partial y}-2M_{yz}^{(1)}\right)+\frac{1}{2}\frac{\partial}{\partial y}\left(\frac{\partial M_{xz}^{(3)}}{\partial x}+\frac{\partial M_{yz}^{(3)}}{\partial y}\right)$$
$$+F_x^{(3)}+\frac{3}{2}c_y^{(2)}+\frac{1}{2}\frac{\partial c_z^{(3)}}{\partial y}=m_3\ddot{u}+m_4\ddot{\theta}_x+m_5\ddot{\phi}_x+m_6\ddot{\psi}_x$$

$$\delta\psi_y:\ \frac{\partial M_{xy}^{(3)}}{\partial x}+\frac{\partial M_{yy}^{(3)}}{\partial y}-3M_{yz}^{(2)}-\frac{3}{2}\left(\frac{\partial M_{xx}^{(2)}}{\partial x}+\frac{\partial M_{xy}^{(2)}}{\partial y}-\frac{\partial M_{zz}^{(2)}}{\partial x}-2M_{xz}^{(1)}\right)-\frac{1}{2}\frac{\partial}{\partial x}\left(\frac{\partial M_{xz}^{(3)}}{\partial x}+\frac{\partial M_{yz}^{(3)}}{\partial y}\right)$$
$$+F_y^{(3)}-\frac{3}{2}c_x^{(2)}-\frac{1}{2}\frac{\partial c_z^{(3)}}{\partial x}=m_3\ddot{v}+m_4\ddot{\theta}_y+m_5\ddot{\phi}_y+m_6\ddot{\psi}_y$$

$$\delta\theta_z:\ \frac{\partial M_{xz}^{(1)}}{\partial x}+\frac{\partial M_{yz}^{(1)}}{\partial y}-M_{zz}^{(0)}-\frac{1}{2}\frac{\partial}{\partial y}\left(\frac{\partial M_{xx}^{(1)}}{\partial x}+\frac{\partial M_{xy}^{(1)}}{\partial y}\right)+\frac{1}{2}\frac{\partial}{\partial x}\left(\frac{\partial M_{yy}^{(1)}}{\partial y}+\frac{\partial M_{xy}^{(1)}}{\partial x}\right)+\frac{1}{2}\left(\frac{\partial M_{xz}^{(0)}}{\partial y}-\frac{\partial M_{yz}^{(0)}}{\partial x}\right)$$

$$+F_z^{(1)}+\frac{1}{2}\left(\frac{\partial c_y^{(1)}}{\partial x}-\frac{\partial c_x^{(1)}}{\partial y}\right)=m_1\ddot{w}+m_2\ddot{\theta}_z+m_3\ddot{\phi}_z$$

$$\delta\phi_z:\ \frac{\partial M_{xz}^{(2)}}{\partial x}+\frac{\partial M_{yz}^{(2)}}{\partial y}-2M_{zz}^{(1)}-\frac{1}{2}\frac{\partial}{\partial y}\left(\frac{\partial M_{xx}^{(2)}}{\partial x}+\frac{\partial M_{xy}^{(2)}}{\partial y}\right)+\frac{1}{2}\frac{\partial}{\partial x}\left(\frac{\partial M_{yy}^{(2)}}{\partial y}+\frac{\partial M_{xy}^{(2)}}{\partial x}\right)+\frac{\partial M_{xz}^{(1)}}{\partial y}-\frac{\partial M_{yz}^{(1)}}{\partial x}$$

$$+F_z^{(2)}+\frac{1}{2}\left(\frac{\partial c_y^{(2)}}{\partial x}-\frac{\partial c_x^{(2)}}{\partial y}\right)=m_2\ddot{w}+m_3\ddot{\theta}_z+m_4\ddot{\phi}_z \quad (19)$$

where the superposed dot on a variable indicates time derivative, for example, $\dot{u}=\partial u/\partial t$, $m_i(i=0, 1, 2, \cdots, 6)$ are the mass moments of inertia

$$m_i=\int_{-\frac{h}{2}}^{\frac{h}{2}}\rho(z)^i\,dz \quad (20)$$

$M_{ij}^{(k)}$ are the stress resultants

$$M_{ij}^{(k)}=\int_{-\frac{h}{2}}^{\frac{h}{2}}(z)^k\sigma_{ij}\,dz,\ M_{ij}^{(k)}=\int_{-\frac{h}{2}}^{\frac{h}{2}}(z)^k m_{ij}\,dz,\ (k=0, 1, 2, 3) \quad (21)$$

and $(\bar{f}_x, \bar{f}_y, \bar{f}_z)$ are the body forces (measured per unit volume), $(\bar{t}_x, \bar{t}_y, \bar{t}_z)$ the surface forces (measured per unit area) on S, and (q_x^t, q_y^t, q_z^t) the distributed forces (measured per unit area) on Ω^+, (q_x^b, q_y^b, q_z^b) the distributed forces (measured per unit area) on Ω^-, and $(\bar{c}_x, \bar{c}_y, \bar{c}_z)$ be the body couples (measured per unit volume) in the (x, y, z) coordinate directions. Additional details can be found in [15].

The general third-order theory developed herein contains all of the existing plate theories but some of them have not been extended to contain the microstructure parameters and the vonKarman nonlinearity. They are summarized in the recent paper by Reddy and Kim [15].

CONCLUSIONS

A general third-order theory of functionally graded plates with microstructure-dependent length scale parameter and the von Karman nonlinearity is presented. The theory accounts for temperature dependent properties of the constituents in the functionally graded material, and modified couple stress theory is used to bring a microstructural length scale parameter. The equations of motions and associated force boundary conditions are derived using Hamilton's principle. The theory developed contains 11 generalized displacements. The existing plate theories, namely, a third-order theory with vanishing surface tractions, the Reddy third-order plate theory [7], the first-order plate theory, and the classical plate theory can be obtained as special cases of the developed general third-order plate theory. Three-dimensional constitutive relations must be used, consistent with the three-dimensional strain field, to develop plate constitutive relations. More complete development is given in the forthcoming paper [15].

The general third-order theory and its special cases developed herein can be used to construct finite element models of functionally graded plates with geometric nonlinearity and microstructure dependent length scale parameter. For the general case, the finite element models allow C^0-approximation of all 11 generalized displacements. The third-order plate theories with vanishing surface tractions require C^0 interpolation of $(u, v, \theta_x, \theta_y)$ and Hermite interpolation of w, θ_z, and ϕ_z. Computational models and their applications of some of the theories presented here are yet to appear. Also, analytical (e. g., Navier) solutions based on the linear theories may be obtained.

ACKNOWLEDGEMENT

The support of this research by the Air Force Office of Scientific Research through Grant FA9550-09-1-0686, P00001 is gratefully acknowledged.

REFERENCES

[1] Yang. F., Chong, A. C. M., Lam, D. C. C., and Tong, P., "Couple stress based strain gradient theory for elasticity." *International Journal of Solids and Structures*, **39**, 2002, 2731-2743.

[2] Park, S.K. and Gao, X-L., "Bernoulli-Euler beam model based on a modified couple stress theory." *Journal of Micromechanics and Microengineering*, **16**, 2006, 2355-2359.

[3] Ma, H.M., Gao, X-L., and Reddy, J.N., "A Microstructure-dependent Timoshenko beam model based on a modified couple stress theory." *Journal of the Mechanics and Physics of Solids*, **56**, 2008, 3379-3391.

[4] Ma, H.M., Gao, X-L., and Reddy, J.N., "A nonclassical Reddy-Levinson beam model based on a modified couple stress theory." *International Journal for Multiscale Computational Engineering*, **8**(2), 2010, 167-180.

[5] Ma, H.M., Gao, X-L., and Reddy, J.N., "A non-classical Mindlin plate model based on a modified couple stress theory." *Acta Mechanica* 2011, DOI 10.1007/s00707-011-0480-4.

[6] Reddy, J. N., "Microstructure-dependent couple stress theories of functionally graded beams." *Journal of Mechanics and Physics of Solids*, 59, 2011, 2382-2399.

[7] Reddy, J.N., "A simple higher-order theory for laminated composite plates." *J Appl Mech*, **51**, 1984, 745-752.

[8] Reddy, J.N., "A refined nonlinear theory of plates with transverse shear deformation." *Int J Solids Struct*, 20, 1984, 881-896.

[9] Reddy, J.N., "A small strain and moderate rotation theory of laminated anisotropic plates." *J Appl Mech*, **54**, 1987, 623-626.

[10] Reddy, J.N., "A general non-linear third-order theory of plates with moderate thickness." *Int J Non-Linear Mech*, **25**(6), 1990, 677-686.

[11] Reddy, J.N., "A general nonlinear third-order theory of functionally graded plates." *Int. J. Aerospace and Lightweight Structures*, **1**(1), 2011, 1-21.

[12] Bose, P. and Reddy, J.N., "Analysis of composite plates using various plate theories, part 1: formulation and analytical results." *Struct Engng Mech*, **6**(6), 1998, 583-612.

[13] Praveen, G.N. and Reddy, J.N., "Nonlinear transient thermoelastic analysis of functionally graded ceramic-metal plates." *Journal of Solids and Structures*, **35**(33), 1998, 4457-4476.

[14] Reddy, J. N., *Energy Principles and Variational Methods in Applied Mechanics*, 2nd ed, John Wiley & Sons, New York, 2002.

[15] Reddy, J.N. and Kim, J., "A nonlinear modified couple stress-based third-order theory of functionally graded plates." *Composite Structures*, **94**, 2012, 1128-1143.

SOME CONCEPTUAL ISSUES IN THE MODELLING OF CRACKED BEAMS FOR LATERAL-TORSIONAL BUCKLING ANALYSIS

* N. Challamel[1], A. Andrade[2], D. Camotim[3]

[1] Université Européenne de Bretagne, University of South Brittany UBS LIMATB,
Centre de Recherche, Rue de Saint Maudé, BP92116_56321 Lorient cedex—France
on sabbatical leave, Mechanics Division, Department of Mathematics
University of Oslo, P.O. Box 1053, Blindern,
NO-0316 Oslo—Norway

[2] Department of Civil Engineering—INESC Coimbra, University of Coimbra
3030-788 Coimbra—Portugal
E-mail: anisio@dec.uc.pt

[3] Department of Civil Engineering and Architecture—ICIST/IST
Technical University of Lisbon, 1049-001 Lisbon—Portugal
E-mail: dcamotim@civil.ist.utl.pt
* Email: noel.challamel@univ-ubs.fr

KEYWORDS

Lateral-torsional buckling, kirchhoff-clebsch theory, connection, crack, non-conservative loading, stability, uniform moment, variational and energy method, spring models

ABSTRACT

This paper is focused on the lateral-torsional buckling of cracked or weakened elastic beams. The crack is modelled with a generalized elastic connection law, whose equivalent stiffness parameters can be derived from fracture mechanics considerations. The same type of generalised spring model can be used for beams with semi-rigid connections, typically in the field of steel or timber engineering. As the basis for the present investigation, we consider a strip beam with fork end supports and exhibiting a single vertical edge crack, subjected to uniform bending in the plane of greatest flexural rigidity. The effect of prebuckling deformation is taken into consideration within the framework of the Kirchhoff-Clebsch theory. First, the three-dimensional elastic connection law adopted is a direct extension of the planar case, but this leads to a paradoxical conclusion: the critical moment is not affected by the presence of the crack, regardless of its location. It is shown that the above paradox is due to the non-conservative nature of the connection model adopted. Simple alternatives to this cracked-section constitutive law are proposed, based on conservative moment-rotation laws (quasi-tangential and semi-tangential) and consistent variational arguments.

INTRODUCTION

The numerous investigations devoted to the buckling of cracked elastic structures have so far focused mainly on the flexural buckling behaviour of columns -*e.g.*, [1, 2]. For such an in-plane analysis, the crack may be reasonably modelled by a simple elastic rotational spring with an "equivalent stiffness", as suggested by Okamura *et al.* [1]. The fundamental constitutive law of the cracked cross-section is therefore expressed as

$$M = k\Delta\theta, \tag{1}$$

where M is the bending moment acting at the cracked section, k is the equivalent stiffness and $\Delta\theta$ is the relative rotation (slope difference) occurring at the cracked cross-section.

In order to tackle out-of-plane buckling problems (*e.g.*, the lateral-torsional buckling of beams— see Figure 1), it is necessary to generalise the above constitutive law. The most straightforward generalisation corresponds to the diagonal relation

$$\begin{pmatrix} M_1 \\ M_2 \\ M_3 \end{pmatrix} = \begin{pmatrix} k_1 & 0 & 0 \\ 0 & k_2 & 0 \\ 0 & 0 & k_3 \end{pmatrix} \begin{pmatrix} \Delta\theta_1 \\ \Delta\theta_2 \\ \Delta\theta_3 \end{pmatrix}, \tag{2}$$

where M_1 is the torsional moment, M_2 is the out-of-plane (minor-axis) bending moment, M_3 is the in-plane (major-axis) bending moment, the $\Delta\theta_i$ ($i=1, 2, 3$) are the relative rotations, associated with each direction, occurring at the cracked cross-section and each k_i is the stiffness relating a relative rotation with the corresponding moment. The cracked cross-section constitutive law can be further generalised by considering off-diagonal (coupling) terms, making it possible to take into account the effects of anisotropy and crack orientation—see, for instance, Wang *et al.* [3] who address the closely related problem of beam vibration.

The amount of work dealing with the lateral-torsional buckling of cracked beams is rather scarce. Carloni *et al.* [4] investigated the lateral-torsional buckling of cracked I-beams under uniform bending, but restricted the constitutive law to the torsional term, *i.e.*,

$$M_1 = k_1 \Delta\theta_1. \tag{3}$$

Karaagac *et al.* [5] studied the lateral-torsional buckling of a cracked cantilever beam submitted to a concentrated force, adopting Eq. (2) to describe the cracked cross-section constitutive behaviour—the buckling problem was solved by means of the finite element method. Finally, the authors are not aware of the publication of any closed-form solution concerning the lateral-torsional buckling behaviour of cracked beams.

The same type of spring model can be used for beams with semi-rigid connections. In the field of steel structures, the effect of semi-rigid connections on the out-of-plane behaviour of I-beams has been numerically assessed by several authors, such as Krenk and Damkilde [6] or more recently Basaglia *et al.* [7]. In order to include the effects of the warping restraint, the constitutive law (2) may be augmented to

$$\begin{Bmatrix} M_1 \\ M_2 \\ M_3 \\ M_\omega \end{Bmatrix} = \begin{bmatrix} k_1 & 0 & 0 & 0 \\ 0 & k_2 & 0 & 0 \\ 0 & 0 & k_3 & 0 \\ 0 & 0 & 0 & k_\omega \end{bmatrix} \begin{Bmatrix} \Delta\theta_1 \\ \Delta\theta_2 \\ \Delta\theta_3 \\ \Delta\theta'_1 \end{Bmatrix}, \tag{4}$$

where M_ω is the bimoment in Vlassov's theory. However, most of these studies deal only with the warping restraint stiffness, which means that Eq. (3) becomes simply

$$M_\omega = k_\omega \Delta\theta'_1. \tag{5}$$

In this paper, the use of spring models for the lateral-torsional buckling analysis of cracked beams (or beams with semi-rigid connections) is investigated. As the basis for this investigation, we consider a strip beam with fork end supports and exhibiting a single vertical edge crack, subjected to uniform bending in the plane of greatest flexural rigidity (see Figure 1)—therefore note that the in-plane constitutive behaviour of the spring plays no role in this problem. The adoption of the constitutive equation (2) for the spring is discussed and shown to lead to a non-conservative model. Simple alternatives to this equation are proposed, based on conservative (quasi-tangential and semi-tangential) moment-rotation laws. The paper focuses on the conceptual issues involved in the modelling of the aforementioned problem and the analyses are always conducted analytically, leading to closed-form characteristic equations for the buckling moments.

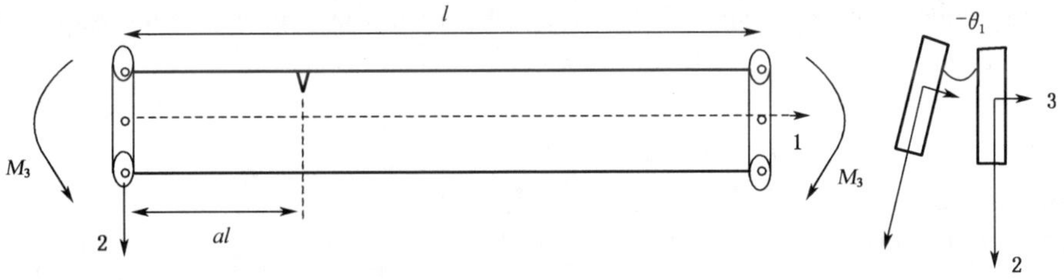

Figure 1 Lateral-torsional buckling of a cracked beam under uniform bending moment

LATERAL-TORSIONAL BUCKLING—A PARADOX?

The beam, with a narrow rectangular cross-section (strip beam) and length l, is subjected to uniform bending in its major-axis flexural plane. It is simply supported in the minor-axis flexural plane and the end cross-sections are prevented from twisting. There are no restraints inducing stretching strain of the beam central line. The location of the crack along the beam length is defined by the dimensionless parameter α, with $0<\alpha<1$. The crack divides the beam into two homogeneous segments, with lengths αl and $(1-\alpha)l$, connected by a rotational spring.

This out-of-plane buckling problem is first examined within the framework of the Kirchhoff-Clebsch rod theory, without making at the onset any specific kinematical assumptions on the curvature-displacement relations. Over each beam segment, the equilibrium equations in an adjacent state to the fundamental equilibrium path read

$$M'_{1i} - \bar{\kappa}_3 M_{2i} + \kappa_{2i}\overline{M}_3 = 0 \tag{6a}$$

$$M'_{2i} + \bar{\kappa}_3 M_{1i} - \kappa_{1i}\overline{M}_3 - F_{3i} = 0 \tag{6b}$$

$$F'_{3i} = 0, \tag{6c}$$

where (ⅰ) M_1 is the torsional moment, (ⅱ) M_2 and M_3 are the minor and major-axis bending moments, (ⅲ) F_3 is a shearing force, (iv) κ_1 is the rate of twist and (v) κ_2 and κ_3 are the components of curvature of the deformed central line. A superposed bar indicates a quantity associated with the fundamental equilibrium path. The subscript i stands for the symbols "−" ($0 \leqslant x \leqslant \alpha l$) or "+" ($\alpha l \leqslant x \leqslant l$) and identifies the beam segment under consideration. Observe that Eq. (6.c) implies that the function F_3 is constant over each beam segment and, for the sake of convenience, the same symbol shall be used to denote the function and its constant value.

Assuming that the torsional and bending moments are related to the rate of twist and curvature components by means of uncoupled linear constitutive equations of the form

$$\begin{pmatrix} M_1 \\ M_2 \\ M_3 \end{pmatrix} = \begin{pmatrix} GJ & 0 & 0 \\ 0 & EI_2 & 0 \\ 0 & 0 & EI_3 \end{pmatrix} \begin{pmatrix} \kappa_1 \\ \kappa_2 \\ \kappa_3 \end{pmatrix}, \tag{7}$$

one obtains from Eqn. 6(a)-6(b)

$$M'_{1i} - M_{2i}\overline{M}_3\left(\frac{1}{EI_3} - \frac{1}{EI_2}\right) = 0 \tag{8a}$$

$$M'_{2i} - M_{1i}\overline{M}_3\left(\frac{1}{GJ} - \frac{1}{EI_3}\right) = F_{3i}. \tag{8b}$$

Differentiating both members of Eqn. (8b) and inserting Eqn. (8a) into the result leads to the linear differential equation

$$\frac{EI_2 GJ}{\left(1-\frac{EI_2}{EI_3}\right)\left(1-\frac{GJ}{EI_3}\right)} M''_{2i} + \overline{M}_3^2 M_{2i} = 0, \tag{9}$$

whose general solution is

$$M_{2i}(x) = A_{2i}\sin\frac{\overline{M}_3 x}{\sqrt{\widetilde{EI}_2\widetilde{GJ}}} + B_{2i}\cos\frac{\overline{M}_3 x}{\sqrt{\widetilde{EI}_2\widetilde{GJ}}}, \text{ where} \tag{10}$$

$$\widetilde{GJ} = \frac{GJ}{1-\frac{GJ}{EI_3}} \qquad \widetilde{EI}_2 = \frac{EI_2}{1-\frac{EI_2}{EI_3}}. \tag{11}$$

The equivalent stiffness values $\widetilde{GJ}$ and $\widetilde{EI}_2$ exhibit the limiting behaviours $\widetilde{GJ} \to GJ$ as $\frac{GJ}{EI_3} \to 0$ and $\widetilde{EI}_2 \to EI_2$ as $\frac{EI_2}{EI_3} \to 0$. The torsional moment is now derived from Eqn. 8(b):

$$M_{1i}(x) = A_{2i}\sqrt{\frac{\widetilde{GJ}}{\widetilde{EI}_2}}\cos\frac{\overline{M}_3 x}{\sqrt{\widetilde{EI}_2\widetilde{GJ}}} - B_{2i}\sqrt{\frac{\widetilde{GJ}}{\widetilde{EI}_2}}\sin\frac{\overline{M}_3 x}{\sqrt{\widetilde{EI}_2\widetilde{GJ}}} - F_{3i}\frac{\widetilde{GJ}}{\overline{M}_3}. \tag{12}$$

Finally, the flexural rotation θ_{2i}, the twist θ_{1i} and the out-of-plane deflection w_i can be obtained by integration (upon the adoption of simplified kinematical relations, in order to keep the analysis tractable):

$$\widetilde{EI}_2\theta_{2i}(x) = -A_{2i}\frac{\sqrt{\widetilde{EI}_2\widetilde{GJ}}}{\overline{M}_3}\cos\frac{\overline{M}_3 x}{\sqrt{\widetilde{EI}_2\widetilde{GJ}}} + B_{2i}\frac{\sqrt{\widetilde{EI}_2\widetilde{GJ}}}{\overline{M}_3}\sin\frac{\overline{M}_3 x}{\sqrt{\widetilde{EI}_2\widetilde{GJ}}} + C_{2i} \tag{13}$$

$$\widetilde{GJ}\theta_{1i}(x) = A_{2i}\frac{\widetilde{GJ}}{\overline{M}_3}\sin\frac{\overline{M}_3 x}{\sqrt{\widetilde{EI}_2\widetilde{GJ}}} + B_{2i}\frac{\widetilde{GJ}}{\overline{M}_3}\cos\frac{\overline{M}_3 x}{\sqrt{\widetilde{EI}_2\widetilde{GJ}}} - F_{3i}\frac{\widetilde{GJ}}{\overline{M}_3}x + C_{1i} \tag{14}$$

$$\widetilde{EI}_2 w_i(x) = A_{2i}\frac{\widetilde{EI}_2\widetilde{GJ}}{\overline{M}_3^2}\sin\frac{\overline{M}_3 x}{\sqrt{\widetilde{EI}_2\widetilde{GJ}}} + B_{2i}\frac{\widetilde{EI}_2\widetilde{GJ}}{\overline{M}_3^2}\cos\frac{\overline{M}_3 x}{\sqrt{\widetilde{EI}_2\widetilde{GJ}}} - C_{2i}x + D_{2i}. \tag{15}$$

The general solution of the lateral-torsional buckling problem thus involves a total of twelve constants: each beam segment accounts for six constants, denoted A_2, B_2, C_2, D_2, C_1 and F_3, with the additional label "$-$" or "$+$" identifying the segment under consideration.

Consideration of the boundary and jump conditions will now lead to the characteristic equation for the buckling moments. The boundary conditions (at $x = 0$ and $x = l$) are

$$w_-(0) = 0 \qquad M_{2-}(0) = 0 \qquad \theta_{1-}(0) = 0 \tag{16}$$

$$w_+(l) = 0 \qquad M_{2+}(l) = 0 \qquad \theta_{1+}(l) = 0 \tag{17}$$

At the cracked cross-section ($x = \alpha l$), the continuity of the out-of-plane deflection, torsional moment, minor-axis bending moment and shear force is enforced and the constitutive law of the spring connecting the two segments is postulated according to Eqn. (2), yielding

$$w_-(\alpha l) = w_+(\alpha l) \qquad M_{1-}(\alpha l) = M_{1+}(\alpha l) \tag{18}$$

$$M_{2-}(\alpha l) = M_{2+}(\alpha l) \tag{19a}$$

$$F_{3-}(\alpha l) = F_{3+}(\alpha l) \tag{19b}$$

$$M_{1+}(\alpha l) = k_1[\theta_{1+}(\alpha l) - \theta_{1-}(\alpha l)] \qquad M_{2+}(\alpha l) = k_2[\theta_{2+}(\alpha l) - \theta_{2-}(\alpha l)] \tag{20}$$

It can be shown that

$$\sin(\overline{\widetilde{m}}_3) = 0, \text{ where } \overline{\widetilde{m}}_3 = \frac{\overline{M}_3 l}{\sqrt{\widetilde{EI}_2\widetilde{GJ}}}, \tag{21}$$

must hold for the twelve constants A_{2-}, …, F_{3-}, A_{2+}, …, F_{3+} not to be all simultaneously zero. Eqn. (21) is the characteristic equation providing the lateral-torsional buckling moments of the cracked beam. Its lowest positive root corresponds to the dimensionless critical moment $\overline{\widetilde{m}}_{3.cr} = \pi$, which is precisely the value obtained by Grober [8] for an uncracked beam. Moreover, when the effect of the pre-buckling curvature is negligible, the cracked beam critical moment tends to Prandtl's solution for an uncracked beam - recall the limiting behaviours of $\widetilde{GJ}$ and $\widetilde{EI}_2$. Hence the paradox: it appears that the crack does not affect the lateral-torsional critical moment of the beam, regardless of its location and size.

To clarify this paradox, a variational approach to the buckling problem is now adopted and, for the sake of simplicity, the effect of the pre-buckling curvature is ignored. Consider the functional

$$V_0 = \frac{1}{2}\int_0^{\alpha l}(GJ\theta_{1-}'^2 + EI_2 w_-''^2 + 2\overline{M}_3\theta_{1-}w''_-)\mathrm{d}x + \frac{1}{2}\int_{\alpha l}^{l}(GJ\theta_{1+}'^2 + EI_2 w_+''^2 + 2\overline{M}_3\theta_{1+}w''_+)\mathrm{d}x$$
$$+ \frac{k_1}{2}[\theta_{1+}(\alpha l) - \theta_{1-}(\alpha l)]^2 + \frac{k_2}{2}[w'_+(\alpha l) - w'_-(\alpha l)]^2, \tag{22}$$

whose admissible functions satisfy the essential conditions $w_-(0) = w_+(l) = 0$, $\theta_{1-}(0) = \theta_{1+}(l) = 0$ and $w_-(\alpha l) = w_+(\alpha l)$. Using standard techniques from the calculus of variations, it is a straightforward matter to see that the variational identity

$$\delta V_0 - [\overline{M}_3\theta_{1-}\delta w'_-]_0^{\alpha l} - [\overline{M}_3\theta_{1+}\delta w'_+]_{\alpha l}^{l} = 0 \tag{23}$$

is a weak form corresponding to the boundary value problem addressed above (recall that the effect of the pre-buckling curvature is now neglected). The paradox of the insensitivity of the critical moment to the presence of the crack can now be explained by observing that there exists no potential which, upon variation, yields the left-hand side of Eqn. (23). Therefore, the buckling problem defined by this equation belongs to the class of non-conservative problems. The non-conservative character is directly attributable to the jump conditions 19(a) and 19(b), which involve the constitutive law (2) postulated for the cracked cross-section.

ON THE USE OF CONSERVATIVE MOMENT-ROTATION LAWS

The beam in Figure 1 is being modelled as two segments connected by a rotational spring at the crack location. In such a physical model, the bending moment $\overline{M}_3$ at the cracked section—*i. e.*, at the connection between segments—performs work in the relative rotation occurring at that connection when the beam goes from a fundamental equilibrium state to an adjacent, buckled one. Accordingly, this bending moment must be regarded as consisting of a pair of opposite conservative moments applied at the connected end of each segment, and not as a "conventional" internal moment (which performs work over a curvature)—see [9]. Moreover, in order to carry out a lateral-torsional buckling analysis, it is necessary to specify the rotation behaviour of these conservative end moments, externally applied to each segment end by the spring device. Unfortunately, the physical model provides no indication on how to select a particular moment-rotation law. Therefore, different possible choices are investigated next and their consequences discussed.

First, the moment $\overline{M}_3$ at the connected end of each segment is assumed to behave as a quasi-tangential moment (see Figure 2a), with $\beta=0$. The appropriate energy functional to investigate the lateral-torsional buckling of the cracked beam is then

$$V_1 = \frac{1}{2}\int_0^{\alpha l}(GJ\theta_{1-}'^2 + EI_2 w_-''^2 - 2\overline{M}_3\theta'_{1-}w'_-)\mathrm{d}x + \frac{1}{2}\int_{\alpha l}^{l}(GJ\theta_{1+}'^2 + EI_2 w_+''^2 - 2\overline{M}_3\theta'_{1+}w'_+)\mathrm{d}x$$
$$+\frac{k_1}{2}[\theta_{1+}(\alpha l)-\theta_{1-}(\alpha l)]^2 + \frac{k_2}{2}[w'_+(\alpha l)-w'_-(\alpha l)]^2 = V_0 - [\overline{M}_3\theta_{1-}w'_-]_0^{\alpha l} - [\overline{M}_3\theta_{1+}w'_+]_{\alpha l}^{l} \quad (24)$$

with the same essential conditions to be satisfied by the admissible functions as in V_0. The vanishing of the first variation of V_1 leads to the jump conditions

$$GJ\theta'_{1-}(\alpha l) - \overline{M}_3 w'_-(\alpha l) = GJ\theta'_{1+}(\alpha l) - \overline{M}_3 w'_+(\alpha l) \quad (25)$$
$$GJ\theta'_{1+}(\alpha l) - \overline{M}_3 w'_+(\alpha l) - k_1[\theta_{1+}(\alpha l) - \theta_{1-}(\alpha l)] = 0 \quad (26)$$

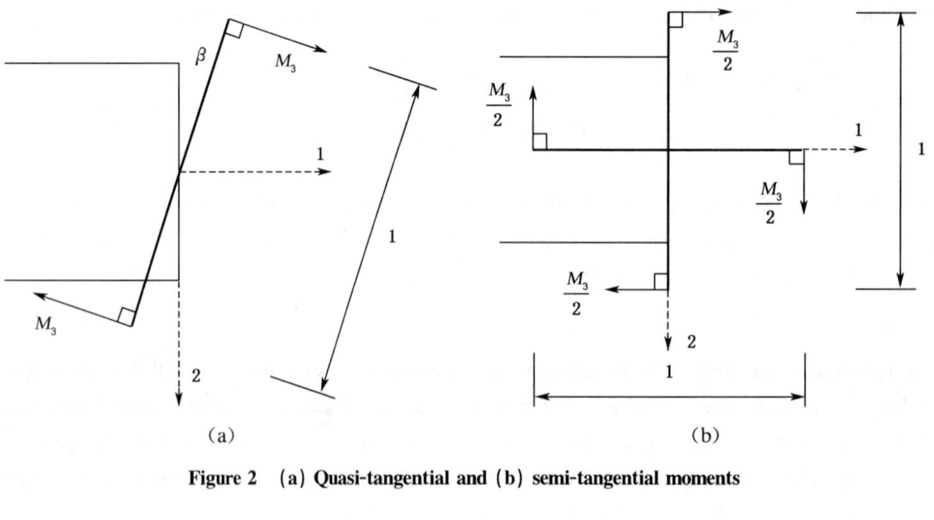

Figure 2 (a) Quasi-tangential and (b) semi-tangential moments

at the cracked section. Thus, it is possible to detect here the emergence of a new constitutive law for the cracked section, namely

$$M_{1-}(\alpha l) + \overline{M}_3\theta_{2-}(\alpha l) = M_{1+}(\alpha l) + \overline{M}_3\theta_{2+}(\alpha l) = k_1\Delta\theta_1(\alpha l) \tag{27}$$

$$M_{2-}(\alpha l) = M_{2+}(\alpha l) = k_2\Delta\theta_2(\alpha l). \tag{28}$$

It can be shown that, in this case, the buckling moments are given by the transcendental equation

$$\gamma_2\sin(\bar{m}_3) - \bar{m}_3\sin(\alpha\bar{m}_3)\sin[(1-\alpha)\bar{m}_3] = 0, \text{ where } \bar{m}_3 = \frac{\overline{M}_3 l}{\sqrt{EI_2GJ}} \text{ and } \gamma_2 = \frac{k_2 l}{EI_2}, \tag{29}$$

which, oddly enough, is independent of the crack torsional stiffness k_1. In fact, it turns out that the twist θ_1 is continuous at the cracked section, that is

$$\Delta\theta_1(\alpha l) = \theta_{1+}(\alpha l) - \theta_{1-}(\alpha l) = 0. \tag{30}$$

Moreover, notice that $\bar{m}_{3.cr} \to \pi$ (Prandtl's solution) as $\gamma_2 \to +\infty$.

The case $\beta = \frac{\pi}{2}$ (see Figure 2a) is discussed next. The appropriate energy functional is the functional V_0 defined in Eq. (22). Proceeding as above leads to a constitutive law for the cracked section of the form

$$M_{1-}(\alpha l) = M_{1+}(\alpha l) = k_1\Delta\theta_1(\alpha l) \tag{31}$$

$$M_{2-}(\alpha l) + \overline{M}_3\theta_{1-}(\alpha l) = M_{2+}(\alpha l) + \overline{M}_3\theta_{1+}(\alpha l) = k_2\Delta\theta_2(\alpha l). \tag{32}$$

Moreover, it can be shown that the out-of-plane rotation θ_2 is continuous at the cracked section, and this renders the buckling moment insensitive to the crack stiffness k_2. In a dimensionless format, the characteristic equation for the buckling moments reads

$$\gamma_1\sin(\bar{m}_3) + \bar{m}_3\cos(\alpha\bar{m}_3)\cos[(1-\alpha)\bar{m}_3] = 0, \text{ where } \gamma_1 = \frac{k_1 l}{GJ}. \tag{33}$$

In the particular case of a crack located at mid-span $\left(\alpha = \frac{1}{2}\right)$, the lateral-torsional buckling moments are insensitive to k_1 as well, and one merely has

$$\alpha = \frac{1}{2} \quad \Rightarrow \quad \bar{m}_{3.cr} = \pi. \tag{34}$$

This insensitivity to the stiffness parameters when the crack is located at mid-span was already noticed by Carloni *et al.* [5] in a numerical study concerning a similar lateral-torsional buckling problem (albeit including the warping term). Clearly, the model considered by these authors falls within the present class of cracked-section constitutive behaviour. The adoption of the two particular types of quasi-tangential moment discussed above (corresponding to $\beta = 0$ and $\beta = \frac{\pi}{2}$ in Figure 2a) does not lead to lateral-torsional buckling moments depending on both the torsional and flexural stiffnesses of the cracked section. However, such a dependence can be achieved by considering a semi-tangential moment (see Figure 2b).

CONCLUSIONS

The flexural buckling of a plane Euler-Bernoulli column with a single edge crack can be investigated by modelling it as two segments connected by a rotational spring located at the cracked cross-section. However, the automatic extension of such a model to the lateral-torsional buckling analysis of a cracked strip beam, with the adoption of the constitutive law (2) for the connecting spring, leads to a paradoxical conclusion: for a beam with standard support conditions and subjected to uniform bending, the critical moment is not affected by the presence of the crack, regardless of its location.

In this paper, it was shown that the above paradox is due to the non-conservative nature of the model adopted. It was argued that the bending moment at the cracked section must be regarded as consisting of a pair of opposite conservative moments applied at the connected end of each segment, which performs work in the relative rotation occurring at the connection when the beam goes from a fundamental equilibrium state to an adjacent, buckled one. Finally, the results obtained in this paper should be equally relevant for the lateral-torsional buckling behaviour of strip beams with semi-rigid connections. In other words, these results do not depend on the physical phenomenon/behaviour that is behind the use of the rotational spring model.

REFERENCES

[1] Okamura H., Liu H. W., Chorn-Shin C. and Liebowitz H., "A cracked column under compression", *Eng. Fract. Mech.*, **1**(3), 547-564, 1969.

[2] Challamel N. and Xiang Y., "On the influence of the unilateral damage behavior in the stability of cracked beam/columns", *Eng. Fract. Mech.*, **77**(9), 1467-1478, 2010.

[3] Wang K., Inman D. J. and Farrar C. R., "Modeling and analysis of a cracked composite cantilever beam vibrating in coupled bending and torsion", *J. Sound Vibr.*, **284**(1-2), 23-49, 2005.

[4] Carloni C., Gentilini C. and Nobile L., "Buckling of thin-walled cracked columns", *Key Engineering Materials*, **324-325**, 1127-1130, 2006.

[5] Karaagac C., Ozturk H. and Sabuncu M., "Free vibration and lateral buckling of a cantilever slender beam with an edge crack: Experimental and numerical results", *J. Sound Vibr.*, **326**(1-2), 235-250, 2009.

[6] Krenk S. and Damkilde L., "Warping of joints in I-beam assemblages", *J. Eng. Mech.—ASCE*, **117**(11), 2457-2474, 1991.

[7] Basaglia C., Camotim D. and Silvestre N., "GBT-based local, distortional and global buckling analysis of thin-walled steel frames", *Thin-Walled Struct.*, **47**(11), 1246-1264, 2009.

[8] Grober M. K., Ein Beispiel der Anwendung der Kirchhoffschen Stabgleichungen, *Physik. Zeitschr.*, **15**, 460-462, 1914.

[9] Izzuddin B. A., "Conceptual issues in geometrically nonlinear analysis of 3D framed structures", *Comput. Meth. Appl. Mech. Eng.*, **191**(8-10), 1029-1053, 2001.

ON THE MECHANICS OF ANGLE COLUMN INSTABILITY

P. B. Dinis, * D. Camotim and N. Silvestre

Department of Civil Engineering and Architecture, ICIST, Instituto Superior Técnico,
Technical University of Lisbon, Av. Rovisco Pais, 1049-001 Lisboa, Portugal
* Email: dcamotim@civil. ist. utl. pt

KEYWORDS

Thin-walled columns, equal-leg angles, buckling and post-buckling mechanics, shell finite element analysis, generalised beam theory (GBT).

ABSTRACT

This paper reports the results of a numerical investigation aimed at providing a fresh insight on the mechanics underlying the buckling and (mostly) post-buckling behaviour of short-to-intermediate equal-leg thin-walled angle steel columns exhibiting fixed and pinned (but with the warping prevented) end supports. Although most of the numerical results presented and discussed were obtained through ABAQUS shell finite element analyses, the paper also includes some GBT-based critical stresses and buckling mode shapes, whose interpretation helps to clarify the distinction between local and global buckling. The shell finite element results displayed consist of (i) elastic post-buckling equilibrium paths and (ii) curves and diagrams providing the evolution, along a given path, of the column deformed configurations and normal stress distributions.

INTRODUCTION

Due to the lack of primary warping resistance, members displaying cross-sections with all their wall mid-lines intersecting at a single point (*e. g.*, angle, T-section or cruciform members) possess a minute torsional stiffness and, therefore, are highly susceptible to buckling phenomena involving torsion. Moreover, in the case of equal-leg angle or cruciform columns, it is very hard to distinguish between torsion and local deformations, due to the fact that their cross-sections consist of two or four identical outstand plate elements (walls). Since these instability phenomena may be associated with clearly different post-critical behaviours (strength reserves), it is fair to say that such distinction may have far-reaching implications on the definition of rational structural models capable of providing accurate ultimate strength estimates for such columns.

The post-buckling behaviour and strength of angle columns has attracted the attention of several researchers in the past (*e. g.*, [1-4]). Recently, numerical investigations carried out by the authors and involving GBT (Generalised Beam Theory) buckling and/or shell finite element post-buckling analyses, shed new light on (i) how to characterise and/or distinguish local and global buckling in angle, T-section and cruciform columns [5], and also on (ii) the post-buckling behaviour and strength of simply supported (locally/globally pinned end sections with free warping) and fixed short-to-intermediate equal-leg angle and cruciform columns [6-8]. The results obtained unveiled some surprising behavioural features concerning fixed short-to-intermediate equal-leg angle columns, namely the fact that, although their critical stresses/loads practically do not vary with the length (*i. e.*, they lie in an almost horizontal "plateau" of the P_{cr} vs. L curve), they exhibit quite different elastic post-buckling behaviours. They are associated with a wide range of post-critical strength reserve levels, varying from quite high (smaller lengths) to rather low (larger lengths)- *i. e.*, akin to post-buckling behaviours ranging from "local" to "global". Moreover, it was also found that the amount of corner flexural displacements plays a key role in separating the various post-buckling behaviours.

This work aims at presenting, mechanically interpreting and discussing a set of numerical results concerning the elastic buckling and post-buckling behaviours of thin-walled steel ($E = 210$ GPa and $\nu = 0.3$) equal-leg angle columns with the end support conditions usually adopted in experimental investigations: fixed (F columns) or pinned (but with warping prevented—P columns) end sections. The equal-leg angle columns analysed (i) exhibit cross-section dimensions 70×70×1.2 mm (the effect of rounded corners is disregarded), (ii) have short-to-intermediate lengths and (iii) contain critical-mode geometrical imperfections with very small amplitudes (10% of the wall thickness t). Almost all the numerical results were obtained through ABAQUS shell finite element analyses, but some buckling results determined by means of GBT analyses (performed with the GBTUL code) are also displayed. They provide invaluable insight on the characteristics of the column critical buckling mode shapes and help define the column length ranges to be considered. The investigation comprises (i) buckling loads and mode shapes, (ii) post-buckling equilibrium paths and curves and/or figures showing the evolution, along those paths, of the column deformed configurations and stress distributions. Insightful mechanical explanations on the similarities and (considerable) differences between the column post-buckling behaviours associated with the support conditions dealt with are also provided.

BUCKLING BEHAVIOUR

The P_{cr} *vs.* L buckling curves depicted in Figure 1(a) concern fixed (F curve) and pin-ended (P curve) columns, and were obtained from GBT analyses including 4 global (**1-4**) and 3 (**5-8**) local deformation modes (see Figure 1(c)). On the other hand, Figure 1(b) displays the GBT modal participation diagrams, providing the contributions of each GBT deformation mode to the column buckling modes. There is clear evidence that the "plateaus" appearing in these buckling curves correspond always to flexural-torsional buckling (see detail in Figure 1(c))—the amount of major axis flexure (mode **2**) is initially minute and becomes increasingly visible as L grows. In order to assess how the column post-buckling behaviour varies as one progresses along the P_{cr} *vs.* L curve plateaus, the following lengths are selected: $L_1 = 53$ cm, $L_2 = 98$ cm, $L_3 = 133$ cm, $L_4 = 182$ cm, $L_5 = 252$ cm, $L_6 = 364$ cm, $L_7 = 420$ cm, $L_8 = 532$ cm, $L_9 = 700$ cm, $L_{10} = 890$ cm—pin-ended columns $P_1 - P_7$ ($23.2 \leqslant \sigma_{cr} \leqslant 27.5$ MPa) and fixed-ended columns $F_1 - F_{10}$ ($21.1 \leqslant \sigma_{cr} \leqslant 27.5$ MPa)—compared to the short-to-intermediate F columns, the P columns only differ in the smaller length range corresponding to the plateau, due to the drop of the minor-axis flexural buckling loads: the transition from flexural-torsional to flexural buckling occurs for $L = 420$ cm (P columns) and $L = 890$ cm (F columns).

POST-BUCKLING BEHAVIOUR

ABAQUS shell finite element analyses were employed to investigate the elastic post-buckling behaviour of

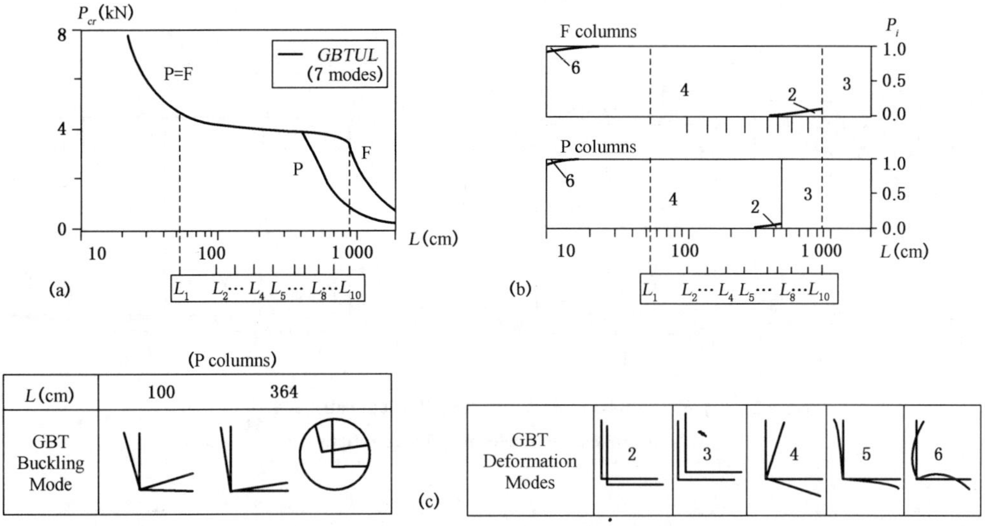

Figure 1 **(a) P_{cr} vs. L curves, (b) GBT modal participation diagrams (F and P columns) and (c) in-plane shapes of 2 buckled mid-span cross-sections (P columns) and the first 5 GBT deformations modes**

angle columns (ⅰ) exhibiting the lengths indicated before and (ⅱ) containing critical-mode initial imperfections with amplitude equal to 10% t (flexural-torsional modes with mid-span torsional rotations $\beta_0 = 0.098$) [6, 8].

Fixed-Ended Columns

Figure 2(a) shows the upper parts of the 10 F column post-buckling equilibrium paths P/P_{cr} *vs*. β, where β is the mid-span web chord rigid-body rotation. To clarify issues raised by the observation of the above curves, Figure 2(b) displays additional post-buckling results, concerning the F_3 and F_9 columns: besides the P/P_{cr} *vs*. β equilibrium paths of Figure 2(a), it shows the evolution, along those paths, of the mid-span (F_3 and F_9 columns) and quarter-span (F_9 column) cross-section deformed configurations. Finally, Figure 3 shows the normalised mid-line longitudinal normal stresses (σ/σ_{cr}) acting on the F_3 and F_9 column mid-span cross-sections at three applied load levels. The observation of these post-buckling results prompts the following comments:

(ⅰ) The equilibrium paths P/P_{cr} *vs*. β become progressively more flexible as L increases and, moreover, the F_1-F_{10} angle column exhibit qualitatively distinct post-buckling behaviours. While those concerning the shorter (F_1-F_7) columns are clearly stable, the ones associated with the longer (F_8-F_{10}) columns exhibit well defined limit points, either ($ⅰ_1$) abrupt and followed by significant torsional rotation reversals (F_8-F_9), or ($ⅰ_2$) smooth and without torsional rotation reversal (F_{10}). In addition, visible corner displacements occur in the longer columns (see Figure 2—F_9 column).

(ⅱ) In the F_8-F_9 columns, the flexural-torsional deformed configuration switches abruptly from a single half-wave to three half-waves soon after the peak load is reached—these deformed configuration switch and peak load occur for gradually smaller β values as the column length grows ($L_8 \rightarrow L_9$). The F_{10} column, corresponding to the transition between major-axis flexural-torsional and minor-axis flexural buckling, has a smooth P/P_{cr} *vs*. β equilibrium path and exhibits a premature well defined limit point.

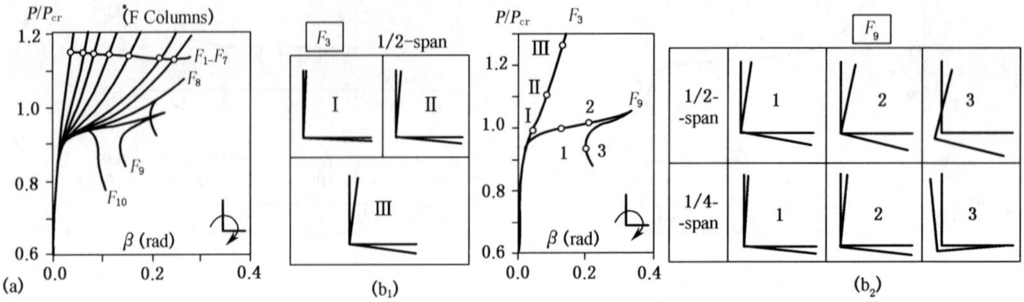

Figure 2 (a) F_1–F_7 column P/P_{cr} *vs*. β equilibrium paths and (b) F_3 and F_9 column cross-section deformed configurations

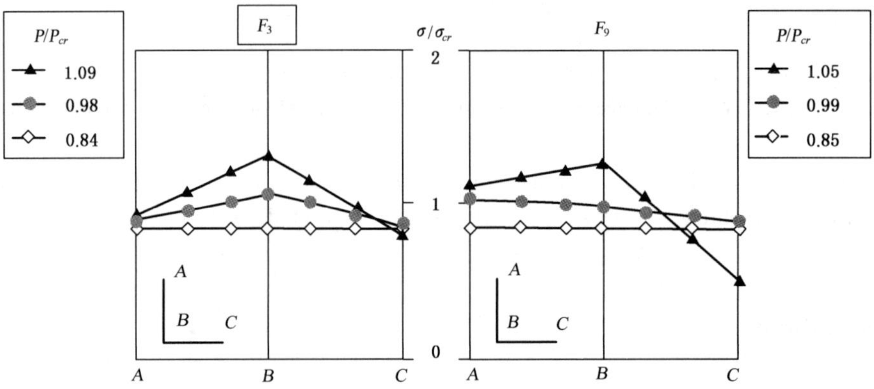

Figure 3 Evolution of the F_3 and F_9 column mid-span normal stress distributions (three load levels)

(ⅲ) The configuration switch is clearly visible in the third F_9 column mid-span and quarter-span cross-section deformed configurations shown in Figure 2(b) (they rotate in opposite senses), which correspond to the post-peak equilibrium state 3. A sizeable amount of (predominantly minor axis) flexure occurs, which is associated with tensile stresses in the cross-section corner regions. Note that the flexural displacements ($ⅲ_1$) are barely visible (but not null) in the remaining F_9 column deformed configurations, as well as in all their F_3 column counterparts, and ($ⅲ_2$) are as relevant as the torsional rotations along the equilibrium path descending branch, which is quite steep and involves significant torsional rotation reversals.

(ⅳ)The mid-line normal stresses of the F columns remain practically uniform up until $P/P_{cr} \approx 0.8$. As the applied load increases further, the stress distribution becomes progressively more non-uniform: the stresses ($ⅳ_1$) are practically linear in both legs and ($ⅳ_2$) "shift" from the leg edges towards the corner, but their distributions differ for the shorter and longer columns. Indeed, they are ($ⅳ_1$) practically symmetric for the sorter column (such as F_3) and ($ⅳ_2$) clearly asymmetric for the longer columns (such as F_9)—Figure 3 illustrates these statements. Moreover, it is worth mentioning that neither of the stress distributions is in agreement with the widespread belief (*e.g.*, [4]) that buckled short-to-intermediate equal-leg angle columns exhibit parabolic normal stress distribution with the higher value at the corner—in other words, that each leg behaves like a pinned-free long plate, as is the case in cruciform columns (*e.*

g., [7]).

The discrepancy between the obtained and expected (widely accepted) equal-leg angle column post-buckling behaviours mentioned before stems from the occurrence of corner (flexural) displacements, particularly in the longer (F_8-F_{10}) columns. In order to confirm and assess the relevance of the flexural displacements, all fixed angle columns were analysed with their internal longitudinal edges *fully restrained* (no corner displacements—FR_1-FR_{10} columns). For comparison purposes, isolated plates with (i) fixed transverse edges, one longitudinal free edge and the other pinned, and (ii) the dimensions of an angle leg (70×1.2 mm and L_1-L_{10} lengths—FP_1-FP_{10} plates) were also analysed. Figure 4(a) displays the upper parts of the P/P_{cr} *vs*. β equilibrium paths of the FP plates and FR columns with lengths L_3 and L_9, while Figure 4(b)

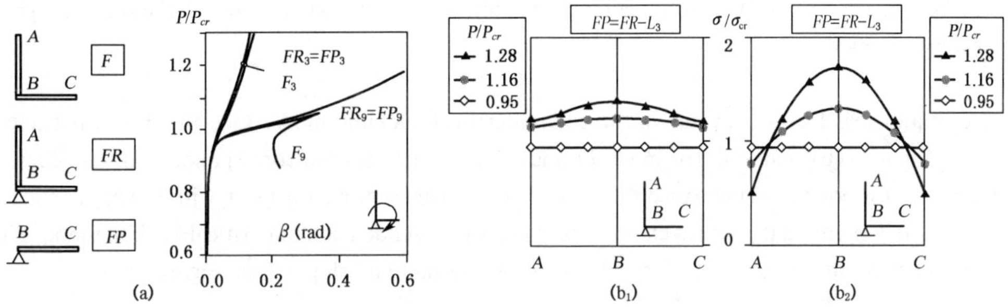

Figure 4 (a) P/P_{cr} *vs*. β equilibrium paths and (b) normal stress distribution evolution at (b_1) the mid and (b_2) quarter-span cross-sections of the *FP* plates, *FR* and *F* columns with lengths L_3 and L_9

shows the stress distribution evolution, at the mid-span and quarter-span cross-sections, of the FP_3 plate and FR_3 column—qualitatively similar results were obtained for all the other lengths. The observation of this new set of results leads to the following remarks:

(i) The FP plates and FR columns share exactly the same (i_1) critical buckling stresses (up to 2.4% higher than those associated with the corresponding F columns) and (i_2) post-buckling behaviours—see the coincident equilibrium paths and normal stress distribution evolutions in Figures 4(a)-(b).

(ii) Restraining the corner displacements, very meaningful in the longer F columns (*e.g.*, see the F_9 column mid-span cross-section deformed configuration in Figure 2(b)), visibly affects the column post-buckling behaviours, particularly for the larger lengths—they are now clearly stable and exhibit no limit points or torsional rotation reversals (*i.e.*, the FR columns "mimic" the corresponding FP plates).

(iii) The corner displacement restraint has also strong impact on the column stress distributions, which now closely resemble those exhibited by pinned-free plates or cruciform columns [7-9]. The mid and quarter-span cross-section normal stress distribution evolutions (see Figure 4(b)) exhibit the well known parabolic shapes.

(iv) Concerning the evolutions of the normal stress distribution along the corner longitudinal edge, discussed in [8], both exhibit a three half-wave pattern with (iv_1) the lower values occurring at the end supports and mid-span, and (iv_2) the higher values acting at the quarter-span and three quarter-span cross-sections. The mechanical grounds for the appearance and development of this three half-wave stress pattern are found in Stowell's work [9]. It stems from the axial extensions caused by the variation of the torsional rotations—their values follow the torsional rotation slope, which explains the characteristics described above.

In order to acquire deeper insight on the angle column post-buckling behaviour, one analyses next the evolution of the corner displacements in the F_3, F_6 and F_9 angle columns. Figures 5(a)-(b) show the upper parts of the corresponding post-buckling equilibrium paths P/P_{cr} *vs.* d_M/t and P/P_{cr} *vs.* d_m/t, where d_M and d_m are the mid-span corner displacements due to major and minor-axis flexure, respectively. As for Figures 5(c)-(d), they provide the longitudinal profiles of those displacements at four equilibrium states located along the corresponding equilibrium paths and associated with increasing P/P_{cr} values. Note that (i) the horizontal coordinate is normalised with respect to the column length (x_3/L) and (ii) the d_M/t and d_m/t scales are quite different for each column (*e.g.*, the L_9 column values are 80 times larger than their L_3 column counterparts). The observation of these results prompts the following comments:

(i) Since major-axis flexure appears in the column critical buckling modes (see the modal participation diagrams in Figure 1(b) and look for mode **2**) and, thus, integrates the corresponding initial geometrical imperfections, it is not surprising to see that (i_1) the d_M values gradually grow with the applied load and that (i_2) their longitudinal profiles exhibit the typical fixed-ended column critical buckling mode shape: one inner half-wave and two outer quarter-waves ensuring null end slopes—see Figure 5(c).

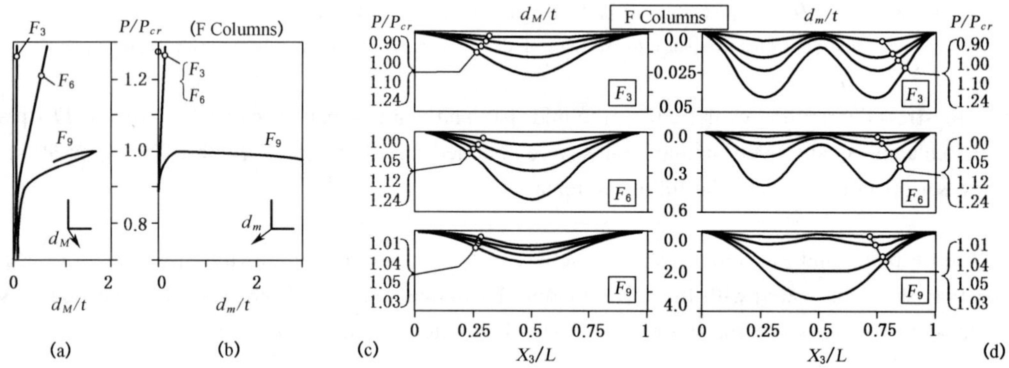

Figure 5 F_3, F_6, F_9 **column (a)** P/P_{cr} *vs.* d_M/t **and (b)** P/P_{cr} *vs.* d_m/t **plots, and (c)** d_M/t **and (d)** d_m/t **profiles**

(ii) For the short (F_3) and intermediate (F_6) columns, the P/P_{cr} *vs.* d_M/t equilibrium paths evolve in a monotonic fashion and the d_M values remain always fairly small, even if they clearly increase with L. This feature is not shared by the long (F_9) column equilibrium path, which evolves monotonically up to $P/P_{cr} \approx 1.05$ and then has a sudden "d_M reversal", coinciding with the equally abrupt growth of minor-axis flexural displacements (d_m).

(iii) As loading progresses, one also observes the (surprising) emergence of minor-axis flexure, which does not participate in the column critical buckling modes—it is associated with pure flexural buckling. The ratios between the column non-critical (flexural) and critical (flexural-torsional) buckling loads are equal to 27.1, 5.5 and 1.6 for the F_3, F_6 and F_9 columns.

(iv) In the (shorter) F_3 and F_6 columns, the d_m values are always small (but increase with L) and their longitudinal profiles exhibit three inner half-waves and two outer quarter-waves (null end slopes)—see Figure 5(c). Note also that all d_m values are *positive* (towards the cross-section convex side) and their maximum values occur around the quarter-span and three quarter-span cross-sections (the mid-span values are quite small).

(ⅴ) For the (longer) F_9 column, the d_m values start growing below $P/P_{cr} = 1.00$ and the growth rate increases dramatically after reaching the peak load ($P/P_{cr} \approx 1.05$). This growth rate increase coincides with the "d_M reversal" mentioned in item (ⅱ). The d_m values are again all *positive* and their profiles gradually switch from ($ⅴ_1$) three inner half-waves ($P/P_{cr} \leqslant 1.04$) to ($ⅴ_2$) a single inner half-wave ($P/P_{cr} > 1.04$, including the post-peak behaviour). Note that the F_9 column d_m values are orders of magnitude larger than their F_3 and F_6 column counterparts.

(ⅵ) The mechanical reasoning behind the emergence and development of the minor-axis displacements d_m is the same invoked earlier to explain the three half-wave stress pattern in the restrained (FR) column, due to Stowell [9]. These displacements, caused by the non-linear cross-section mid-line longitudinal stress distributions due to the torsional rotation variation, produce effective centroid shifts (towards the cross-section corner) following that same three half-wave longitudinal pattern. These effective centroid shifts are responsible for the (positive) d_m values that have a meaningful impact on the angle column response.

(ⅶ) In the (longer) F_7-F_9 columns, the impact mentioned in the previous item is particularly strong and eventually brings about the minor-axis flexural *buckling mode* (recall that the flexural-torsional and flexural buckling loads are not too far apart in these columns), leading to a *mode interaction* phenomenon involving the two "close enough" (competing) buckling modes. Figure 5(d) confirms this assertion—indeed, it is possible to observe how the F_9 column d_m pattern switches from three inner half-waves to a single inner half-wave.

(ⅷ) The occurrence of d_M and d_m displacements, together with the ensuing fixed-end bending moments, explains the practically linear longitudinal normal stress distributions shown in Figure 3. Moreover, the symmetry (F_3 column) or asymmetry (F_9 column) of these stress distributions just reflects the dominance of major-axis and minor-axis bending, respectively—note also that the stress distribution diagram varies along the column length (according to the d_M and d_m variations).

Pin-Ended Columns

Next, one assesses how releasing the end section minor-axis flexural rotations affects the post-buckling behaviour of the equal-leg angle columns with short-to-intermediate lengths. Figures 6(a)-(c) show the upper parts ($P/P_{cr} > 0.6$) of the P_1-P_7 column (ⅰ) P/P_{cr} *vs.* β, (ⅱ) P/P_{cr} *vs.* d_M/t and (ⅲ) P/P_{cr} *vs.* d_m/t curves, where β, d_M and d_m retain the previous meanings. As for Figures 7(a)-(b), they display the d_M/t and d_m/t longitudinal profiles at three P_3 and P_6 column equilibrium states (increasing P/P_{cr} values). Finally, Figure 8 depicts the normalised mid-line longitudinal normal stresses (σ/σ_{cr}) acting on the P_3 and P_6 columns. Observing these pin-ended column post-buckling results prompts the following comments:

(ⅰ) Like in the fixed-ended columns, there are two distinct post-buckling behaviours: while ($ⅰ_1$) the

P_1-P_2 columns are clearly stable and exhibit minute flexural displacements, (i$_2$) the P_3-P_7 columns are barely stable, involve much larger d_m values and exhibit limit points, either (i$_1$) abrupt and followed by significant torsional rotation reversals (P_3-P_4 columns), or (i$_2$) smooth and exhibiting no torsional rotation reversal (P_5-P_7 columns). Once more, the amount of corner d_m displacements plays a key role in separating the various behaviours.

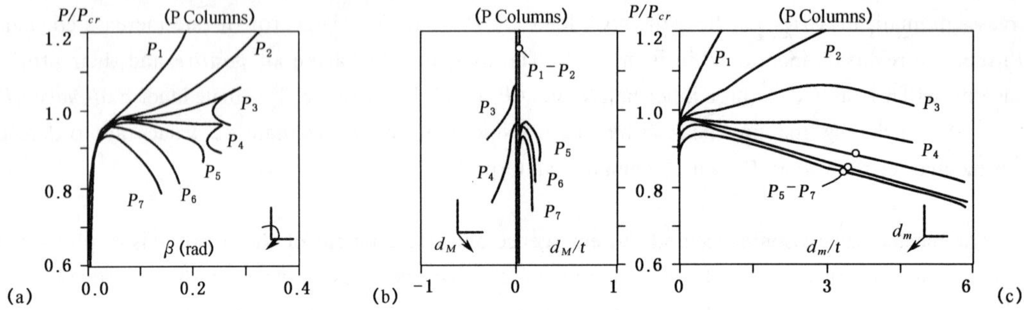

Figure 6 (a) P/P_{cr} vs. β, (b) P/P_{cr} vs. d_M/t and (c) P/P_{cr} vs. d_m/t equilibrium paths (P_1-P_7 columns)

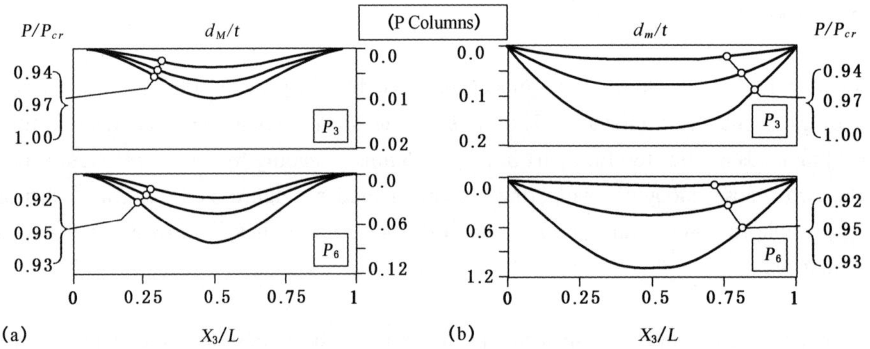

Figure 7 (a) d_M/t and (b) d_m/t longitudinal profiles of the P_3 and P_6 columns

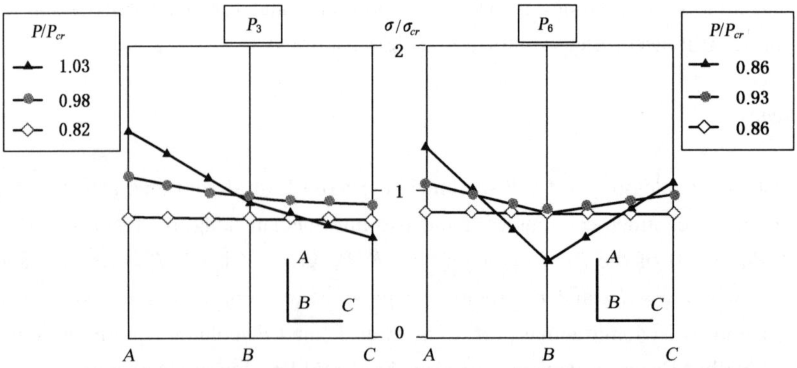

Figure 8 Mid-span cross-section longitudinal normal stress distributions of the P_3 and P_6 columns

(ii) The F and P column equilibrium paths share a few common features, namely (ii$_1$) rotation reversals (P_3-P_4 columns), involving switches from a single to three inner half-waves, (ii$_2$) small d_M values for all applied load levels (they grow with L and exhibit the fixed-ended column critical buckling mode shape) and (ii$_3$) non-parabolic mid-span longitudinal normal stress distributions—they are virtually linear, either practically symmetric (*e. g.*, the P_6 column) or clearly asymmetric (*e. g.*, the P_3 column).

(iii) However, there are significant differences making it possible to separate the pin-ended and fixed-ended column post-buckling behaviours: (iii$_1$) naturally, the d_m/t longitudinal profiles now exhibit a single half-wave (the columns are pin-ended for minor-axis flexure), (iii$_2$) the d_m values are now much higher (about ten times) than their d_M counterparts for all the applied load levels (in the F columns both displacement magnitudes were similar until the emergence of minor-axis flexure), and (iii$_3$) the elastic limit loads are now lower and, in addition, they decrease faster with L—indeed, one has $1.06 \geqslant P_u/P_{cr} \geqslant 0.93$ ($1.16 \geqslant P_u/P_{cr} \geqslant 1.01$ for the F columns).

(iv) The above differences stem from removing the minor-axis flexural rotation restraints. Although the mechanical reasoning for the appearance and development of the three half-wave d_m profile remains valid in the P columns, the absence of the end moments makes it impossible to oppose the minor-axis bending due to the effective centroid shifts caused by the normal stress redistribution (see [8])—this explains the dominance of the "well curved" half-wave longitudinal profile component over its three half-wave counterpart (the switch from a three half-wave profile to a single half-wave one is no longer visible). Such dominance is much more "suffocating" for the longer columns (*e. g.*, the P_6 column), due to the interaction with the minor-axis flexural *buckling mode*, which is brought about by the proximity between the flexural-torsional and flexural buckling loads.

(v) The considerably higher relevance of the minor-axis flexural displacements in the pin-ended columns, when compared with their fixed-ended counterparts, is then responsible for (v$_1$) the more flexible post-buckling behaviours and (v$_2$) the substantially lower (normalised) ultimate strength values.

CONCLUSION

This work reported and discussed the results of a numerical investigation on the elastic buckling and post-buckling behaviours of short-to-intermediate fixed-ended and pin-ended (but with warping prevented) thin-walled equal-leg angle columns. Although a few GBT-based buckling results were presented, in order to (i) clarify the distinction between local and global buckling and (ii) justify the column length selection, the vast majority of the results displayed were obtained through ABAQUS shell finite element analyses. They consisted of (i) elastic post-buckling equilibrium paths, and (ii) curves and diagrams providing the evolution, along a given path, of column deformed configurations and longitudinal normal stresses.

Amongst the various conclusions drawn from behavioural features unveiled during this in-depth study, the following ones deserve to be specially mentioned:

(i) Both the fixed-ended and pin-ended columns exhibit well defined critical stress plateaus associated with buckling modes combining torsion with major-axis flexure—the amount of major-axis flexure grows as the column length increases.

(ii) For each length falling inside the plateaus, both the fixed-ended and pin-ended column post-buckling behaviours are characterised by the simultaneous occurrence of cross-section torsional rotations and transverse translations (corner displacements), stemming from the major and minor-axis flexures. The relevance of the minor-axis flexural displacements has a very strong impact on the column post-buckling response, namely on its post-critical strength reserve and longitudinal normal stress distributions. This impact is more severe in the pin-ended columns that in the fixed-ended ones.

(iii) Due to the relevance of the corner displacements (mostly those due to the minor-axis flexure), the behaviour of both fixed-ended and pin-ended equal-leg angle columns cannot be viewed as the "sum" of two pinned-free long plates. In particular, the cross-section longitudinal normal stress distributions become far from parabolic as post-buckling progresses. If the corner displacements (*i.e.*, the major and minor-axis flexures) are prevented along the column, it is possible to recover the pinned-free long plate post-buckling behaviour.

Some of the behavioural features unveiled in this work are bound to have far-reaching implications in the design of cold-formed steel equal-leg angle columns, namely on the development of rational models able to provide accurate ultimate strength estimates for such columns. At this stage, it appears that different design rules are required to estimate efficiently the ultimate loads of fixed vended and pin-ended columns—this subject has also been recently addressed by the authors [10, 11].

REFERENCES

[1] Kitipornchai S and Chan SL, "Nonlinear finite-element analysis of angle and tee beam-columns", *Journal of Structural Engineering* (ASCE), **113**(4), 721-739, 1987.

[2] Popovic D, Hancock, GJ and Rasmussen, KJR, "Axial compression tests of cold-formed angles", *Journal of Structural Engineering* (ASCE), **125**(5), 515-523, 1999.

[3] Young B, "Tests and design of fixed-ended cold-formed steel plain angle columns", *Journal of Structural Engineering* (ASCE), **130**(12), 1931-1940, 2004.

[4] Rasmussen KJR, "Design of angle columns with locally unstable legs", *Journal of Structural Engineering* (ASCE), **131**(10), 1553-1560, 2005.

[5] Dinis PB, Camotim D and Silvestre N, "On the local and global buckling behaviour of angle, T-section and cruciform thin-walled members", *Thin-Walled Structures*, **48**(10-11), 786-797, 2010.

[6] Dinis P. B., Camotim D. and Silvestre N., "Post-buckling behaviour and strength of angle columns", *Proceedings of International Colloquium on Stability and Ductility of Steel Structures*, Rio de Janeiro, 2010, pp. 1141-1150 (vol. 2).

[7] Dinis PB and Camotim D, "Buckling, post-buckling and strength of equal-leg angle and cruciform columns: similarities and differences", *Proceedings of* 6th *European Conference on Steel and Composite Structures* (Eurosteel 2011—Budapest, 31/8-2/9), L. Dunai *et al*. (eds.), 105-110 (vol. A), 2011.

[8] Dinis PB, Camotim D and Silvestre N, "On the mechanics of thin-walled angle column instability", *submitted*, 2011.

[9] Stowell EZ, *Compressive Strength of Flanges*, National Advisory Committee for Aeronautics Report 1029, 1951.

[10] Silvestre N; Dinis PB, Camotim D, "Design of Cold-Formed Steel Angles: A Novel Approach", *submitted*, 2011.

[11] Dinis PB, Camotim D and Silvestre N, "Behaviour and design of cold-formed steel angles", *Proceedings of* 7th *International Conference on Advances in Steel Structures* (ICASS 2012—Nanjing, 6-8/4), 2012.

FINITE ELEMENT METHOD TO DETERMINE CRITICAL WEIGHT OF FLEXIBLE PIPE CONVEYING FLUID SUBJECTED TO END MOMENTS

* C. Athisakul[1], B. Phungpaingam[2], W. Chatanin[3], S. Chucheepsakul[1]

[1] Department of Civil Engineering, Faculty of Engineering,
King Mongkut's University of Technology Thonburi, Bangkok, Thailand
[2] Department of Civil Engineering, Rajamangala University of Technology Thanyaburi,
Thanyaburi, Pathum-thani, Thailand
[3] Department of Mathematics, Faculty of Science, King Mongkut's University of
Technology Thonburi, Bangkok, Thailand
* Email: Athisakul@gmail.com

KEYWORDS

Finite element method, flexible pipe, critical weights, variational approach, end moments, nonlinear.

ABSTRACT

This paper aims to evaluate the critical weight of flexible pipe subjected to applied end moments at fixed support locations. The pipe is hinged at one end, while the other end is free to slide over a frictionless support. The horizontal distance between the two supports is fixed. The model formulation is developed by the variational approach, and the finite element method is employed to obtain the numerical solutions. The critical weights are evaluated for various values of end moments and the proportional parameter of the end moments.

INTRODUCTION

The failure of flexible pipes used in offshore engineering operations causes severe environmental pollution. In order to ensure the strength and stability of flexible pipe, accurate determination of the wall thickness of pipe is necessary. There has been a considerable amount of research works dealing with the stability of pipes conveying fluids such as Chen [1], and Païdoussis and Issid [2]. Thompson and Lunn [3] presented the static elastic theory for nonlinear analysis of pipe conveying fluid. They found that the internal flow velocity can induce the buckling-type or fluttering-type instabilities. The divergence instability of a variable-arc-length elastica pipe due to steady flow velocity of internal fluid was presented by Chucheepsakul and Monprapussorn [4]. They used the elliptic integral method to obtain analytical solutions. However, this work is focused only on the effect of internal flow velocity by neglecting the weight of the pipe and the internal fluid. A more recent investigation on nonlinear buckling of marine elastic pipes transporting fluid was presented by Chucheepsakul and Monprapussorn [5]. They concluded that the nonlinear buckling of the marine elastica pipe can occur due to insufficient stiffness and overloading. The critical weights of pipes for a particular example were also presented in their works. Athisakul and Chucheepsakul [6] used the finite element method to evaluate the critical loads of the variable-arc-length beam for various inclinations. Their results can be applied as benchmarks for the analysis of free hanging marine pipes/risers. In order to reduce the stress at both the touchdown point and at the platform connection, a subsea buoy is added to produce the S, Wave, and Camel configurations. According to the subsea buoy system, the additional bending moment may occur at the ends of the pipes. Consequently, flexible pipes have to resist the double curvature bending. Chucheepsakul et al. [7] published a paper dealing with the double curvature bending of variable-arc-length elasticas under two applied moments. The elliptic integral method was utilized to obtain the closed-form solutions. However, the weight of structure is neglected.

This paper continues in this line of investigation by considering the combination of its uniform self-weight and two applied moments at both ends in the same direction. Since the elliptic integrals method cannot be applied to this kind of problems, the finite element method (FEM) is an alternative method to determine the numerical solutions. The variational approach is employed to develop the model formulation. The first variation of the total potential energy is derived to establish the system of nonlinear finite element equations. The numerical solutions are obtained by an iterative procedure. The second variation of the total potential energy is evaluated to form the tangent stiffness matrix of the pipes. The critical uniform self-weights of the pipe are the maximum value, which the determinant of tangent stiffness matrix is equal to zero. In practice, the critical uniform self-weight can be defined from changing the sign of tangent stiffness matrix from positive to negative.

VARIATIONAL FORMULATION

Consider a flexible pipe of uniform flexural rigidity EI as shown in Figure 1(a). The pipe is supported by a pin support at end A and by a frictionless support at end B. The constant distance between ends A and B is L. The pipe is subjected to a clockwise moment $M_A = (1-\beta)M_0$ at the end A and a clockwise moment $M_B = \beta M_0$ at the end B. The scalar parameter β represents the proportion of the moment at end B to the total moment M_0. The uniform self-weight of the pipe per unit arc-length is equal to w. The internal fluid of density ρ_i is transported from end A to end B with a uniform and steady flow speed U. The internal area of pipe is represented by A_i. Figure 1(b) shows the deformed configuration of the flexible pipe. The total arc-length S_t is an unknown parameter. The overhang length l is small compared with the total arc-length S_t. Therefore, the loads in the portion of overhang length l can be neglected.

Figure 1 (a) Undeformed configuration of flexible pipe; (b) Deformed configurations of flexible pipe.

According to variational principle [8, 9], the total virtual work of the flexible pipe can be expressed as

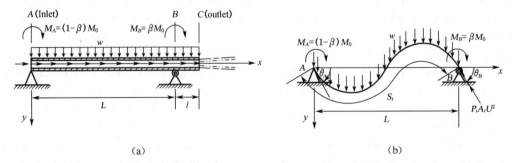

$$\delta\pi = \int_0^L \left\{ \frac{EIy''}{(1+y'^2)^{5/2}}\delta y'' - \frac{2EIy''^2 y'}{(1+y'^2)^{7/2}}\delta y' + \frac{Ny'}{\sqrt{1+y'^2}}\delta y' - w\sqrt{1+y'^2}\delta y + \rho_i A_i \kappa U^2 \delta y \right\} dx$$
$$- \frac{M_A}{(1+y'^2)}\delta y' \bigg|^{x=0} - \frac{M_B}{(1+y'^2)}\delta y' \bigg|^{x=L} \tag{1}$$

The first two terms represent the virtual strain energy due to bending. The third term represents the virtual strain energy due to axial deformation, where N is the axial force in the pipe section. The fourth term is the virtual work done by pipe's weight. The fifth term is the virtual work done by the internal flow inside the pipe. The last two terms are the virtual work done by the applied end moments.

According to differential geometry of a plane curve, one obtains

$$\frac{dx}{ds} = \cos\theta,\ \frac{dy}{ds} = \sin\theta,\ ds = \sqrt{1+y'^2}dx,\ \kappa = \frac{\theta'}{s'} = \frac{y''}{s'^3} \tag{2 a-d}$$

The prime represents the derivatives with respect to x. Since the beam material is linear elastic, the moment curvature relation becomes

$$M = -EI\kappa = -EI\frac{d\theta}{ds} \tag{3}$$

FINITE ELEMENT METHOD

The span length L is divided into n equally spaced regions or elements. Each of these elements has a length l. The displacement of the beam segment is approximated by

$$y(x) = [N]\{q\} \tag{4}$$

where $[N]$ is the row of fifth-order polynomials shape functions, and $\{q\}$ is the vector containing the values of y and its first and second derivatives at both ends of the element. Consequently, the system of element equations can be expressed as follows

$$\left\{\frac{\partial\pi}{\partial q_i}\right\} = \int_0^L \left\{ [N'']^T \frac{EIy''}{(1+y'^2)^{5/2}} + [N']^T\left[\frac{Ny'}{\sqrt{1+y'^2}} - \frac{2EIy''^2 y'}{(1+y'^2)^{7/2}}\right] - [N]^T[w\sqrt{1+y'^2} - \rho_i A_i \kappa U^2] \right\} dx \tag{5}$$

The contribution from the applied moments is

$$\left\{\frac{\partial\pi}{\partial q}\right\}_{x=0,\ x=L} = -[N']^T \frac{M_A}{(1+y'^2)}\bigg|^{x=0} - [N']^T \frac{M_B}{(1+y'^2)}\bigg|^{x=L} \tag{6}$$

For equilibrium, the total virtual of the system is zero ($\delta\pi = 0$). Therefore, the nonlinear global equilibrium equation $\left\{\frac{\partial\pi}{\partial Q}\right\} = \{0\}$ can be obtained by assembling the element equations. The iterative procedure is used to obtain the numerical solutions of the global degree of freedom Q.

In order to find the critical configuration of the pipe, the second variation of the total potential energy is derived into the matrix form [6] as

$$\delta^2\pi = \{q\}^T[K_T]\{q\} \tag{7}$$

The critical weight is evaluated by optimizing for an increment of load step by step until the determinant of tangent stiffness matrix $[K_T]$ is equal to zero or it changes sign from positive to negative [6]. The critical weight is the maximum value of pipe weight, which the equilibrium of pipe is still satisfied.

NUMERICAL RESULTS

The following non-dimensional parameters are introduced for the sake of generality.

$$s^* = s/s_t,\ x = x/L,\ y = y/L,\ s = s/L, \qquad (8\ \text{a-d})$$

$$M = ML/EI,\ \hat{Q} = QL^2/EI,\ N = NL^2/EI,\ \hat{w} = wL^3/EI,\ U = UL\ \sqrt{\rho_i A_i/EI} \qquad (8\ \text{e-i})$$

The parameters ρ_i and A_i represent density of internal fluid and internal cross sectional area of the pipe, respectively.

In order to validate the numerical results obtained from this study, some numerical solutions for double curvature bending of the variable-arc-length elastica are evaluated as shown in Table 1. It is clearly found that the numerical solutions obtained from this study are in very good agreement with the exact solutions computed from the elliptic integral method [7].

The particular problems of flexible pipes with the total end moment $\hat{M}$ from -2 to 2, and the internal flow speed $\hat{U}$ of 0, 0.5, and 1.5 are considered. Table 2 shows the critical weights of the pipe for various values of the proportional parameter β of the end moments. As shown in Table 2, the critical weights depend on the value of total end moments, the direction of end moments, and the internal flow speed. For $\hat{M}=0$ and $\hat{U}=0$, the critical weight of 8.252 7 is identical to the value suggested by Athisakul and Chucheepsakul [6]. In case where $\hat{M}$ and β are specified, the critical weight of the pipe decreases as the internal flow speed increases. The critical configurations of the pipe for $\hat{U}=1$ are illustrated in Figure 2. Considering the case of negative end moments, the critical weight decreases as the parameters β increases (see Table 2). According to the positive sign convention of the applied end moments shown in Figure 1, the negative end moment at end A resists the deflection induced by the pipe weight while the negative end moment at end B enlarges the deflection of the pipe. Therefore, the deflection of the pipe at a critical state increases as the parameter β of the negative end moment increases as shown in Figures 2(a) and 2 (b). On the contrary, the positive end moment at end A enlarges the deflection of pipe while the positive end moment at end B resists the weight of pipe. Consequently, the deflection of the pipe at critical state decreases as the parameter β of the positive end moment increases as shown in Figures 2(c) and 2(d). For a given value of positive end moment, the critical weight of pipe increases as the parameter β increases (see Table 2). Table 2 also shows that the equilibrium of the pipe with the positive value of pipe weight may not exist for the case of large end moments.

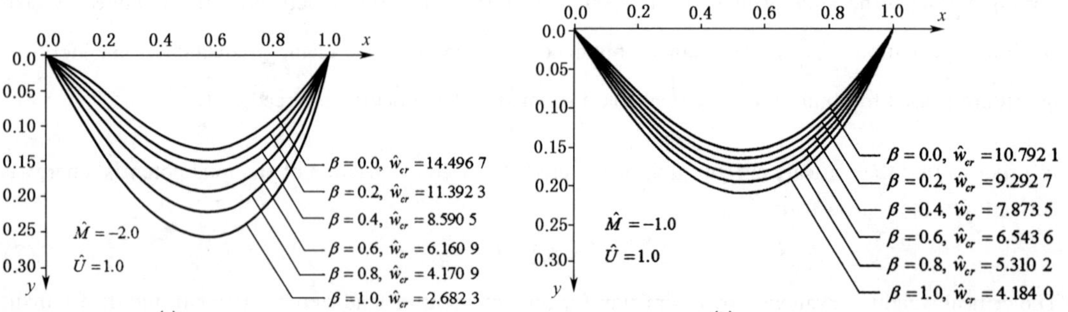

(a) (b)

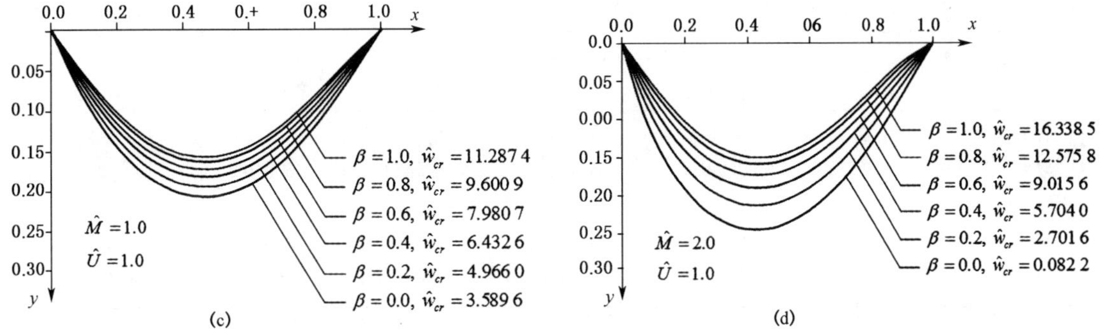

Figure 2 Critical configurations of the pipe for $\hat{U} = 1$.

TABLE 1 VALIDATION OF NUMERICAL RESULTS FOR THE PROBLEM OF ELASTICAS WITH DOUBLE CURVATURE BENDING $\hat{M} = 3$

	Rotation at end A θ_A (rad)		Rotation at end B θ_B (rad)		Total arc-length $\hat{s}_t$	
β	EIM [7]	FEM (this study)	EIM [7]	FEM (this study)	EIM [7]	FEM (this study)
0.2	0.809 178	0.809 161	−0.250 095	−0.250 097	1.057 947	1.057 950
0.4	0.404 192	0.404 200	0.101 105	0.101 101	1.010 185	1.010 183
0.6	0.107 839	0.107 836	0.389 108	0.389 121	1.009 265	1.009 258
0.8	−0.144 274	−0.144 305	0.652 506	0.652 565	1.031 259	1.031 254
1.0	−0.357 892	−0.358 003	0.887 168	0.887 392	1.066 052	1.066 079

EIM = Elliptic integral method
FEM = Finite element method

TABLE 2 CRITICAL WEIGHTS OF THE FLEXIBLE PIPE FOR $\hat{M} = -2, -1, 0, 1, 2$, AND $\hat{U} = 0, 0.5, 1.0$

	Internal Flow Speed $\hat{U} = 0.0$					
	$\beta=0.0$	$\beta=0.2$	$\beta=0.4$	$\beta=0.6$	$\beta=0.8$	$\beta=1.0$
$\hat{M} = -2$	15.397 3	12.374 2	9.662 1	7.320 3	5.413 6	3.995 3
$\hat{M} = -1$	11.791 0	10.332 9	8.955 3	7.666 5	6.475 1	5.388 9
$\hat{M} = 0$			8.252 7			
$\hat{M} = 1$	4.784 6	6.123 6	7.552 8	9.064 4	10.651 1	12.304 0
$\hat{M} = 2$	1.377 4	3.924 2	6.854 1	10.099	13.600 7	17.315 4
	Internal Flow Speed $\hat{U} = 0.5$					
	$\beta=0.0$	$\beta=0.2$	$\beta=0.4$	$\beta=0.6$	$\beta=0.8$	$\beta=1.0$
$\hat{M} = -2$	15.167 8	12.125 8	9.390 9	7.028 8	5.101 5	3.666 3
$\hat{M} = -1$	11.538 5	10.069 2	8.681 6	7.383 1	6.182 0	5.086 3
$\hat{M} = 0$			7.975 7			
$\hat{M} = 1$	4.483 6	5.831 6	7.270 6	8.791 2	10.385 7	12.047 1
$\hat{M} = 2$	1.052 4	3.616 6	6.564 1	9.825 0	13.342 0	17.067 5

continued

Internal Flow Speed $\hat{U}=1.0$						
	$\beta=0.0$	$\beta=0.2$	$\beta=0.4$	$\beta=0.6$	$\beta=0.8$	$\beta=1.0$
$\hat{M}=-2$	14.496 7	11.392 3	8.590 5	6.160 9	4.170 9	2.682 3
$\hat{M}=-1$	10.792 1	9.292 7	7.873 5	6.543 6	5.310 2	4.184 0
$\hat{M}=0$			7.155 4			
$\hat{M}=1$	3.589 6	4.966 0	6.432 6	7.980 7	9.600 9	11.287 4
$\hat{M}=2$	0.082 2	2.701 6	5.704 0	9.015 6	12.575 8	16.338 5

CONCLUSIONS

The finite element method for determining the critical uniform weight of the flexible pipe subjected to two end moments is presented. The end moments are applied at both ends of the pipe in the same direction. The critical weight of the pipe decreases as the internal flow speed increases. As the parameter β increases the critical weight increases when the positive end moments are applied. The results are opposite when the negative end moments are applied, as the parameter β increases the critical weight decreases. It is also found that the equilibrium of the pipe with the positive value of pipe weight may not exist when the absolute value of moment becomes large.

ACKNOWLEDGEMENTS

The authors gratefully acknowledge the financial support by the Thailand Research Fund (TRF) under Contract No. MRG5380034.

REFERENCES

[1] Chen, S.S., "Flow-induced instability of an elastic tube", *ASME* Paper No. 71-Vibr. -39, 1971.

[2] Païdoussis, M.P. and Issid, N.T., "Dynamic stability of pipes conveying fluid", *Journal of Sound and Vibration*, 1974, 33, pp. 267-294.

[3] Thompson, J.M.T. and Lunn, T.S., "Static elastica formulations of a pipe conveying fluid", *Journal of Sound and Vibration*, 1981, 77, pp. 127-132.

[4] Chucheepsakul, S. and Monprapussorn, T., "Divergence instability of variable-arc-length elastica pipes transporting fluid", *Journal of Fluids and Structures*, 2000, 14, pp. 895-916.

[5] Chucheepsakul, S. and Monprapussorn, T., "Nonlinear buckling of marine elastica pipes transporting fluid", *International Journal of Structural Stability and Dynamics*, 2001, 1(3), pp. 333-365.

[6] Athisakul, C., and Chucheepsakul, S., "Effect of inclination on bending of variable-arc-length beams subjected to uniform self-weight", *Engineering Structures*, 2008, 30, pp. 902-908.

[7] Chucheepsakul, S., Wang, C.M., He, X.Q. and Monprapussorn, T., "Double curvature bending of variable-arc-length elasticas", *Journal of Applied Mechanics*, *ASME*, 1999, 66, pp. 87-94.

[8] Chucheepsakul, S., Monprapussorn, T. and Huang T., "Large strain formulations of extensible flexible marine pipes transporting fluid", *Journal of Fluids and Structures*, 2003, 17, pp. 185-224.

[9] Athisakul, C., Monprapussorn, T. and Chucheepsakul, S., "A variational formulation for three-dimensional analysis of extensible marine riser transporting fluid", *Ocean Engineering*, 2011, 38, 609-620.

INSTABILITY OF VARIABLE-ARC-LENGTH ELASTICA SUBJECTED TO END MOMENT

* B. Phungpaingam[1], C. Athisakul[2] and S. Chucheepsakul[2]

[1] Department of Civil Engineering, Rajamangala University of Technology Thanyaburi
Patumthani, Thailand 12110

[2] Department of Civil Engineering, King Mongkut's University of Technology Thonburi
Bangkok, Thailand 10140

* Email: phungpaigram_b@yahoo.com

KEYWORDS

Variable-arc-length elastica, instability, concentrated moment, shooting method.

ABSTRACT

This paper presents the instability of a variable-arc-length elastica where one end is attached on the hinged joint and the other end is placed on the sleeve support. The friction is also introduced at the sleeve end. At the hinged end, the concentrated moment is applied to turn the elastica around the joint. The system of governing differential equations is obtained from equilibrium equations, constitutive equation and nonlinear geometric relations. To extract the behaviour of this problem, the system of differential equations needs to be integrated from one end to the other and satisfied boundary conditions. In this problem, we utilize the Runge-Kutta scheme as an integration tool and the shooting method plays a vital role to compute the results. The results are interpreted by using the load-deflection diagram where the stiffness of the system can be observed. From the results, it can be found that there are two equilibrium points for a given end moment that is less than its critical value. This results from the non-monotonic load-deflection curves. There is a limit load point (buckling load) for each value of frictional coefficient. Beyond the buckling load, the elastica can lose its stability. Moreover, we discovered that, with presence of the friction (i.e., dry friction), the elastica shows more stable behaviour than its counterpart without friction since the buckling load increases when the frictional coefficient increases.

INTRODUCTION

An elastica often loses its stability at a bifurcation point or at a limit load point. The instability resulting at the bifurcation point occurs for an elastica under compression [1, 2]. On the other hand, the instability of an elastica at the limit load point occurs when the elastica is subjected to a moment couple [3, 4]. The elastica problems presented in Refs [3, 4] involve elasticas with a constant arc-length. In the case of the variable-arc-length (VAL) elastica [5, 6, 7], the instability of the VAL elastica under an end moment or a moment gradient is also influenced by the limit load point. It is interesting that, for the VAL elastica subjected to a moment gradient [7], the stability of the elastica can be enhanced by a distribution factor β which proportions the magnitude of the moment at the near end and the remote end. The friction at the support is another effect that influences the instability of the VAL elastica. In this paper, we investigate the instability of the VAL elastica subjected to the end moment by allowing for the effect of dry friction at the sleeve end.

We first develop the governing equation for the aforementioned elastica problem by considering the equilibrium of an elastica segment, the constitutive equation and the nonlinear geometric relations. To obtain the whole behaviour of the elastica, one needs to integrate the governing equation from one end to the other. This constitutes a two-point boundary value problem. The powerful tool for investigation is the shooting method. The method is able to solve the nonlinear problem without any difficulty. From the results, we found several interesting aspects such as stability and instability of the VAL elastica, critical load, effect of friction and equilibrium shapes.

STATEMENT OF THE PROBLEM

In this problem, an elastica of the span length L and the flexural rigidity EI is placed on the hinged support at end A and the sleeve support at end B (see Figure 1a). At the sleeve support, friction is introduced in terms of coefficient of the friction μ.

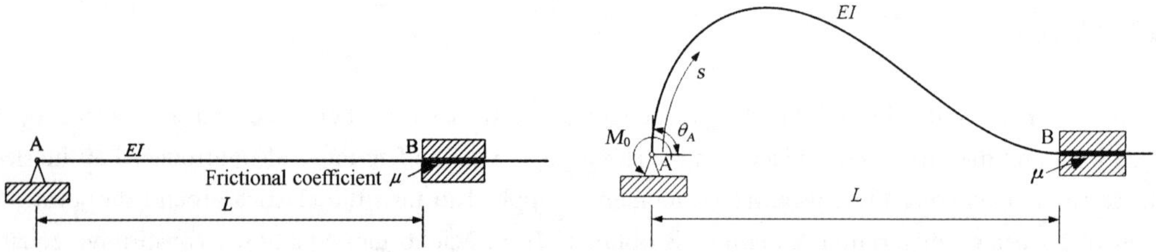

(a) Undeformed shape of the VAL elastica (b) Deformed shape of the VAL elastica

Figure 1 Undeformed and Deformed shapes of VAL elastica: a) undeformed shape; b) deformed shape

Atthe end A, a couple M_0 is applied in a counter clockwise direction that results in the deflections of the elastica (see Figure 1b). As it can be seen in Figure 1b, after deformation, the span length L is not affected by the applied moment but by the total arc-length s_t (the arc-length of point A to B). The increase in arc-length is the key characteristic of the VAL elastica.

PROBLEM FORMULATION

The governing differential equations of the problem can be obtained from equilibrium equations, the constitutive equation and the geometric relations. The free body diagrams of the elastic are shown in Figures 2a and 2b.

From Figure 2, the moment curvature relation may be deduced as

$$\frac{d\theta}{ds}=\frac{V(0)x+H(0)y-M_0}{EI} \tag{1}$$

where $V(0)$ is an unknown vertical force at end A, $H(0)$ the horizontal force at end A and it relates to vertical force by $H(0)=\mu V(0)$.

The nonlinear geometric relations of the inextensible and unshearable elastica can be described in the

following equations

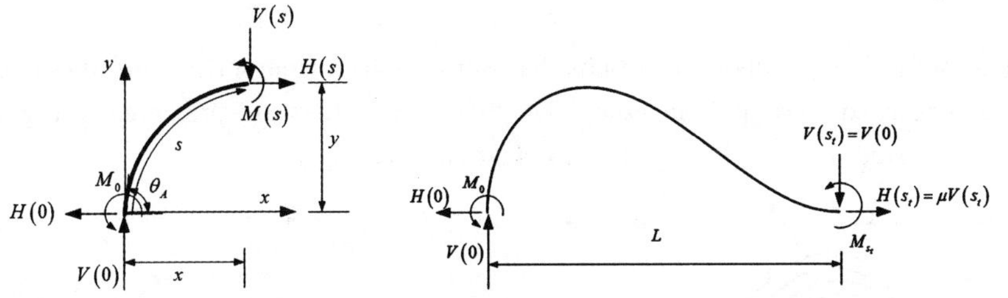

(a) Elastica segment (b) Entire length of the elastica

Figure 2 Equilibrium of elastica: a) elastica segment; b) entire length of the elastica

$$\frac{dx}{ds} = \cos\theta; \ \frac{dy}{ds} = \sin\theta \qquad (2, 3)$$

For convenience in computation, the following non-dimensional terms are introduced

$$\overline{H}_0 = \frac{H(0)L^2}{EI}; \ \overline{V}_0 = \frac{V(0)L^2}{EI}; \ \overline{M}_0 = \frac{M_0 L}{EI};$$

$$\overline{s} = \frac{s}{L}; \ \overline{s}_t = \frac{s_t}{L}; \ \overline{x} = \frac{x}{L}; \ \overline{y} = \frac{y}{L} \qquad (4a\text{-}g)$$

In view of the abovementioned non-dimensional terms, Eqns. (1) to (3) may be re-written in non-dimensional forms as

$$\frac{d\theta}{d\overline{s}} = \overline{V}_0\overline{x} + \mu\overline{V}_0\overline{y} - \overline{M}_0; \ \frac{d\overline{x}}{d\overline{s}} = \cos\theta; \ \frac{d\overline{y}}{d\overline{s}} = \sin\theta \qquad (5\text{-}7)$$

Equations (5) to (7) form the system of differential equations for explaining the behaviour of the elastica in an infinitesimal length $d\overline{s}$. To describe global behaviour of the elastica, the system of differential equations in Eqns (5)-(7) must be integrated so that the boundary conditions of the problem are satisfied. The boundary conditions of the problem are summarized in Table 1.

TABLE 1 BOUNDARY CONDITIONS

variables	$\overline{s}=0$	$\overline{s}=\overline{s}_t$
$\overline{x}$	0	1
$\overline{y}$	0	0
θ	θ_A	0

For each value of frictional coefficient μ, there are four unknown parameters (i.e., $\overline{V}_0$, $\overline{M}_0$, θ_A, $\overline{s}_t$). One of these parameters is set as a control parameter where, with regard to the limit load point, we choose the angle θ_A as the control variable. Thus, there are three unknown parameters to be solved which are $\overline{V}_0$, $\overline{M}_0$ and $\overline{s}_t$. According to the three unknown parameters, the three constraint equations need to be supplied.

$$\overline{x}(\overline{s}_t) - 1 = 0; \ \overline{y}(\overline{s}_t) = 0; \ \theta(\overline{s}_t) = 0 \qquad (8\text{-}10)$$

In order to compute the unknown parameters, the constraint equations (Eqns (8)-(10)) established from end conditions should be minimized by using the Newton-Raphson iterative scheme.

RESULTS AND DISCUSSION

In this section, we will discuss about the results that were obtained from solving differential Eqns (5)-(7) that satisfy the end conditions given in Eqns (8)-(10). The results are presented in a graphical form where we have referred to as the load-deflection diagrams (see Figure 3).

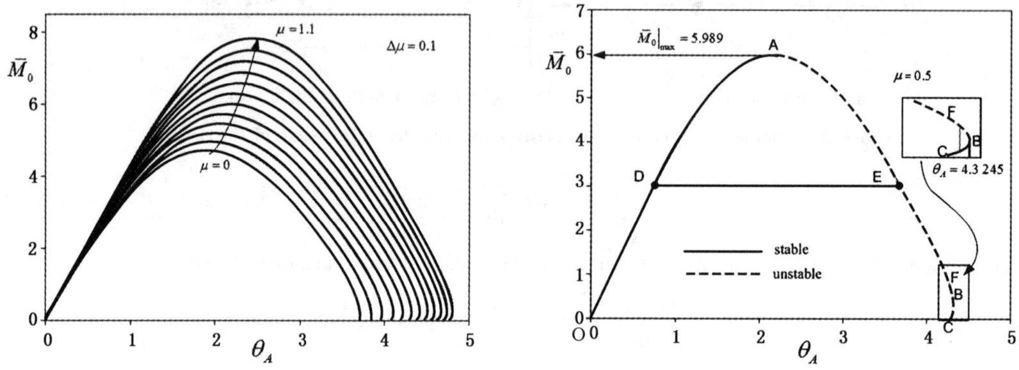

(a) Load-deflection curves (b) Characteristics of a load-deflection curve for $\mu = 0.5$

Figure 3 Load-deflection curves with various values of frictional coefficient μ from 0 to 1.1 : a) load-deflection curves; b) characteristics of a load-deflection curve for $\mu = 0.5$

From the curves in Figure 3. the behaviour of the VAL elastica can be interpreted. We begin by discussing the stability of the elastica with the aid of Figure 3b ($\mu = 0.5$) as an example. From the load-deflection curve in Figure 3b, we can see that the curve begins at point O and it moves on the stable path OA (solid line) in which the slope on OA is positive (i. e., positive stiffness). As the couple $\overline{M}_0$ increases, the deflection of the elastica increases as well. When the couple magnitude reaches point A (i. e. the critical point), the stiffness of the system becomes zero. At this point, the stability of the elastica can be changed from a stable state to an unstable state and vice versa. If we attempt to further increase the couple $\overline{M}_0$, no static equilibrium exists and the system would undergo motion without further moment action. Beyond the point A, the equilibrium path becomes unstable on path AB (broken line) where the stiffness of the system becomes negative. However, the instability of the elastica ceases at point B where it is a turning point and the stability of the system can be switched from unstable to stable. Afterwards, the elastica returns to the stable state by a decrease in both the couple $\overline{M}_0$ and the angle θ_A until the couple $\overline{M}_0 \to 0$. Based on the characteristics of load-deflection curves, there are two possible equilibrium configurations (i. e., points D and E) for a given value of the couple $\overline{M}_0$ ranging between limit load point A and limit point of displacement B. One is a stable shape while the other belongs to an unstable shape. It should be noted that, on the path FBC, there are also two possible equilibrium shapes for a given value of the angle θ_A at end A. One is referred to the stable shape whereas the other is considered as the unstable shape. The equilibrium shapes of the elastica are shown in Figure 4. The stable equilibrium configurations are represented by the black line while the unstable configurations are represented by the red line.

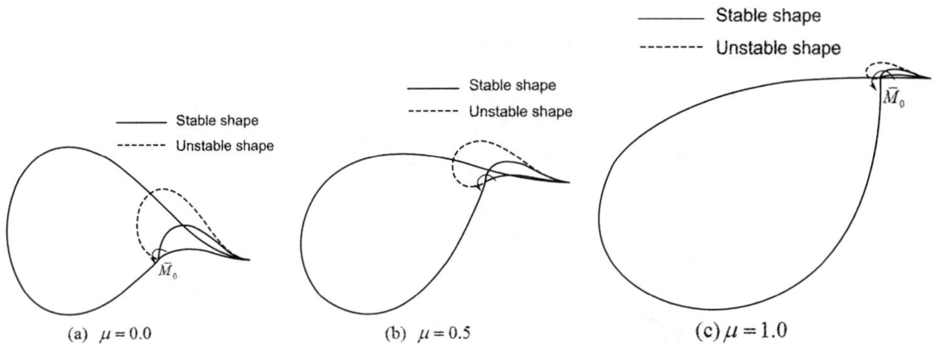

Figure 4 Equilibrium configurations for various values of coefficients of friction: a) μ=0. 0; b) μ=0. 5; c) μ=1. 0

As it can be seen, the friction helps to make the elastica stable by increasing the critical load and by increasing the range of rotation. Moreover, the slope of the load-deflection curves on the stable path becomes steeper in its inclination when the coefficient of friction is increased. Thus, we may conclude that the increase in the coefficient of friction results in an increase of the stiffness of the system (and hence increase in the stability of the system). On the unstable paths (e. g. , path AB), the slopes of load-deflection curves are similar and we may conclude that, on unstable paths, the coefficient of friction μ affects marginally the degree of instability of the elastica. The effect of the friction on the elastica shapes can be observed in Figure 5. The higher value of frictional coefficient yields a lower deflection at a given value of rotation at end A.

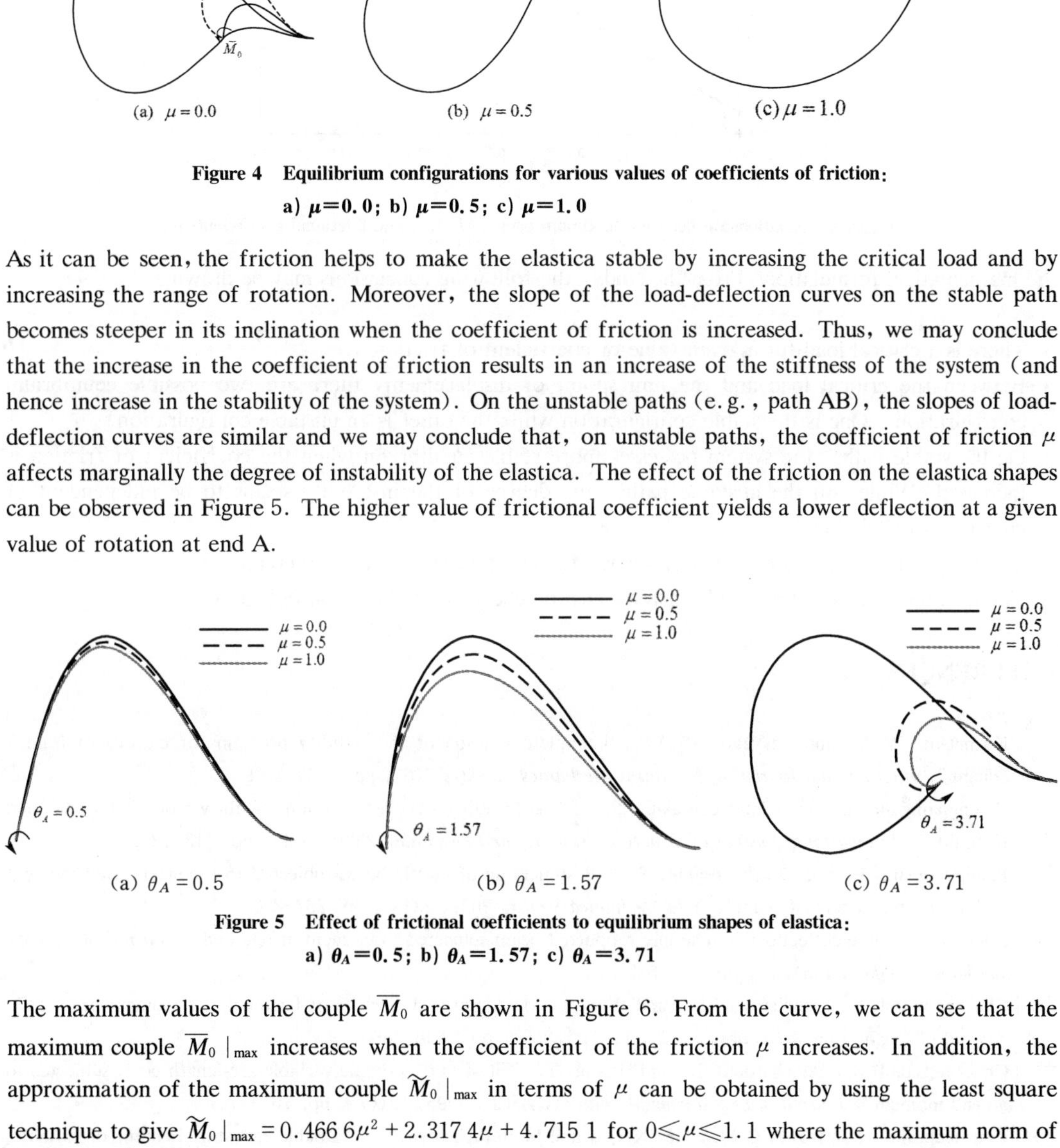

Figure 5 Effect of frictional coefficients to equilibrium shapes of elastica: a) θ_A=0. 5; b) θ_A=1. 57; c) θ_A=3. 71

The maximum values of the couple $\overline{M}_0$ are shown in Figure 6. From the curve, we can see that the maximum couple $\overline{M}_0\,|_{\max}$ increases when the coefficient of the friction μ increases. In addition, the approximation of the maximum couple $\widetilde{M}_0\,|_{\max}$ in terms of μ can be obtained by using the least square technique to give $\widetilde{M}_0\,|_{\max} = 0.466\,6\mu^2 + 2.317\,4\mu + 4.715\,1$ for $0 \leqslant \mu \leqslant 1.1$ where the maximum norm of error $\|\,\overline{M}_0\,|_{\max} - \widetilde{M}_0\,|_{\max}\,\|$ is 0. 005 13 (compared with the results from the numerical experiment).

CONCLUSIONS

The stability and deformed shape of a variable-arc-length elastica under an end couple are investigated numerically by using the shooting method. The friction at the support is also taken into consideration in

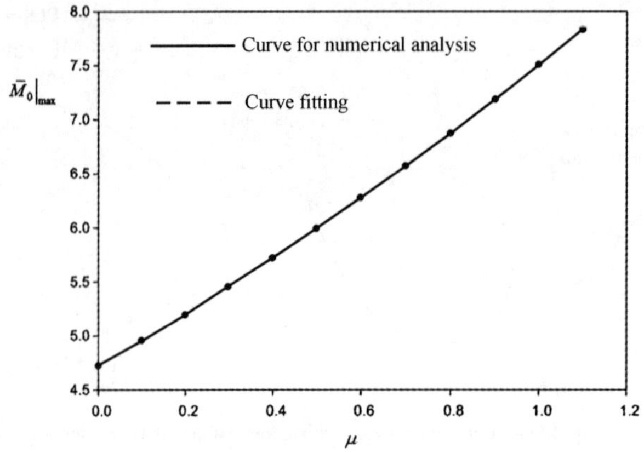

Figure 6 Relationship between maximum couple $\overline{M}_0 \,|_{max}$ and frictional coefficients μ

the mathematical formulation. From the study, the following conclusions may be drawn

- There is a critical load for a given value of coefficient of friction
- Between the critical load and the limit point of displacement, there are two possible equilibrium configurations. One is the stable configuration while the other is an unstable configuration.
- On the stable paths, the system possesses more stable equilibrium when the coefficient of friction is increased. While, on the unstable paths, the degree of the instability seems to be independent of coefficient of friction.
- The flexibility of the system decreases when the coefficient of friction increases.
- The maximum value ofthe couple increases as the coefficient of friction increases.

REFERENCES

[1] Kuznetsov, V. V. and Levyakov, S. V., "Complete solution of the stability problem for elastica of Euler's column", *International Journal of Nonlinear Mechanics*, 2002, 37(6), pp. 1003-1009.

[2] Phungpaingam, B. and Chucheepsakul, S., "Post-buckling of elastic column with various rotational end restraints", *International Journal of Structural Stability and Dynamics*, 2005, 5(1), pp. 113-123.

[3] Phungpaingam, B. and Chucheepsakul, S., "Postbuckling of elastic beam subjected to a concentrated moment within the span length of beam", *Acta Mechanica Sinica*, 2007, 23(3), pp. 287-296.

[4] Seide, P., "Large deflections of a simply supported beam subjected to moment at one end", *Journal of Applied Mechanics, ASME*, 1984, 51, pp. 519-525.

[5] Chucheepsakul, S., Buncharoen, S., and Wang C. M., "Large deflection of beams under moment gradient", *Journal of Engineering Mechanics, ASCE*, 1994, 120(9), pp. 1848-1860.

[6] Chucheepsakul, S., Buncharoen, S., and Huang, T., "Elastica of a simple variable-arc-length beam subjected to an end moment", *Journal of Engineering Mechanics, ASCE*, 1995, 121(7), pp. 767-772.

[7] Chucheepsakul, S., Wang, C.M., He, X.Q., and Monprapussorn, T., "Double curvature bending of variable-arc-length elasticas", *Journal of Applied Mechanics, ASME*, 1999, 66(1), pp. 87-94.

DYNAMICS OF A DUFFING-VAN DER POL OSCILLATOR WITH TIME DELAYED POSITION FEEDBACK

*A. Y. T. Leung[1, *], Z. J. Guo[1, 2] and H. X. Yang[1]

[1] Department of Civil and Architectural Engineering, City University of Hong Kong, Kowloon, Hong Kong, China
[2] School of Mathematics, Taishan University, Taian, 271021, China
* Email: BCALEUNG@cityu.edu.hk

KEYWORDS

Duffing-van der Pol, delayed position feedback, residue harmonic balance, Hopf bifurcation, saddle-node bifurcation.

ABSTRACT

The two main objectives of this paper are (1) to determine the bifurcating periodic responses accurately and to study the effects of time delay and feedback gain on the steady state response in autonomous oscillators by means of the residue harmonic balance method; and (2) to study the dynamics of both the autonomous and non-autonomous Duffing-van der Pol oscillators with delayed position feedback as examples. The results of the steady state response to frequency and amplitude excitation versus the time delay, positive and negative feedback gains are studied. Both saddle-node and subcritical Hopf bifurcations are found and discussed. The obtained second-order solutions are in excellent agreement with numerical integration results. By combining both harmonic balance and polynomial homotopy continuation techniques, the superharmonic resonance responses of the non-autonomous oscillator are investigated in order to obtain all possible steady state solutions. The effects of the feedback gains and time delay on the steady state response are found analytically. It is shown that the fundamental and superharmonic amplitudes decrease with increasing feedback gain. Furthermore, the multiple symmetric solutions of the response amplitudes versus those of external excitation are detected when time delay becomes smaller.

INTRODUCTION

Over the past decades, extensive attention has been paid to the study of delay differential equations in science and engineering [1-3]. Owing to the fact that the solutions of delay differential equations are history dependent, the difficulty lies in getting an analytical method to solve the highly transcendental equations [4-5]. As a result, many efforts have been made to find the approximate solutions [6-8] instead. In recent years, the homotopy perturbation [6], perturbation incremental [7], and pseudo-oscillator analysis [8] are the most widely applied methods for the purpose.

In this study, an attempt is made to gain an insight into the dynamics of a Duffing-van der Pol oscillator with time delayed position feedback. The relevant oscillator being considered is defined by

$$\ddot{u}+\Omega_0^2 u-(\alpha-\beta u^2)\dot{u}+\gamma u^3=\delta[u(t-\tau)-u]+q\cos(\omega t) \tag{1}$$

where $q \geqslant 0$, $\omega > 0$ and $\tau \geqslant 0$ are positive parameters. The position feedback is regarded as negative if the feedback gain $\delta < 0$ and positive if $\delta > 0$. The first objective of this work is to accurately determine the bifurcating periodic responses to frequency and amplitude excitations and to study the effect of time delay and feedback gain on the steady state response for the autonomous oscillator by the residue harmonic balance. The second objective is to focus on the dynamics of the non-autonomous oscillator and study the effect of time delay and feedback gain on the steady state response by combining both harmonic balance and homotopy continuation techniques.

AUTONOMOUS DUFFING-VAN DER POL OSCILLATOR

In this section, we consider the response and dynamics of the autonomous Duffing -van der Pol oscillator with time delayed position feedback for $q=0$ in Eq. (1), i.e.

$$\ddot{u}+\Omega_0^2 u-(\alpha-\beta u^2)\dot{u}+\gamma u^3=\delta[u(t-\tau)-u] \tag{2}$$

Hopf Bifurcation and Residue Harmonic Balance Method

The linearized equation of Eq. (2) and the characteristic equation are, respectively, given by

$$\ddot{u}+\Omega_0^2 u-\alpha\dot{u}-\delta[u(t-\tau)-u]=0 \tag{3}$$

$$D(\lambda):=\lambda^2+\Omega_0^2-\alpha\lambda+\delta-\delta e^{-\lambda\tau}=0 \tag{4}$$

We assume that the delayed vibration system undergoes a Hopf bifurcation at a $\tau=\tau_0$, where exactly one pair of pure imaginary roots $\pm i\omega$ of $D(\lambda)=0$ stay on the imaginary axis and all other characteristic roots remain in the open left-half complex plane and the transversality condition holds at the critical point. The corresponding vibration frequency ω_0 is determined from solving two transcendental equations, i.e.

$$\mathrm{Re}D(i\omega):=-\omega^2+\Omega_0^2+\delta-\delta\cos(\omega\tau)=0,\quad \mathrm{Im}D(i\omega):=-\alpha\omega+\delta\sin(\omega\tau)=0.$$

Our aim is to accurately obtain the bifurcation and response subject to various frequency and amplitude excitations by using the residue harmonic balance, a new technique developed by the authors recently. The residue harmonic balance has been proven to be a powerful tool for various kinds of nonlinear dynamic systems. The method does not depend on a small parameter. The solution procedure is mainly solving linear equations and a few iterations gives highly accurate solutions that cover a large solution domain. The general solution procedure can be found in Refs [9-10]. Further, the zeroth-, first- and second-orders harmonic balance equations are given in the Appendix in general forms.

Results and Discussions

In this section, the steady state solutions of Eq. (2) are computed and the results obtained by the residue harmonic balance method are compared with those of numerical simulations. Numerical integration results are computed by the fourth order Runge-Kutta scheme with a fixed step-size of 0.005 for the selected parameters $\alpha=0.2$, $\beta=0.2$, $\gamma=0.1$, $\Omega_0=0.316\,23$. Numerical simulation fails at bifurcation so that comparisons are made at points away from there.

Steady state response versus time delay and Hopf bifurcation

In this subsection, various the steady state responses of an autonomous oscillator are investigated. The parameter $\delta=0.2$ is selected and time delay τ is taken as control parameter. The relation of the steady state response subject to frequency and the amplitude excitation against time delay are shown in Figs. 1 and 2 respectively. The dashed, dash-dotted, solid and dotted lines denote the stable zeroth-order, unstable zeroth-order, stable second-order and numerical solutions, respectively. From Figs. 1 and 2, one observes that a saddle-node bifurcation occurs at point $\tau \approx 3.432\,2$. With increasing time delay τ, there are a stable upper and an unstable lower limit cycles coexisting. Further, a subcritical Hopf bifurcation occurs at $\tau \approx 3.395$. The stable second-order solutions are in excellent agreement with numerical results and the amplitude increases with increasing time delay τ. A comparison about the frequency and amplitude of current various-order solutions as well as numerical results are tabulated in Table 1.

TABLE 1 CURRENT VARIOUS-ORDER FREQUENCIES AND AMPLITUDES AFTER HOPF BIFURCATION

τ	$\omega_{(0)}$	$\omega_{(1)}$	$\omega_{(2)}$	$\omega_{(N)}$	$A_{(0)}$	$A_{(1)}$	$A_{(2)}$	$A_{(N)}$
3.56	0.914 1	0.889 2	0.886 8	0.886 2	2.119 5	2.004 9	1.988 0	1.983 7
4.00	0.978 4	0.954 6	0.953 5	0.953 3	2.617 5	2.532 0	2.518 5	2.516 0
4.25	0.982 2	0.958 9	0.958 1	0.957 9	2.738 1	2.661 9	2.647 6	2.645 0
4.50	0.976 8	0.953 8	0.953 3	0.953 1	2.809 1	2.741 2	2.725 8	2.723 1
4.75	0.965 9	0.943 3	0.943 0	0.942 8	2.847 7	2.787 8	2.771 3	2.768 4
5.00	0.951 7	0.929 5	0.929 3	0.929 2	2.863 3	2.811 8	2.794 0	2.791 1

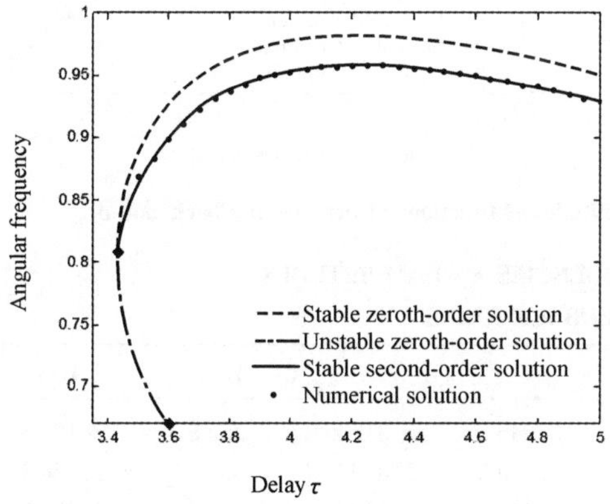

Figure 1 Frequency as a function of time delay τ.

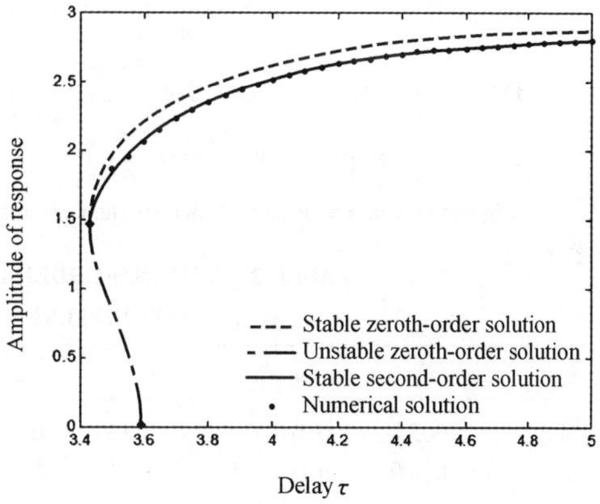

Figure 2 Amplitude as a function of time delay τ.

Accurate response versus feedback gain

In this subsection, we take the feedback gain δ as a bifurcation parameter and fix the delay $\tau=4$. Figures 3 and 4 depict the relations of the steady state responses versus positive and negative feedback gains respectively. The dash-dotted, dashed, solid and dotted lines denote the zeroth-, first-, second and numerical solutions respectively. From Figs. 3 and 4, the steady responses to angular frequency and amplitude are shown. A difference is observed under positive and negative feedback gains. A comparison about the frequency and amplitude of the various-order solutions are tabulated in Table 2. It is evident that the present second-order solutions are accurate enough when comparing with numerical integration results.

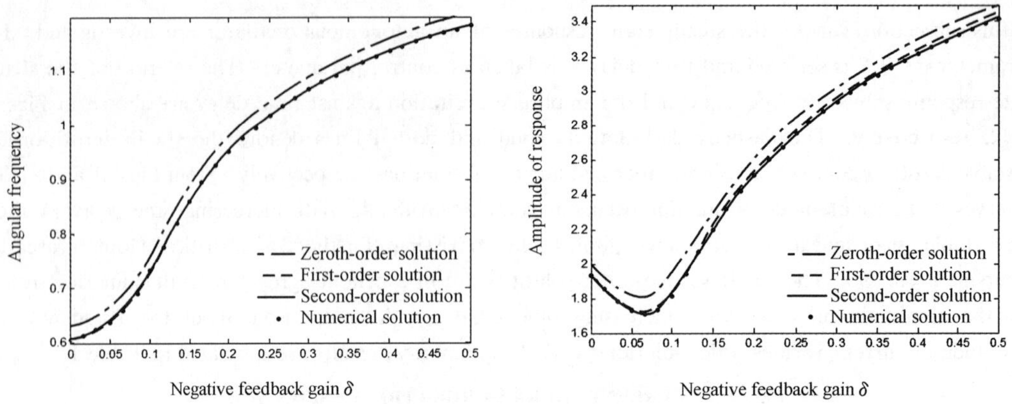

Figure 3 Current various-order frequencies and amplitudes as functions of positive feedback gain δ.

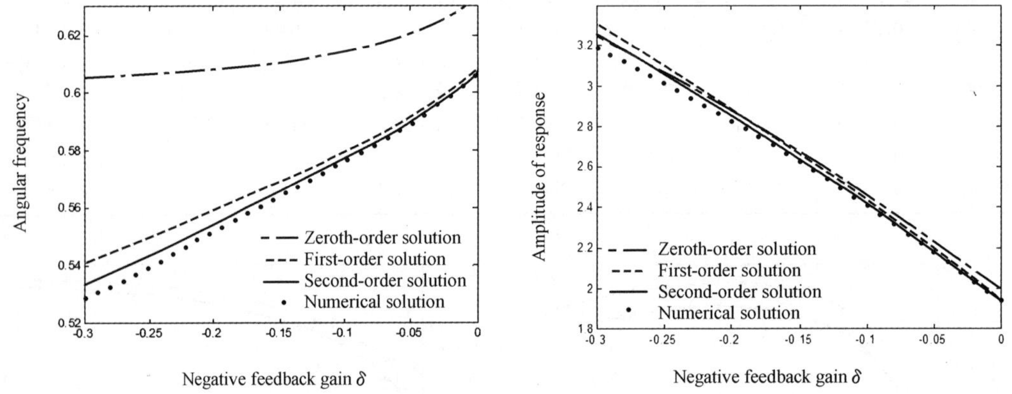

Figure 4 Current various-order frequencies and amplitudes as functions of negative feedback gain δ

TABLE 2 VARIOUS-ORDER FREQUENCIES AND AMPLITUDES FOR DIFFERENT FEEDBACK GAINS

δ	$\omega_{(0)}$	$\omega_{(1)}$	$\omega_{(2)}$	$\omega_{(N)}$	$A_{(0)}$	$A_{(1)}$	$A_{(2)}$	$A_{(N)}$
-0.3	0.605 0	0.540 6	0.533 1	0.528 8	3.248 2	3.312 2	3.259 5	3.190 8
-0.2	0.608 0	0.559 0	0.554 3	0.551 8	2.878 5	2.886 0	2.855 0	2.826 4
-0.1	0.614 1	0.579 1	0.576 9	0.576 1	2.461 9	2.433 7	2.418	2.433 7
0.0	0.632 5	0.607 5	0.606 4	0.606 1	2.000	1.945 2	1.938 2	1.936 3
0.1	0.762 8	0.736 1	0.733 4	0.732 8	1.939 8	1.835 6	1.820 3	1.816 5
0.2	0.978 4	0.954 6	0.953 5	0.953 3	2.617 5	2.532 0	2.518 5	2.516 0
0.3	1.090 9	1.068 2	1.067 8	1.067 7	3.028 0	2.958 7	2.941 4	2.938 6
0.4	1.159 8	1.137 2	1.137 2	1.137 0	3.298 4	3.242 7	3.219 6	3.216 2
0.5	1.207 5	1.184 7	1.185 0	1.184 8	3.496 3	3.454 8	3.424 8	3.420 9

NON-AUTONOMOUS DUFFING-VAN DER POL OSCILLATOR

In this section, we consider the effects of feedback gain and time delay on the steady state response to the non-autonomous Eq. (1). The approximate solution corresponding to the superharmonic resonance steady state is assumed to be

$$u = \frac{a_0}{2} + \sum_{i=1}^{3}[a_i \cos(i\omega t) + b_i \sin(i\omega t)] \tag{5}$$

By substituting Eq. (5) into Eq. (1) and using the Galerkin procedure, one obtains a set of harmonic balance equations. Unfortunately, it is rather difficult to obtain all the possible solutions of the above harmonic balance equations. In this section, the polynomial homotopy continuation technique [11] which can obtain all possible solutions is presented and applied.

Effect of Feedback Gain

This subsection studies the effect of feedback gain on the response amplitudes subject to external frequency excitation. The parameters $\alpha = 0.1$, $\beta = 0.2$, $\gamma = 0.1$, $q = 1$ and $\Omega_0 = 0.316\,23$ are selected. There exists symmetric response which is shown in Figure 5. It is observed that (i) the fundamental amplitude increases first and then decreases for feedback gains $\delta = 0$, 0.2 and 0.4, (ii) the fundamental and superharmonic amplitudes decrease with increasing feedback gains, (iii) the stable periodic solution becomes unsteady when the frequency of external excitation $\omega > 0.95$ for feedback gain $\delta = 0$.

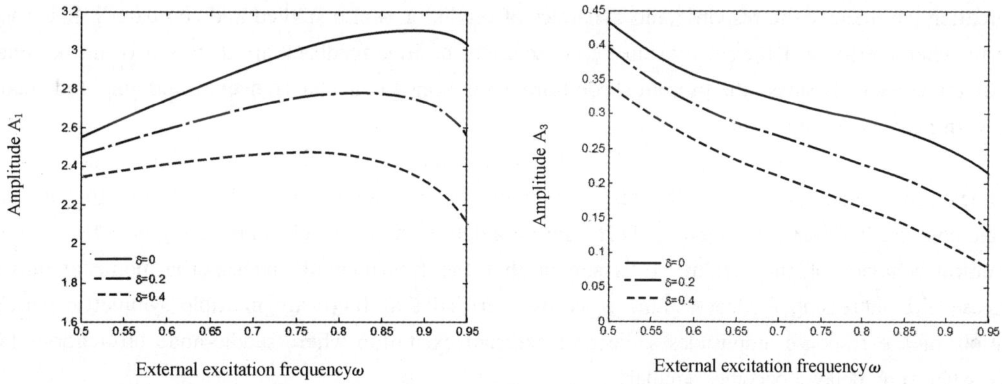

Figure 5 Influence of feedback gain on response amplitudes versus frequency of external excitation with $\tau = 1$.

Effect of Time Delay

This subsection investigates the effect of time delay on the response amplitudes subject to external amplitude excitation. In Figure 6, the solid and dashed lines denote stable and unstable responses, respectively. From Figure 6, it is observed that (i) when time delay is larger ($\tau \geqslant 1$), the fundamental and superharmonic amplitudes increase with the increasing amplitude of external excitation and (ii) for time delay $\tau = 0.1$, a saddle-node bifurcation occurs at $q = 0.405\,5$, and there are three symmetric periodic solutions in which only the upper branch is stable for $0.405\,5 < q < 0.557\,1$.

CONCLUSIONS

A delay position feedback has a dramatic influence on the dynamics of nonlinear oscillators. So time delay existing in the Duffing-van der Pol oscillator is an important factor for one to consider. When time delay and feedback gain are chosen as the bifurcation parameters, the Hopf and saddle-node bifurcations can be studied by the modified harmonic balance method.

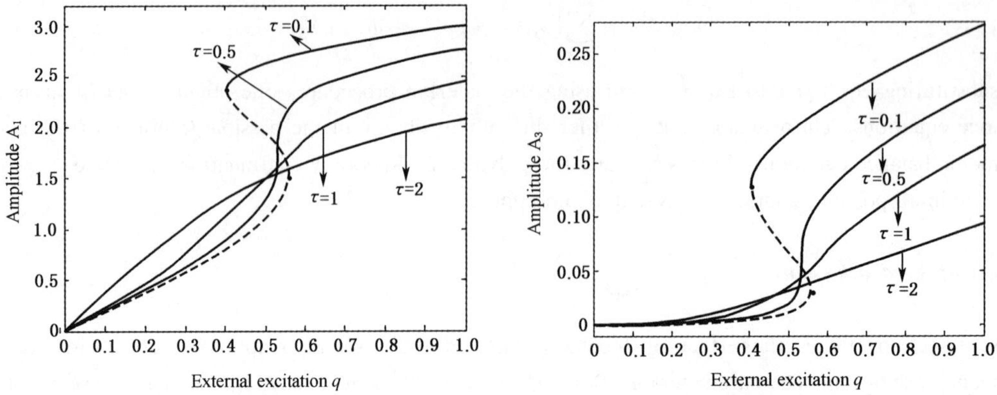

Figure 6 Influence of time delay on excitation amplitude-response amplitude with ω=0.8.

By using the residue harmonic balance technique, the response to angular frequency and amplitude of autonomous oscillator are obtained accurately and verified. The subcritical Hopf bifurcation is investigated when time delay is chosen as a bifurcation parameter. When feedback gain δ is chosen as a bifurcation parameter, the response and dynamics of oscillator is also studied and compared. It is found that the characteristics of the steady state response under positive feedback are different from that under negative feedback. Besides, the present second-order solutions to angular frequency and amplitude match well with numerical simulation.

By combining the harmonic balance and the polynomial homotopy continuation techniques, the effects of time delay and feedback gain on the response amplitudes subject to external excitation for the non-autonomous oscillator are investigated. Different time delays and feedback gains have great effects on the dynamical behavior of the system. It is shown that the fundamental and superharmonic amplitudes decrease with increasing feedback gain. It is also verified that there are multiple symmetric periodic solutions of the response amplitudes subject to external excitation where saddle-node bifurcations take place when time delay τ becomes smaller.

ACKNOWLEDGEMENTS

We wish to thank the reviewers for their valuable suggestions. This work is supported by Shan Dong Provincial Natural Science Foundation Grant No. ZR2011AQ022.

APPENDIX

The zeroth-, first- and second-order residue equations of autonomous oscillator are given by:

- Zeroth-order harmonic equation:

$$p^0:\ \omega_0^2 u''_0 + (\Omega_0^2+\delta)u_0 - \omega_0(\alpha-\beta u_0^2)u'_0 + \gamma u_0^3 - \delta u_0(x-\tau\omega_0)$$

- First-order harmonic equation:

$$p:\ \omega_0^2 u''_1 + 2\omega_0\omega_1 u''_0 + (\Omega_0^2+\delta)u_1 - \omega_0(\alpha-\beta u_0^2)u'_1 - \omega_1\alpha - \beta u_0^2)u'_0 + 2\beta\omega_0 u_0 u_1 u'_0$$
$$+ 3\gamma x_0^2 u_1 - \delta u_1(x-\tau\omega_0) + \tau\omega_1\delta u'_0(x-\tau\omega_0) + 1/4\gamma a_0^3\cos(3x) - 1/4\omega_0\beta a_0^3\sin(3x)$$

- Second-order harmonic equation:

$$
\begin{aligned}
p^2:\ &\omega_0^2 u''_2 + 2\omega_0\omega_1 u'_1 + (\omega_1^2 + 2\omega_0\omega_2)u''_0 + (\Omega_0^2 + \delta)u_2 - \omega_0(\alpha - \beta u_0^2)u'_2 + 2\omega_0\beta u_0 u_1 u'_1 \\
&+ \beta\omega_0(2u_0u_2 - u_1^2)u'_0 - \omega_1(\alpha - \beta u_0^2)u'_1 + 2\omega_1\beta u_0 u_1 u'_0 - \omega_2(\alpha - \beta u_0^2)u'_0 + 3\gamma u_0^2 u_2 \\
&+ 3\gamma u_0 u_1^2 - \delta u_2(x - \tau\omega_0) + \tau\omega_1\delta u'_1(x - \tau\omega_0) + \tau\omega_2\delta u'_0(x - \tau\omega_0) - \frac{1}{2}\tau^2\omega_1^2\delta u''_0(x - \tau\omega_0) \\
&+ (5/4\beta\omega_0 a_0^2 b_{31} + 3/4\gamma a_0^2 a_{31})\cos(5x) + (-5/4\beta\omega_0 a_0^2 a_{31} + 3/4\gamma a_0^2 b_{31})\sin(5x)
\end{aligned}
$$

REFERENCES

[1] Kuang, Y., *Delay Differential Equations with Applications in Population Dynamics*, Academic Press, Boston, 1993.

[2] Hu, H. Y. and Wang, Z. H., *Dynamics of Controlled Mechanical Systems with Delayed Feedback*, Springer-Verlag, New York, 2002.

[3] Xu, J. and Chung, K. W., "Effects of time delayed position feedback on a van der Pol-Duffing oscillator", *Physica D-Nonlinear Phenomena*, 2003, 180(1-2), pp. 17-39.

[4] Fofana, M.S. and Ryba, P.B., "Parametric stability of nonlinear time delay equations", *International Journal of Nonlinear Mechanics*, 2004, 39, pp. 79-91.

[5] Asl, F.M. and Ulsoy, A.G., "Analysis of a system of linear delay differential equations", *Journal of Dynamics Systems, Measurement and Control*, 2003, 125, pp. 215-223.

[6] Shakeri, F. and Dehghan, M., "Solution of delay differential equations via a homotopy perturbation method", *Mathematical and Computer Modelling*, 2008, 48(3-4), pp.486-498.

[7] Chung, K.W., Chan, C.L. and Xu, J., "A perturbation-incremental method for delay differential equations", *International Journal of Bifurcation and Chaos*, 2006, 16(9), pp. 2529-2544.

[8] Zhang, L.L., Huang, L.H. and Zhang, Z.Z., "Hopf bifurcation of the maglev time-delay feedback system via pseudo-oscillator analysis", *Mathematical and Computer Modelling*, 2010, 52(5-6), pp. 667-673.

[9] Leung, A. Y. T. and Guo, Z. J., "Residue harmonic balance approach to limit cycles of non-linear jerk equations", *International Journal of Non-linear Mechanics*, 2011, 46(6), pp. 898-906.

[10] Leung, A. Y. T. and Guo, Z. J., "Forward residue harmonic balance for autonomous and non-autonomous systems with fractional derivative damping", *Communications in Nonlinear Science and Numerical Simulation*, 2011, 16(4), pp. 2169-2183.

[11] Andrew, J.S. and Charles, W.W. II., *The Numerical Solution of Systems of Polynomials: Arising in Engineering and Science*, World Scientific Press, Singapore, 2005.

IJSSD Symposium on Progress in Structural Stability and Dynamics
14-16 April 2012, Nanjing, China

NONLINEAR VIBRATION OF FUNCTIONALLY GRADED PIEZOELECTRIC ACTUACTORS

*J. Yang[1], Y.J. Hu[2], S. Kitipornchai[3] and T. Yan[3]

[1] School of Aerospace, Mechanical and Manufacturing Engineering, RMIT University, PO Box 71, Bundoora, Victoria, 3083 Australia
[2] College of Mechanical Engineering, University of Shanghai for Science and Technology, Shanghai, 200093, China
[3] Department of Civil and Architectural Engineering, City University of Hong Kong, Kowloon, Hong Kong, China
* Email: j.yang@rmit.edu.au

KEYWORDS

Functionally graded materials, piezoelectric materials, actuator, nonlinear vibration, initial imperfection, Timoshenko beam theory, differential quadrature method.

ABSTRACT

Piezoelectric actuators have been extensively used in a variety of engineering applications such as microelectromechanical systems, active vibration control, acoustic and pressure sensing. This paper investigates the nonlinear free vibration of a novel class of monomorph and bimorph actuators made of functionally graded piezoelectric materials (FGPM) with piezo-elastic constants varying smoothly along the layer thickness according to a power law distribution in terms of the volume fraction of the constituents. The actuator is modeled as a Timoshenko beam to account for the effects of transverse shear deformation, axial and rotary inertia. The partial differential equations of motion which include the effect of local and global geometric imperfections due to fabrication process are derived by employing Hamilton's principle and von Karman type of nonlinear kinematics. The differential quadrature method is used to obtain the nonlinear vibration frequencies for FGPM monomorph and bimorph actuators. Numerical results are presented in tabular form showing the influences of material composition, slenderness ratio and initial geometric imperfection on both the linear and nonlinear frequency parameters.

INTRODUCTION

Typical piezoelectric bending actuators involve multilayer stacks and make use of the flexural deformation mode to produce larger deflections than extensional mode actuators. Conventional layered piezoelectric actuators suffer heavily from high stress concentrations near the interlayer surfaces because of the abrupt changes in both material composition and thermo-electro- elastic properties. This can cause severe deterioration in both interlayer bonding strength and overall response performance. Under repeated strain reversals, the high stress concentration is also very likely to be the cause of premature failure. The use of functionally graded piezoelectric materials (FGPM) with thermo-electro-elastic constants varying smoothly along the thickness direction in one or more layers or sandwiched between distinct piezoelectric layers provides a promising remedy for this problem (Hauke et al. [1]). It has been proven both theoretically and experimentally that the durability, reliability and actuation performance of actuators can be greatly improved by the use of FGPM. Hauke et al. [1] reported experimental results and gave a linear regression formula to calculate the tip deflection of an FGPM cantilever actuator with a linearly variable piezoelectric coefficient d31. Kruusing [2] presented an analytical solution for tip deflection and blocking force of a cantilever actuator with a graded elastic constant Q11 under a transverse point load at the tip. Yang and Xiang [3] studied the thermo-electro-mechanical characteristics, including the static bending, free vibration, and dynamic response, of various FGPM monomorph, bimorph, and multimorph actuator.

This paper investigates the nonlinear free vibration of FGPM monomorph and bimorph actuators whose electro-elastic properties varying along the thickness direction within the framework of Timoshenko beam theory and von Karman geometrical nonlinearity. The effect of geometric imperfection due to fabrication process is also included. The differential quadrature method is used to obtain the nonlinear vibration frequencies for FGPM monomorph and bimorph actuators. The effects of material composition, slenderness ratio and initial geometric imperfection on the frequency parameters are discussed in detail through a parametric study.

THE ORETICAL FORMULATIONS

Figure 1 shows a N_L-layer functionally graded piezoelectric actuator with length L, width b, and total thickness h. The electric field is generated by top and bottom electrodes. The actuator is polarized in the thickness direction and subjected to an applied voltage $V(t)$ that satisfies Maxwell equation

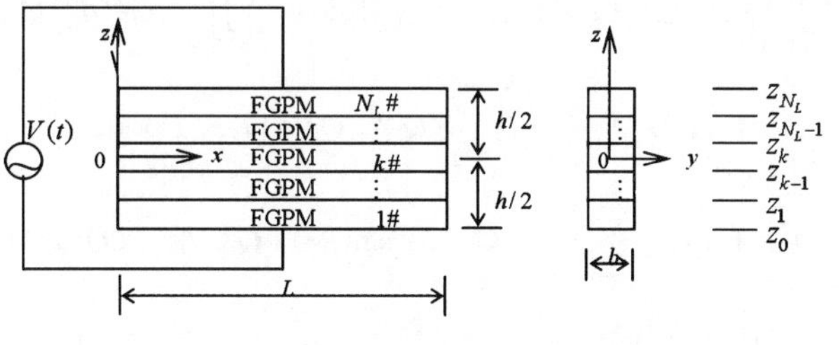

Figure 1 Schematic configuration of a multilayer FGPM actuator

$$V(x, z, t) = -\sin(\beta z)\phi(x, t) + \frac{V_0}{h} z\cos(\Omega_{Vt}) \quad (1)$$

where $\beta = 2\pi/h$, $\phi(x, t)$ is the electric potential, V_0 is the voltage, amplitude and Ω_V is the excitation frequency. The electric fields appropriate for this electric voltage are

$$E_x = -\frac{\partial V}{\partial x} = \sin(\beta z)\frac{\partial \phi}{\partial x}, \quad E_z = -\frac{\partial V}{\partial z} = \beta\cos(\beta z)\phi - E_0, \quad E_0 = \frac{V_0}{h}\cos(\Omega_V t) \quad (2)$$

The material composition follows a power law distribution along the thickness direction. Hence, the effective material properties P_k of the k-th layer can be determined from

$$P^{(k)} = P_U^{(k)} V_U^{(k)} + P_L^{(k)} V_L^{(k)} \tag{3}$$

where $P_U^{(k)}$ ($P_L^{(k)}$), $V_U^{(k)}$ ($V_L^{(k)}$) are the material properties and volume fraction at the upper (lower) surface of the k- th layer, respectively. The actuator is modeled as an elastic beam. According to Timoshenko beam theory, the strain-displacement relationship gives

$$\varepsilon_x = \frac{\partial u}{\partial x} + z\frac{\partial \varphi}{\partial x} + \frac{1}{2}\left(\frac{\partial w}{\partial x}\right)^2 + \frac{\partial w}{\partial x}\frac{\partial w_0}{\partial x}, \quad \gamma_{xz} = \varphi + \frac{\partial w}{\partial x} + \frac{\partial w_0}{\partial x} \tag{4}$$

where u and w are the mid-plane displacements of an arbitrary point along the x- and z-axes, φ is the cross-sectional rotation. The initial imperfection w_0 is described by (Yang and Huang [5])

$$w_0 = \eta \sec h[\delta(x-\xi)]\cos[\mu\pi(x-\xi)] \tag{5}$$

where η, δ, μ, and ξ are imperfection parameters.

The nonlinear governing differential equations can be derived by employing Hamilton's principle as

$$A_{11}\frac{\partial}{\partial x}\left(\frac{\partial u}{\partial x} + \frac{1}{2}\left(\frac{\partial w}{\partial x}\right)^2 + \frac{\partial w}{\partial x}\frac{\partial w_0}{\partial x}\right) + B_{11}\frac{\partial^2 \varphi}{\partial x^2} - A_{31}\frac{\partial \phi}{\partial x} = I_0\frac{\partial^2 u}{\partial t^2} + I_1\frac{\partial^2 \varphi}{\partial t^2} \tag{6}$$

$$K_s A_{55}\frac{\partial}{\partial x}\left(\varphi + \frac{\partial w}{\partial x} + \frac{\partial w_0}{\partial x}\right) - K_s D_{15}\frac{\partial^2 \phi}{\partial x^2} + A_{11}\frac{\partial}{\partial x}\left[\left(\frac{\partial u}{\partial x} + \frac{1}{2}\left(\frac{\partial w}{\partial x}\right)^2 + \frac{\partial w}{\partial x}\frac{\partial w_0}{\partial x}\right)\left(\frac{\partial w}{\partial x} + \frac{\partial w_0}{\partial x}\right)\right] +$$
$$+ B_{11}\frac{\partial}{\partial x}\left(\frac{\partial \varphi}{\partial x}\left(\frac{\partial w}{\partial x} + \frac{\partial w_0}{\partial x}\right)\right) - A_{31}\frac{\partial}{\partial x}\left(\phi\left(\frac{\partial w}{\partial x} + \frac{\partial w_0}{\partial x}\right)\right) = I_0\frac{\partial^2 w}{\partial t^2} \tag{7}$$

$$B_{11}\frac{\partial}{\partial x}\left(\frac{\partial u}{\partial x} + \frac{1}{2}\left(\frac{\partial w}{\partial x}\right)^2 + \frac{\partial w}{\partial x}\frac{\partial w_0}{\partial x}\right) + D_{11}\frac{\partial^2 \varphi}{\partial x^2} + K_s D_{15}\frac{\partial \phi}{\partial x} - B_{31}\frac{\partial \phi}{\partial x} - K_s A_{55}\left(\varphi + \frac{\partial w}{\partial x} + \frac{\partial w_0}{\partial x}\right)$$
$$= I_1\frac{\partial^2 u}{\partial t^2} + I_2\frac{\partial^2 \varphi}{\partial t^2} \tag{8}$$

$$A_{31}\left(\frac{\partial u}{\partial x} + \frac{1}{2}\left(\frac{\partial w}{\partial x}\right)^2 + \frac{\partial w}{\partial x}\frac{\partial w_0}{\partial x}\right) + B_{31}\frac{\partial \varphi}{\partial x} - D_{15}\frac{\partial}{\partial x}\left(\varphi + \frac{\partial w}{\partial x} + \frac{\partial w_0}{\partial x}\right) - F_{11}\frac{\partial^2 \phi}{\partial x^2} + F_{33}\phi - H_{33}E_0 = 0 \tag{9}$$

where

$$(I_0, I_1, I_2) = \sum_{k=1}^{N_l}\int_{z_{k-1}}^{z_k} \rho^{(k)}(1, z, z^2)dz, \quad (A_{31}, B_{31}) = \sum_{k=1}^{N_l}\int_{z_{k-1}}^{z_k} Q_{11}^{(k)} d_{31}^{(k)}(1, z)\beta\cos(\beta z)dz,$$

$$(A_{55}, A_{11}, B_{11}, D_{11}) = \sum_{k=1}^{N_l}\int_{z_{k-1}}^{z_k} [Q_{55}^{(k)}, Q_{11}^{(k)}(1, z, z^2)]dz,$$

$$(D_{15}, D_{31}, E_{31}) = \sum_{k=1}^{N_l}\int_{z_{k-1}}^{z_k} [Q_{11}^{(k)} d_{15}^{(k)}\sin(\beta z), Q_{11}^{(k)} d_{15}^{(k)}, Q_{11}^{(k)} d_{15}^{(k)} z]dz,$$

$$(F_{11}, F_{33}, H_{33}) = \sum_{k=1}^{N_l}\int_{z_{k-1}}^{z_k} [k_{11}^{(k)}\sin^2(\beta z), k_{33}^{(k)}\beta^2\cos^2(\beta z), k_{33}^{(k)}\beta\cos(\beta z)]dz \tag{10}$$

In Eqs (6)-(10), shear correction factor $K_s = \pi^2/12$, $\rho^{(k)}$ is the mass density, $Q_{11}^{(k)}$ and $Q_{55}^{(k)}$ are elastic stiffness constants, $d_{31}^{(k)}$, $d_{15}^{(k)}$ are piezoelectric strain coefficients, $k_{11}^{(k)}$, $k_{33}^{(k)}$ can be expressed in terms of $d_{31}^{(k)}$, $d_{15}^{(k)}$ and dielectric permittivity coefficients $\in_{11}^{(k)}$, $\in_{33}^{(k)}$ as $k_{11}^{(k)} = \in_{11}^{(k)} - Q_{55}^{(k)}(d_{15}^{(k)})^2$, $k_{33}^{(k)} = \in_{33}^{(k)} - Q_{11}^{(k)}(d_{13}^{(k)})^2$. Subscript "$k$" refers to the k-th layer.

The beam may be clamped (C), free (F), or hinged (H) at its ends ($x = 0$, l) with the following boundary conditions and associated electrical boundary condition

Free (F): $$N = 0,\ Q + N\frac{\partial w}{\partial x} = 0,\ M = 0 \tag{11a}$$

Clamped (C): $u = w = \varphi = 0$ (11b)

Hinged (H): $u = w = M = 0$ (11c)

$$D_{15}\left(\varphi + \frac{\partial w}{\partial x} + \frac{\partial w_0}{\partial x}\right) + F_{11}\frac{\partial \phi}{\partial x} = 0 \tag{12}$$

Here, N, Q and M are the stress resultant, shear force and bending moment.

DIFFERENTIAL QUADRATURE SOLUTION

According to the DQM rule (Bert et al. [6]), the unknown functions u, w, φ, and ϕ are approximated in terms of their function values at a number of pre-selected sampling points by

$$\{u, w, \varphi, \phi\} = \sum_{j=1}^{N_p} l_j(x)\{u_j, w_j, \varphi_j, \phi_j\} \tag{13}$$

and their m-th partial derivatives with respect to x are given by

$$\frac{\partial^m}{\partial x^m}(u, w, \varphi, \phi)\bigg|_{x=x_i} = \sum_{j=1}^{N_p} C_{ij}^{(m)}(u_j, w_j, \varphi_j, \phi_j) \tag{14}$$

where N_p is the total number of sampling points, $l_j(x)$ is Lagrange interpolation polynomials, the weighting coefficients $C_{ij}^{(m)}$ can be calculated through a recursive formula and are dependent on the coordinates of the sampling points which are determined by

$$x_1 = 0.0,\quad x_2 = 0.0001,\quad x_j = \frac{1}{2}\left[1 - \cos\frac{\pi(j-2)}{N_p - 3}\right],\quad x_{N_p - 1} = 0.9999,\quad x_{N_p} = 1.0 \tag{15}$$

Application of the above approximation to Eqns (6)-(9), the associated boundary conditions in Eq. (11) and the electrical condition in Eq. (12) yields a nonlinear algebraic equation system. By dropping the dynamic load term in the equation and employing an iterative algorithm, the nonlinear natural frequencies can be obtained.

NUMERICAL RESULTS AND DISCUSSIONS

This section presents the numerical results for functionally graded monomorph and bimorph actuators whose cross-section are shown in Figure 2. PZT-4 and PZT-5H are selected as the material constituents of each layer with the following material constants

$$Q_{11} = 81.3\ \text{GPa},\ Q_{55} = 25.6\ \text{GPa},\ d_{31} = -123 \times 10^{-12}\ \text{C/N},\ d_{31} = 496 \times 10^{-12}\ \text{C/N}$$
$$\in_{11} = 1470 \in_0,\ \in_{33} = 1300 \in_0,\ p_3 = 2.5 \times 10^{-5}\ \text{C/(m}^2\text{K)},\ a_1 = 0.2 \times 10^{-5}\ \text{K}^{-1} \rho = 7500\ \text{kg/m}^3, \quad \text{(PZT-4)}$$

$$Q_{11} = 60.6\ \text{GPa},\ Q_{55} = 23.0\ \text{GPa},\ d_{31} = -274 \times 10^{-12}\ \text{C/N},\ d_{31} = 741 \times 10^{-12}\ \text{C/N}$$
$$\in_{11} = 3130 \in_0,\ \in_{33} = 3400 \in_0,\ p_3 = 0.548 \times 10^{-5}\ C/(\text{m}^2\text{K}),\ a_1 = 1 \times 10^{-5}\ \text{K}^{-1} \rho = 7500\ \text{kg/m}^3 \quad \text{(PZT-5H)}$$

It is assumed that the upper and lower layers in the bimorph Ⅰ and bimorph Ⅱ actuators are of the same thickness and have opposite poling directions. The volume fractions follow a power law distribution

Monomorph: $V_U = 1 - (1 - 2\bar{z})^\lambda$, $V_L = (1 - 2\bar{z})^\lambda$ (16)

Bimorph: $V_U = (2\bar{z})^\lambda$, $V_L = 1 - (2\bar{z})^\lambda$ (upper layer) (17a)

$$V_U = 1 - (-2\bar{z})^{\lambda}, \quad V_L = (-2\bar{z})^{\lambda} \quad \text{(lower layer)} \tag{17b}$$

where $\bar{z} = z/h$, λ is the volume fraction index.

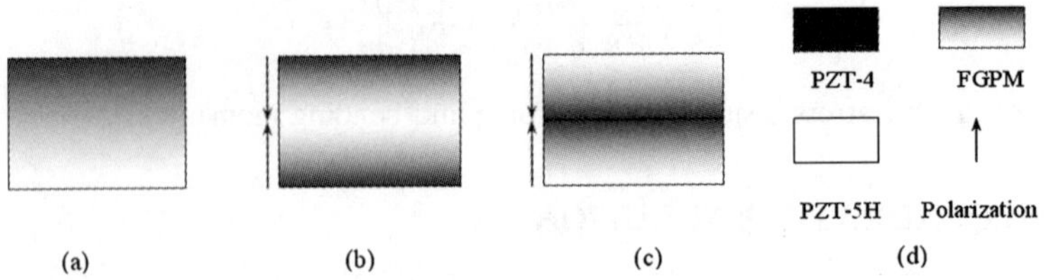

Figure 2 Configurations of FGPM actuators: (a) monomorph; (b) bimorph Ⅰ; (c) bimorph Ⅱ; (d) notations

To validate the present analysis, Table 1 gives the first four dimensionless linear frequencies $\omega = \Omega h \sqrt{(\rho/Q_{66})}$ for fully clamped (C-C) and cantilever (C-F) functionally graded beams with 100% material A ($Q_{11} = 138.4$ GPa, $Q_{55} = 162.5$ GPa, $\rho = 7\,500$ kg/m^3) at the bottom surface and 100% material B ($Q_{11} = 157.3$ GPa, $Q_{55} = 42.5$ GPa, $\rho = 5\,676$ kg/m^3) at the top surface. The volume fraction index is $\lambda = 1$. Excellent agreement is achieved between our results and the 2-D elasticity solutions (Lu and Chen [7]). In this example and following calculations, the total thickness is $h = 0.001$ m.

TABLE 1 THE FIRST FOUR LINEAR FREQUENCIES OF FGM BEAMS

Mode	C-F				C-C			
	$l/h = 15$		$l/h = 7$		$l/h = 15$		$l/h = 7$	
	Present	Lu & Chen [7]	Present	Lu & Chen [7]	Present	Lu & Chen [7]	Present	Lu & Chen [7]
1	0.004 6	0.004 7	0.020 8	0.021 3	0.028 7	0.029 2	0.124 1	0.126 4
2	0.028 3	0.028 8	0.123 1	0.012 53	0.077 2	0.078 6	0.313 3	0.318 4
3	0.077 5	0.079 0	0.228 4	0.235 8	0.146 8	0.149 5	0.456 7	0.471 9
4	0.106 6	0.109 9	0.318 3	0.323 4	0.213 1	0.220 5	0.558 2	0.567 5

To facilitate a direct comparison of the free vibration characteristics of different FGPM actuators, the fundamental frequency Ω_1 is normalized as $\omega_1 = \Omega_1 l \sqrt{(I_0/A_{11})_{PZT-4}}$, where $(I_0/A_{11})_{PZT-4}$ is the ratio between the I_0 and A_{11} of an isotropic homogeneous PZT-4 beam. Table 2 compares the nonlinear fundamental frequencies ω_{1nl} of perfect and imperfect C-C and H-H monomorph actuators ($L/h = 15$) where ω_{1l} denotes linear frequencies. The dimensionless central deflection is $W_c = w_c/h = 0.4$ for nonlinear frequencies. The geometrical imperfection considered herein include Type Ⅰ ($\delta = 0$, $\mu = 1$, $\xi = 0.5$, $\eta = 0.002$) which is a highly localized imperfection at the centre of the beam and Type Ⅱ ($\delta = 25$, $\mu = 2$, $\xi = 0.5$, $\eta = 0.002$) which is a global imperfection. The effects of the imperfection and the nonlinear deformation are observed to increase the fundamental frequency of the actuator. Compared with its counterpart with a Type I imperfection, the actuator with a Type 2 imperfection has a slightly higher fundamental frequency. It is also seen that for monomorph actuator, its fundamental frequency increases monotonically as the volume fraction index λ increases.

TABLE 2 FUNDAMENTAL FREQUENCY OF PERFECT AND IMPERFECT MONOMORPH ACTUATORS

	λ	ω_{1l} (Perfect)	ω_{1nl} (Perfect)	ω_{1nl} (Type Ⅰ)	ω_{1nl} (Type Ⅱ)
C-C	0.2	0.372 9	0.394 0	0.394 8	0.398 4
	1.0	0.388 8	0.411 2	0.412 1	0.415 9
	6.0	0.402 7	0.426 4	0.427 4	0.431 5

continued

	λ	ω_{1l}(Perfect)	ω_{1nl}(Perfect)	ω_{1nl}(Type I)	ω_{1nl}(Type II)
H-H	0.2	0.260 4	0.293 3	0.295 1	0.295 4
	1.0	0.271 8	0.308 5	0.310 3	0.310 8
	6.0	0.281 5	0.318 9	0.320 8	0.321 2

TABLE 3 FUNDAMENTAL FREQUENCY OF PERFECT BIMORPH I ACTUATORS

		$L/h = 15$				$L/h = 30$			
	λ	$W_c = 0.0$	$W_c = 0.2$	$W_c = 0.4$	$W_c = 0.6$	$W_c = 0.0$	$W_c = 0.2$	$W_c = 0.4$	$W_c = 0.6$
C-C	0.2	0.402 5	0.408 1	0.424 4	0.450 0	0.205 2	0.208 0	0.216 1	0.228 9
	1.0	0.409 1	0.414 6	0.430 6	0.455 8	0.208 8	0.211 5	0.219 5	0.232 1
	6.0	0.404 0	0.409 7	0.426 0	0.451 4	0.206 1	0.208 9	0.217 0	0.229 7
H-H	0.2	0.280 7	0.287 1	0.309 1	0.342 9	0.141 9	0.145 0	0.156 1	0.173 0
	1.0	0.285 4	0.293 6	0.317 1	0.351 5	0.144 3	0.148 5	0.160 2	0.177 5
	6.0	0.281 9	0.292 5	0.317 9	0.353 7	0.142 5	0.147 8	0.160 6	0.178 5

TABLE 4 FUNDAMENTAL FREQUENCY OF PERFECT BIMORPH II ACTUATORS

		$L/h = 15$				$L/h = 30$			
	λ	$W_c = 0.0$	$W_c = 0.2$	$W_c = 0.4$	$W_c = 0.6$	$W_c = 0.0$	$W_c = 0.2$	$W_c = 0.4$	$W_c = 0.6$
C-C	0.2	0.393 7	0.399 4	0.416 1	0.442 1	0.200 3	0.203 1	0.211 5	0.224 5
	1.0	0.390 0	0.395 8	0.412 6	0.438 9	0.198 2	0.201 1	0.209 5	0.222 6
	6.0	0.391 5	0.397 3	0.414 0	0.440 2	0.199 1	0.202 0	0.210 3	0.223 4
H-H	0.2	0.274 3	0.285 0	0.311 0	0.347 4	0.138 5	0.143 9	0.156 9	0.175 2
	1.0	0.271 4	0.280 1	0.304 6	0.340 3	0.136 9	0.141 3	0.153 6	0.171 5
	6.0	0.272 7	0.279 0	0.301 4	0.335 8	0.137 6	0.140 8	0.152 0	0.169 3

Tables 3 and 4 list the linear and nonlinear fundamental frequencies of perfect C-C and H-H bimorph I and II actuators ($L/h = 15, 30$) where the results with $W_c = 0.0$ are linear frequencies. As can be observed, bimorph I actuator has a higher fundamental frequency than bimorph II actuator. When fully claimed and with $\lambda = 1$, bimorph I actuator has the largest while bimorph II actuator has the lowest fundamental frequency. For bimorph actuators hinged at both ends, however, the relationship between the fundamental frequency and the volume faction index λ is more complicated. An increase in λ leads to a lower fundamental frequency for slender bimorph II actuators ($L/h = 30$).

CONCLUSIONS

Nonlinear free vibration of functionally graded piezoelectric monomorph and bimorph actuators with electro-elastic properties varying continuously in thickness direction is studied herein by employing the Timoshenko beam theory, von Karman geometrical nonlinearity, and the differential quadrature method. The effect of both localized and global geometric imperfections is taken into account in the analysis. Numerical results show that the fundamental frequency not only increases due to the geometrically nonlinear deformation and initial imperfection but is also significantly influenced by material composition, slenderness ratio and boundary conditions of the actuators.

ACKNOWLEDGEMENT

The work described in this paper was fully supported by a grant from the Research Grants Council of the Hong Kong Special Administrative Region, China [Project no. 9041668 (CityU 114911)]. The authors are grateful for this support.

REFERENCES

[1] Hauke, T., Kouvatov, A., Steinhausen, R., Seifert, W., Beige, H., Langhammer, H. T. and Abicht, H. P., "Bending behavior of functionally gradient materials", *Ferroelectrics* 2000, 238, pp. 195-202.

[2] Kruusing, A., "Analytical and optimization of loaded cantilever beam microactuators", *Smart Materials & Structures*, 2000, 9, pp. 186-196.

[3] Lee, H. J., "Layerwise laminate analysis of functionally graded piezoelectric bimorph beams", *Journal of Intelligent Material Systems and Structures*, 2005, 16, pp. 365-371.

[4] Yang, J. and Xiang, H. J., "Thermo-electro-mechanical characteristics of functionally graded piezoelectric actuators", *Smart Materials & Structures*, 2007, 16 (3), pp. 784-797.

[5] Yang, J. and Huang, X. L., "Nonlinear transient response of functionally graded plates with general imperfections in thermal environments", *Computer Methods in Applied Mechanics and Engineering*, 2007, 196(25-28), pp. 2619-2630.

[6] Bert, C. W., Wang, X. and Striz, A. G., "Differential quadrature for static and free vibration analysis of anisotropic plates", *International Journal of Solids and Structures*, 1993, 30, pp. 1737-1744.

[7] Lu, C. F. and Chen, W. Q., "Free vibration of orthotropic functionally graded beams with various end conditions", *Structural Engineering and Mechanics*, 2005, 20, pp. 465-476.

ANTI-SEISMIC RELIABILITY ANALYSIS OF CONTINUOUS RIGID-FRAME BRIDGE BASED ON ANSYS

* Y. L. Jin

State Key Laboratory of Mechanical System and Vibration, Shanghai Jiao Tong University
Shanghai 200240, China
* Email: jyl19820525@sjtu.edu.cn

KEYWORDS

Anti-seismic reliability, continuous rigid-frame bridge, random variation, Monte-Carlo method, finite element analysis.

ABSTRACT

The continuous rigid-frame bridge with major bending rigidity and torsion rigidity is a continuous structure of the frusta-beam permanent connection. The complex shape of this structure, the special geographic and environmental conditions present one of the greatest challenges for structural engineers. In particular, its dynamic performance under earthquake or wind actions still requires more intensive research. The main objective of this paper is to investigate the earthquake-resistance performance of a continuous rigid-frame bridge under different seismic loads. To this end, an ideal finite element mechanical model of a continuous rigid-frame bridge was first constructed through the use of ANSYS codes. By using this model, the seismic spectrum analysis and the time history analysis were successively carried out to obtain the seismic behaviours of this continuous rigid-frame bridge. Based on the foregoing analysis results and the Monte-Carlo numerical method, the earthquake-resistance reliability was calculated in detail. From the results obtained, it is found that the large-scale continuous rigid-frame bridge has a relatively high seismic-resistance capacity and could satisfy the design requirements under the action of earthquake excitations. The findings of this paper are useful in providing some guidelines on seismic analysis for other similar large-scale structures.

INTRODUCTION

The large continuous rigid-frame bridge [1], with major bending rigidity and torsion rigidity, is a continuous structure of the frusta-beam permanent connection. The typical continuous rigid-frame bridge is a symmetrical structure, and is constructed by adopting the method of cantilever construction. It has satisfactory loading capability, large-scale span capability and slinky outline.

Owing to the randomicity and uncertainty of the earthquake actions, the traditional strength computational method cannot be used to compute the strength of the structure under stochastic loads. As a result, the structure reliability method based on probability theory is presented herein. First, an ideal finite element mechanical model of a continuous rigid-frame bridge was constructed. By using this model, the seismic spectrum analysis and seismic time history analysis were successively carried out to obtain the seismic behaviours of this structure. Finally, based on the computed results and the Monte-Carlo numerical method, the earthquake-resistance reliability was calculated.

FINITE ELEMENT MODEL OF CONTINUOUS RIGID-FRAME BRIDGE

Consider a continuous rigid-frame bridge (V-shape pier) with an overall length of 170 m and height 55 m as shown in Figure 1. Its cross-sectional parameters for the box sections are also shown in Figure 1. According to the structural drawings of the continuous rigid-frame bridge, the corresponding finite element model (FEM) was built up by using the preprocessor module in ANSYS codes, as shown in Figure 2. The FEM considered herein is an idealized, completely symmetrical model.

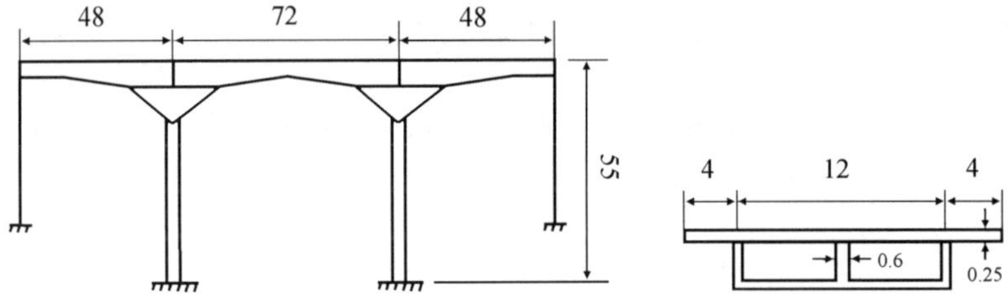

(a) Model of bride (b) Box section

Figure 1 Model of the continuous rigid-frame bridge and its box section

Figure 2 Finite element model of continuous rigid-frame bridge

The top plate, bottom plate, web plate and V-shape bracing of box beams are all represented by the Shell63 element. Prestressing steel reinforcements are characterized by Link8 element. The degrees of freedom at the bottom of V-shape bracing are all restricted. The material properties for concrete are: Young's Modulus $E = 3.25 \times 10^{10}$ N/m², Poisson's ratio $\mu = 0.2$, and density $\rho = 2\,600$ kg/m³ and for the steel reinforcements—Young's Modulus $E = 2.07 \times 10^{11}$ N/m², Poisson's ratio $\mu = 0.3$, and density $\rho = 7\,850$ kg/m³.

SEISMIC RESPONSE SPECTRUM ANALYSIS

Modal analysis is the basis of the response spectrum analysis. By using the Subspace method of modal analysis in ANSYS software, the natural frequencies and the corresponding mode shapes for the first four modes are shown in Table 1 and Figure 3, respectively.

TABLE 1 NATURAL FREQUENCIES AND MAIN VIBRATION CHARACTERISTICS FOR THE FIRST FOUR MODES

Order	Frequency /Hz	Vibration characteristics
1	1.936 7	Anti-symmetric bending of main girder along Y direction
2	3.507 0	Symmetric bending of main girder along Y direction
3	3.769 5	Anti-symmetric torsion of main girder along Z direction
4	3.964 4	Anti-symmetric torsion of main girder along Z direction

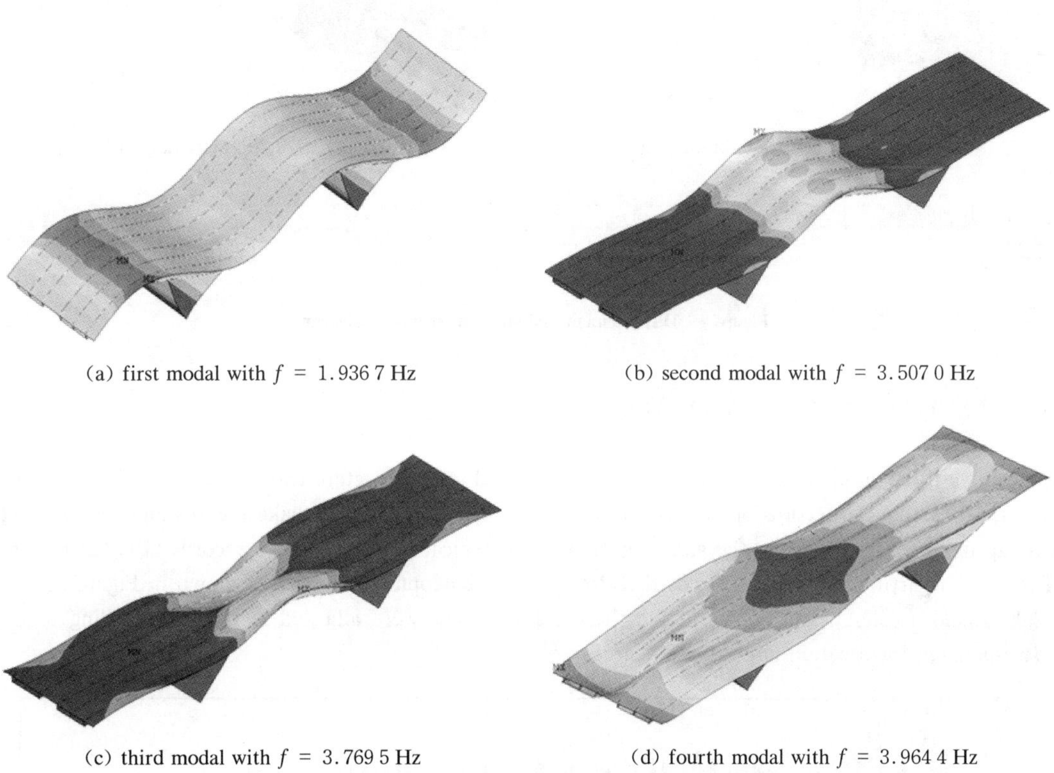

(a) first modal with f = 1.936 7 Hz (b) second modal with f = 3.507 0 Hz

(c) third modal with f = 3.769 5 Hz (d) fourth modal with f = 3.964 4 Hz

Figure 3 The first four mode shapes of continuous rigid-frame bridge

According to the relevent documents and criterions, the site properties for continuous rigid-frame bridge are represented by seismic fortification intensity, 8 degree, II type site condition, characteristic period $T_g = 0.35$ s, maximum value of seismic influence coefficient $\alpha_{max} = 0.24$ and damping ratio $\zeta = 0.05$, respectively [2]. Based on the aforementioned basic parameters, the curve for seismic influence coefficient can be obtained through following equations:

$$\begin{cases} \alpha = 1.32T + 0.108 (0 < T \leqslant 0.1) \\ \alpha = \alpha_{max} = 0.24 (0.1 < T \leqslant 0.35) \\ \alpha = (0.35/T)^{0.9} \times 0.24 (0.35 < T \leqslant 3) \end{cases} \tag{1}$$

Only the seismic behaviors of the continuous rigid-frame bridge under horizontal seismic excitations (longitudinal seismic excitation + transverse seismic excitation) are studied. Figures 4(a) and 4(b) show the displacement distribution diagram and the stress distribution diagram, respectively. It can be seen from Figure 4 that the maximum displacement value occurs at the upper margin 2 of the top plate (the location is indicated by 2), and the maximum stress value is located in the joint between the web plate and the V-shape bracing (the location is indicated by 7).

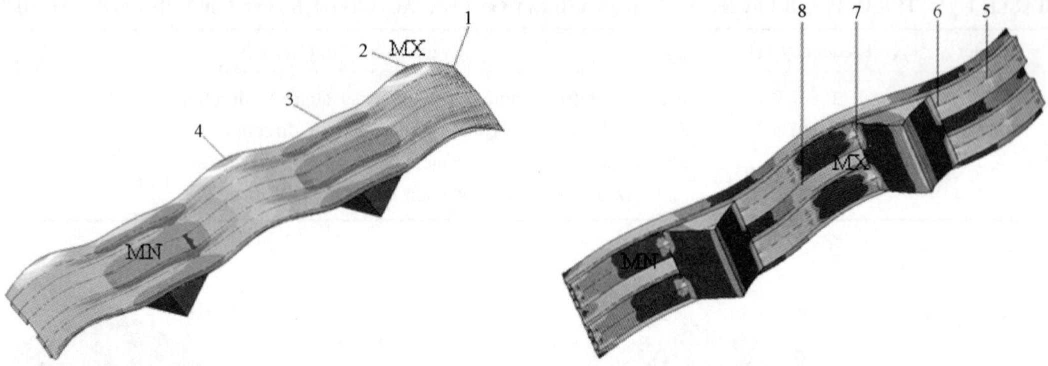

(a) displacement distribution diagram (b) stress distribution diagram

MN MX

Figure 4 Displacement and stress distribution diagrams

SEISMIC TIME-HISTORY ANALYSIS

Generally, three groups of earthquake records are applied to the structure transient time-history analysis: the actual earthquake records of the proposed site, the typical past earthquake records and the artificial earthquake records. Based on the site conditions, two typical past earthquake records (EL Centro and Taft) and an artificial earthquake record (SWNN2) [2] are adopted as input, as shown in Figure 5. The peak ground accelerations (PGA) in the three seismic loads were adjusted to 0.2 g according to the corresponding fortification intensity (8 degrees) [2].

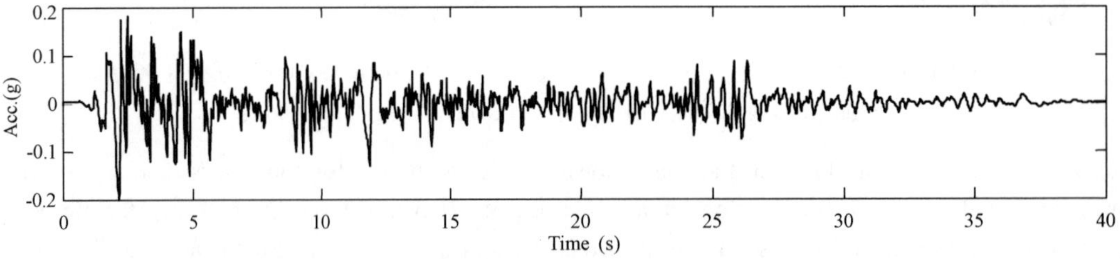

(a) EL-Centro, PGA = 0.2 g

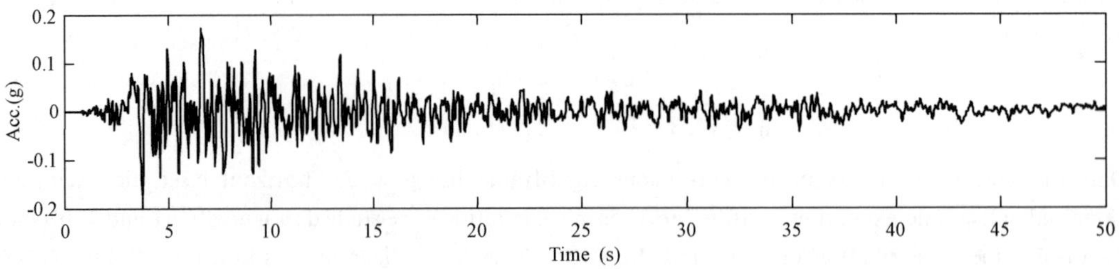

(b) Taft, PGA = 0.2 g

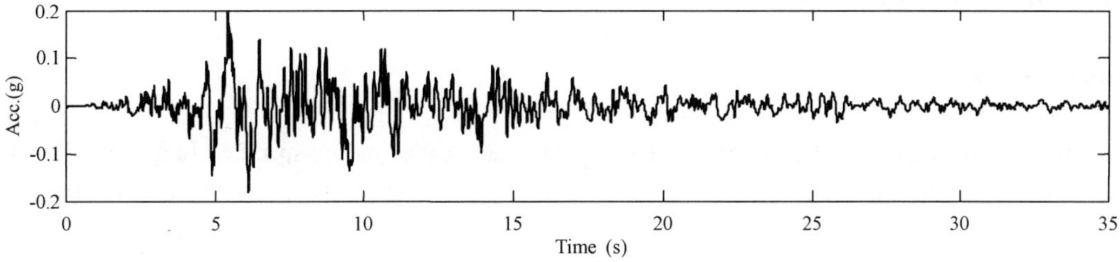

(c) SHWN2, PGA = 0.2 g

Figure 5 Earthquake records

Since it is not possible to furnish the dynamic characteritic of each node due to the heavy workload, it is necessary to select some key nodes as analysis objects during time-history analysis. The concrete positions and the codes of these key nodes are shown in Figure 4 and Table 2.

TABLE 2 CODES AND CONCRETE POITIONS OF KEY NODES

Code	Key nodes	Concrete position
1	Node 3 169	The upper margin 1 of the top plat
2	Node 3 170	The upper margin 2 of the top plat
3	Node 2 299	The upper margin 3 of the top plat
4	Node 1 317	The upper margin 4 of the top plat
5	Node 4 760	Nearby the margin of the top plat
6	Node 4 511	Nearby the joint between the web plate and V-shape bracing
7	Node 4 301	Nearby the joint between the web plate and V-shape bracing
8	Node 4 119	Nearby the middle of the top plate

Table 3 lists the maximum stress and displacement response of some key nodes of continuous rigid-frame bridge under three seismic excitations. It can be seen from Table 3 that the maximum stress value is located at key node 7, and the maximum displacement value occurs at key node 2, which are in agreement with the results of the spectrum analysis.

TABLE 3 MAXIMAL STRESS AND DISPLACEMENT RESPONSE OF CONTINUOUS RIGID-FRAME BRIDGE

Code	Maximum dynamic responses					
	EL Centro		Taft		SHWN2	
	Stress / MPa	**Displ.** / mm	Stress / MPa	**Displ.** / mm	Stress / MPa	**Displ.** / mm
1	—	133.18	—	124.29	—	146.47
2	—	171.24	—	155.47	—	190.36
3	—	76.18	—	50.33	—	89.20
4	—	166.29	—	150.21	—	186.35
5	32.86	—	28.55	—	40.66	—
6	40.62	—	35.69	—	50.25	—
7	42.15	—	38.02	—	53.37	—
8	20.13	—	17.11	—	25.25	

Note that the **Displ.** is the abbreviation of the displacement.

RELIABILITY ANALYSIS OF MONTE-CARLO

Reliability Theory

The structure reliability[3] refers to the probability that can realize its pre-specified function within the prescribed time and conditions. In generally, two factors can affect the reliability of the structure: loading effect S and resisting force R. Here,

$$Z = g(R, S) = R - S \tag{2}$$

Since both the loading effect S and the resisting force R are random variables, Z is also a random variable. This leads to three different results: $Z>0$ (the structure reliability), $Z<0$ (the structure failure) and $Z=0$ (the limited state). Herein, Eq. (3) is called the limited state equation

$$Z = R - S = 0 \tag{3}$$

According to the reliability theory [3], the structure reliability actually solves the probability of the limited state function $Z \geqslant 0$, which can be calculated by using the probability design function (PDF) in the ANSYS codes.

Reliability Index and Failure Probability

The structure failure probability refers to the probability that could not realize its pre-specified function within prescribed time and conditions (the probability of the limited state function $Z<0$). The expression is shown as follows

$$P_f = p(Z<0) = \int_{r-s} f_R(r) f_S(s) dr ds \tag{4}$$

where $f_R(r)$ and $f_s(s)$ are the probability density functions of random variable R and S, respectively.

The structure reliability index β is defined as

$$\beta = \frac{\mu_Z}{\sigma_Z} \tag{5}$$

where, μ_z and σ_z are the mean and the standard deviation of the limited state equation $Z = 0$, respectively. According to the theory of the first-order second-moment method (FOSM), the relationship between structure failure reliability p_f and the structure reliability index β is approximated by

$$P_f = 1 - \Phi(\beta) \tag{6}$$

where $\Phi(\beta)$ is the standard normal distribution function.

Limited State Equation of Carrying Capacity of Continuous Rigid-frame Bridge

The limited state equation of carrying capacity of the continuous rigid-frame bridge is defined as

$$Z = \frac{\sigma_s}{r_{RE}} - \sigma_{\max} \tag{7}$$

where σ_s is the resisting force and r_{RE} the seismic adjustment coefficient of the structure. It is assumed that $\sigma_s = 345$ MPa and $r_{RE} = 1.10$. $\sigma_{\max}$ is the maximum loading effect (maximum stress value) and could be calculated by ANSYS software.

Fitting of The Ground Acceleration

According to the descriptions in Ref. [4], the ground acceleration approximately complies with the extreme value I type distribution. However, there is no corresponding distribution function in the ANSYS codes. Therefore, we can obtain the extreme value I type distribution curves through fitting several standard normal distribution curves in this paper, as shown in Figure 6. It can be observed from Figure 6 that the fitted curves agree fairly well with the initial curves. The means of the three seismic loads after fitting are −0.145 70, −0.049 80 and −0.037 08, respectively. The corresponding standard deviations are 1.034 09, 0.762 29 and 0.543 157, respectively.

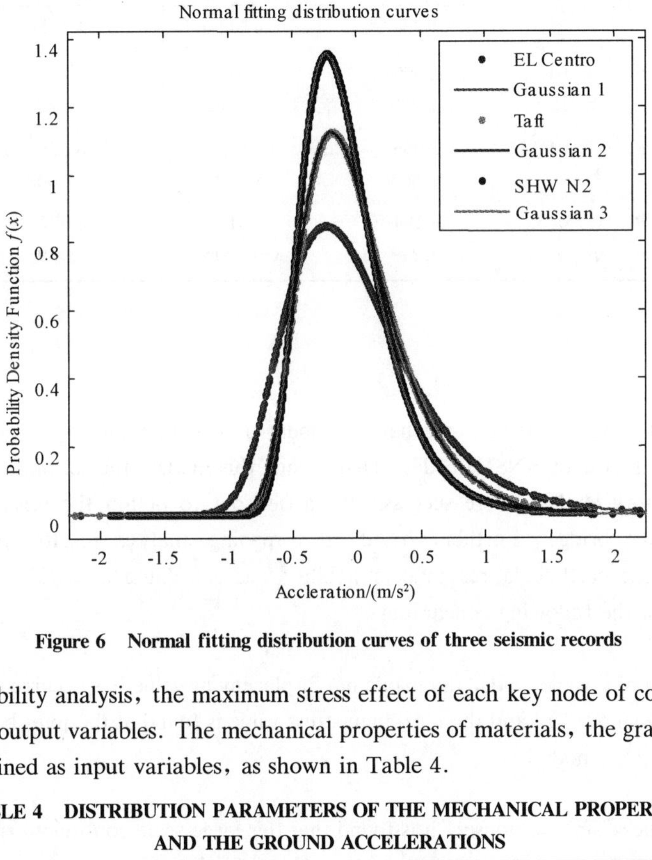

Figure 6 Normal fitting distribution curves of three seismic records

During seismic reliability analysis, the maximum stress effect of each key node of continuous rigid-frame bridge is defined as output variables. The mechanical properties of materials, the gravity, and the ground acceleration are defined as input variables, as shown in Table 4.

TABLE 4 DISTRIBUTION PARAMETERS OF THE MECHANICAL PROPERTIES AND THE GROUND ACCELERATIONS

Item	Mean	S-D	Distribution pattern
Young's Modulus E1	3.25e10	2.55e9	Normal distribution
Density D1	2 700	32.58	Normal distribution
Young's Modulus E2	2.07e11	1.40e10	Normal distribution
Density D2	7 800	52.595	Normal distribution
Yield strength S	345e6	2.45e10	Lognormal distribution
Gravity G	1.73e7	1.25e6	Normal distribution
El Centro seismic load	−0.145 70	1.034 09	Normal distribution
Taft seismic load	−0.049 80	0.762 29	Normal distribution
SHWN2 seismic load	−0.037 08	0.543 157	Normal distribution

Note that the **S-D** is the abbreviation of the standard deviation.

Reliability Results

Based on the Monte-Carlo method, the simulation times are 1 000. The structure failure probability p_f and the reliability index β of these key nodes are shown in Table 5. It can be seen from Table 5 that (1) the maximum failure probability of continuous rigid-frame bridge is at the joint between the web plate and V-shape bracing (i.e. key node 7) and (2) the reliability indexes β of the several main components have appeared to take on negative values which means that this structure has a failure possibility.

TABLE 5 FAILURE PROBABILITY PF AND RELIABILITY INDEX β

Seismic records	P_f and β	The P_f and β value of key nodes (only the stress response)			
		4 760	4 511	4 301	4 119
EL Centro	P_f (E-04)	2 055	5 025	5 925	1 233
	β	0.823	-0.062	-0.235	1.156
Taft	P_f (E-04)	3 056	0.097 2	5 028	1 047
	β	0.509	0.323	-0.008	1.255
SHWN2	P_f (E-04)	4 124	5 816	6 537	1 568
	β	0.219	-0.206	-0.396	1.009

CONCLUSIONS

In this study, an ideal finite element mechanical model of a continuous rigid-frame bridge was first constructed through the use of ANSYS codes. Then, using this model, the seismic spectrum analysis and the seismic time history analysis were successively carried out to obtain the seismic behaviors of this continuous rigid-frame bridge. Further, based on foregoing analysis results and the Monte-Carlo numerical method, the earthquake-resistance reliability was calculated in detail. From the computed results, one can draw the following conclusions:

(1) Under the action of seismic loads, the maximum displacement value has occurred at the upper margin 2 of the top plate (key node 2), and the maximum stress value is found at the joint between the web plate and V-shape bracing (key node 7).

(2) Form the dynamic seismic responses, it is found that this large-scale continuous rigid-frame bridge has relatively high earthquake-resistance capacity and could satisfy the design requirements under the action of earthquake excitations.

(3) The reliability indexes β of the several main components have taken on negative values which means that this structure has a failure possibility.

REFERENCES

[1] Liu Z. Y., Li Q., Zhao C. H. and Jiang J. S., "Analysis of seismic responses of continuous rigid-frame bridge in deep water", *Journal of Earthquake Engineering and Engineering Vibration*, 2009, 29(4), pp. 119-124 (in Chinese).

[2] China Ministry of Transport, *Code of Earthquake Resistance Design for Water Transport Engineering* (JTJ 225-98), Xiamen University Press, 2004, pp. 50-55.

[3] Palle T. C. and Baker M. J., *Structural Reliability Theory and its Applications*, Springer, 1982, pp. 267.

[4] Ou J. P., Duan Y. B. and Liu H. Y., "Structural Random Earthquake Action and Its Statistical Parameters", *Journal of Harbin University of Civil Engineering and Architecture*, 1994, 27(5), pp. 1-10 (in Chinese).

[5] Metropolis N., " The Monte Carlo Method", *Journal of the American Statistical Association*, 1949, 44(247), pp. 335-341.

IJSSD Symposium on Progress in Structural Stability and Dynamics
14-16 April 2012, Nanjing, China

FRACTURE MECHANICS ANALYSIS OF STEEL CONNECTIONS UNDER COMBINED ACTIONS

H. B. Liu and * X. L. Zhao

Department of Civil Engineering, Monash University, Melbourne, VIC 3800, Australia
* Email: ZXL@monash.edu

KEYWORDS

Fracture mechanics, fatigue life, steel connections, combined actions, girth weld.

ABSTRACT

In this paper an analytical method is presented for estimating the fatigue crack growth and fatigue life of steel connections under combined actions. The steel connections are made of two steel plates joined together by single-sided girth welds. The combined actions that applied on the connections include a set of constant amplitude cyclic mode I load and a set of perpendicular static load. The analytical model is developed on the basis of the method of linear elastic fracture mechanics. This model is verified by the experimental results. The comparison between the theoretical and experimental results discloses that the analytical method is able to predict the fatigue crack growth life with acceptable engineering accuracy.

INTRODUCTION

Subsea pipelines have been used to carry oil or gas produced from fields that lie under the sea to shore. Free spans may form due to the irregularities of sea bed or the scouring of underlying soil. Fatigue is a key issue for the girth welded connections in such subsea pipelines (e.g. Busby[1], Langley [2], Macdonald & Maddox [3], Xiao and Zhao [4], [5]). These connections are subjected to combined stresses in most of their service life, which include the fluctuating axial stress normal to the girth welds and the hoop stress that can viewed as static.

The numerical studies based on boundary element method have been carried out to predict the fatigue life, crack growth rate and crack-tip stress intensity factor ranges in the steel connections under combine actions by the authors (Liu and Zhao [6]). The steel connections consist of two steel plates joined together by single-sided girth welds as an approximation of connections in subsea pipelines. However, it is impractical to rely solely on complex and time-consuming boundary element computations for routine engineering designs. It is therefore necessary to develop theoretical models to predict the fatigue life of steel connections under combined actions. In this paper an efficient analytical method is proposed for predicting the fatigue behavior of steel connections under cyclic mode I load and perpendicular static tension. It was proved to be an efficient method by the reasonable agreement between the theoretical results and corresponding experimental results.

EXPERIMENTAL STUDIES

The experimental studies were conducted by the authors (Liu et al. [7]) to investigate the fatigue behavior of steel connections under combined loads. The information related to the current theoretical study was summarized as below.

A total of five specimens were included in this experimental program. Each specimen is made of two steel plates connected by a complete penetrated girth welding. All the steel plates have a uniform thickness of 10 mm. To prevent premature failure, an initial notch was machined on each specimen, consisting of a 5 mm hole and two initial slots. The slots were wire cut with dimensions of 1 mm long and 0.3 mm wide. Two additional holes with diameter of 21 mm were fabricated for the application of perpendicular static load. Details of their geometries, girth welds and initial notch are shown in Figure 1. Six strain gauges were also mounted on each steel plate, which distributions were illustrated in Figure 2.

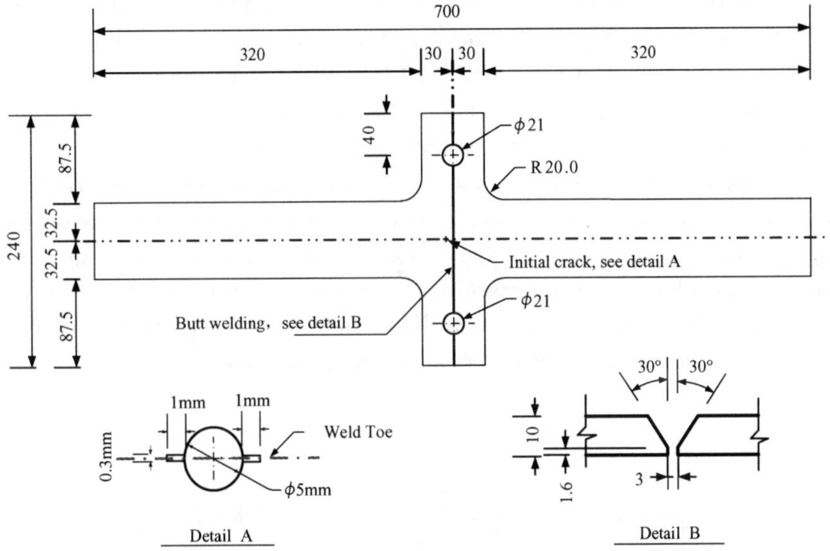

Figure 1 Dimensions and details of the specimens in the preliminary tests

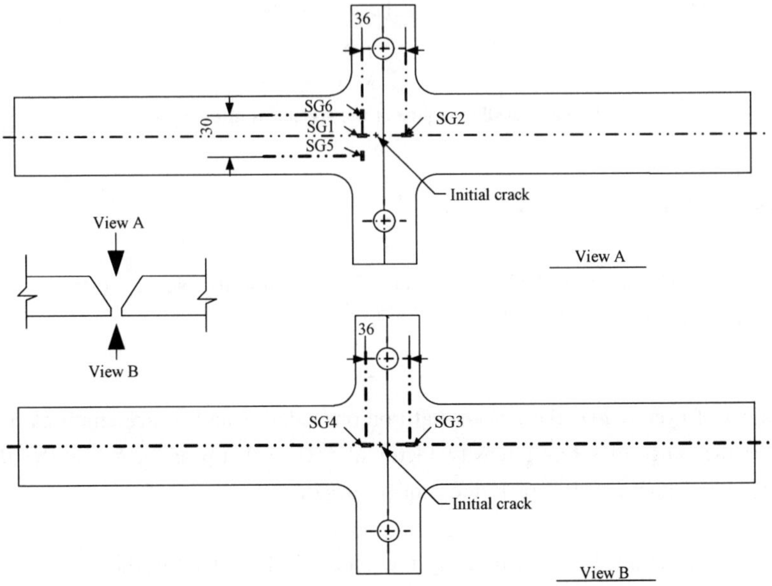

Figure 2 Distribution of strain gauges

The five specimens are made of two batches of steels. Each specimen is given a name according to its steel and the value of perpendicular static force. The specimens of S1H0 and S1H80 are made of the same type of steel with a grade of 300. They were applied 0 kN and 80 kN of perpendicular static load, respectively. The tensile coupon tests were performed using Baldwin Universal testing machine according to Australian Standard 1 391 [8]. Its mean elastic modulus, yield stress and tensile strength are 213 GPa, 385 MPa and 513 MPa, respectively. The specimens of S2H0, S2H40 and S2H80 are made of another batch of steel, which has a mean elastic modulus of 203 GPa, yield stress of 224 MPa and tensile strength of 346 MPa. They were tested by perpendicular static load of 0 kN, 40 kN and 80 kN, respectively.

The specimens were tested to failure under cyclic mode I load and perpendicular static tension load. The loading regime is shown in Figure 3, in which, F_h represents perpendicular static tension load, F_1 represents cyclic loading in the longitudinal direction. The static load was applied using the hydraulic of Enerpac RCH 120, which has a load cell capacity of 120 kN. The load from hydraulic was transferred to the specimen through a loading frame, which was designed to apply the force on two sides of the specimen equally. In the fatigue tests all the specimens were subjected to a uniform cyclic loading with a constant frequency of 15 HZ and a stress ratio of 0.1 until fracture failure occurred. The constant amplitude sine waves range from 6.5 kN to 65 kN. In this experimental program three different values of perpendicular force were tested: 0 kN, 40 kN and 80 kN. Before fatigue test, each specimen was firstly tested by static load on Baldwin Universal machine. The tensile static load was applied at two ends of the steel plate in longitudinal direction. It ran three rounds from 0 kN to 30 kN. During this procedure, strain gauge reading was collected for the calibration of the applied load.

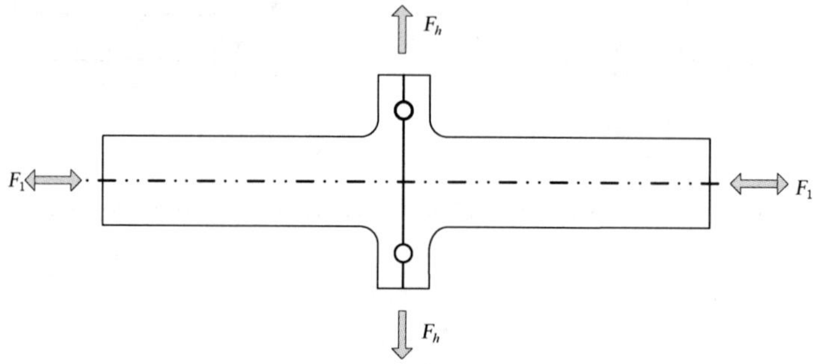

Figure 3 Loading regime of the specimens in the tests

FATIGUE ANALYSIS

Linear Elastic Fracture Mechanics (LEFM) is a widely used method to calculate fatigue crack propagation life, which is mainly based on the Paris Law:

$$da/dN = C \cdot (\Delta K^m - \Delta K_{th}^m) \tag{1}$$

where N is number of cycles, a is the crack length on one side, C and m are empirical material related constants, ΔK is the range of stress intensity factor at the crack tip and ΔK_{th} is the threshold stress intensity factor below which fatigue crack does not propagate.

The LEFM theory was adopted in constructing the theoretical model for the steel connections under combined loads. Figure 4 illustrates the cracks emanating from a circular hole in an infinite plate

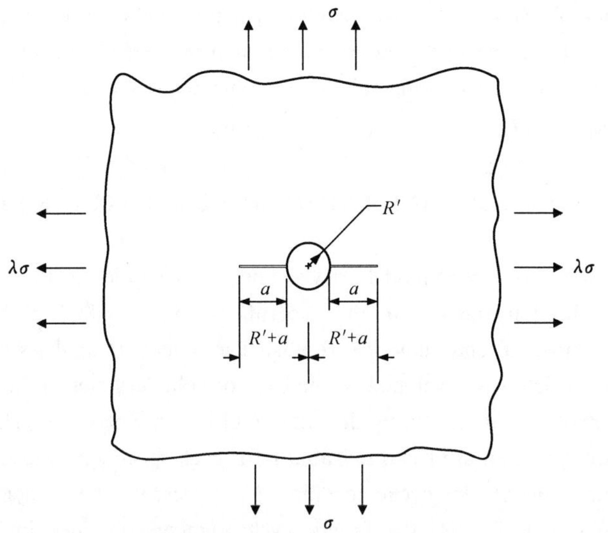

Figure 4 Cracks emanating from a circular hole in an infinite plate subjected to biaxial stress

subjected to biaxial stress. Tada et al. [9] proposed the formulae for its stress intensity factor as below,

$$K_I = \sigma \sqrt{\pi \cdot a} \cdot F_\lambda(s) \tag{2}$$

in which,

$$F_\lambda(s) = (1-\lambda) \cdot F_0(s) + \lambda F_1(s) \tag{3}$$

$$F_0(s) = 0.5 \cdot (3-s)[1+1.243 \cdot (1-s)^3] \tag{4}$$

$$F_1(s) = 1+(1-s)[0.5+0.743 \cdot (1-s)^2] \tag{5}$$

$$s = \frac{a}{R' + a} \tag{6}$$

where λ was defined as the ratio between the stress longitudinal to the crack path and the one in normal direction. However in the current study, the force applied in perpendicular direction is a fluctuating one. Therefore the definition of λ was specified, in the present calculation, as the ratio between the stress longitudinal to the crack path and the maximum value of the fluctuating axial stress. As shown Figure 4, these formulas were proposed under the condition that the plate width is infinite. If they are applied to the steel specimens introduced above, some modification is needed. Albrecht and Yamada [10] had ever proposed a correction factor of F_w for a central crack in a plate with finite width. This factor is able to account for the effects of finite width on stress intensity factor:

$$F_w = \sqrt{\sec \frac{\pi \cdot (R' + a)}{2W}} \tag{7}$$

Therefore, the stress intensity factor, K, for the crack emanating from a circle hole in a finite width plate under combined stress, was modified as below,

$$K = \sigma \sqrt{\pi \cdot a} \cdot F_\lambda(s) \cdot F_w \tag{8}$$

The predicted fatigue life (N_t) is calculated by integrating Eqn. 1 as:

$$N_t = \int_{a_i}^{a_f} \frac{1}{C \cdot (\Delta K^m - \Delta K_{th}^m)} da \tag{9}$$

where a_i and a_f are the initial and final size of crack. The initial crack of a_i is equal to 2.5 mm, which is the radius of its the initial crack. The final crack of a_f is the last value of crack size before the crack growths beyond the edge. For steels operating in marine environments at temperatures up to 20 °C, with or without cathodic protection the following values are recommended in BS7910 [11]: $m = 3$, $C = 2.3 \times 10^{-12}$, $\Delta K_{th} = 2.0$ MPa $\sqrt{m}$.

COMPARISON OF NUMERICAL AND EXPERIMENTAL FATIGUE LIVES

To verify the accuracy of the proposed model, it was applied in predicting the fatigue life of the steel connections, which had been introduced in the experimental program. The theoretical results were compared with their experimental and numerical results. The numerical analysis was conducted by using the software of BEASY, which was developed on the basis of boundary element analysis method. Details are able to be found in another publication by the authors (Liu and Zhao [6]). The comparison between these results and the load condition are listed in Table 1 for each specimen. F_{1max} and F_{1min} represent the maximum and minimum value of the cyclic loading. N_e represents the fatigue life obtained in the experimental testing. N_{2D} and N_{3D} are the fatigue cycle numbers obtained in 2D and 3D numerical analysis, respectively. The symbol "n" is a correction factor. The cyclic load adopted in the theoretical model was adjusted for each specimen by multiplying a factor of "n". This factor was obtained through strain gauge reading. It accounts for the occurrence of secondary shell bending stresses introduced by material shrinkage of girth welding.

As shown in Table 1, the ratios (N_t/N_e) between the theoretical and experimental fatigue lives are varied from 0.76 to 1.08. For the specimen of S1H80, the greatest difference was observed. Its estimated fatigue life is 24% less than its experimental result with a faster theoretical crack growth. Aside from S1H80, the theoretical fatigue lives are in excellent accordance with their experimental value. The difference is no more than 8%. When they are compared with the numerical results, the ratios of N_t/N_{2D} are ranged from 0.80 to 0.95 and the ratios of N_t/N_{3D} are ranged from 0.83 to 1.03. It disclosed that the theoretical predictions tally well with the numerical results.

Figure 5 shows the relationships of the theoretical results with their experiential and numerical results. It is clearly stated that the fracture mechanics model is able to provide a close estimation for the steel connections under combined actions. A much closer estimation is able to be achieved for the steel connections with shorter fatigue lives.

TABLE 1 COMPARISONS OF THE THEORETICAL RESULTS WITH THEIR NUMERICAL AND EXPERIMENTAL RESULTS

Items/No.	S1H0	S1H80	S2H0	S2H40	S2H80
F_{1max} (kN)	65	65	65	65	65
F_{1min} (kN)	6.5	6.5	6.5	6.5	6.5
F_h (kN)	0	80	0	40	80
n	1.10	1.04	1.05	1.05	1.06
N_e (million)	1.36	2.55	1.45	1.64	1.93
N_t (million)	1.36	1.93	1.56	1.71	1.82
Ratio (N_t/N_e)	1.00	0.76	1.08	1.04	0.94
N_{2D} (million)	1.43	2.40	1.66	1.96	2.28
Ratio (N_t/N_{2D})	0.95	0.81	0.94	0.87	0.80
N_{3D} (million)	1.32	2.28	1.53	1.83	2.20
Ratio (N_t/N_{3D})	1.03	0.85	1.02	0.94	0.83

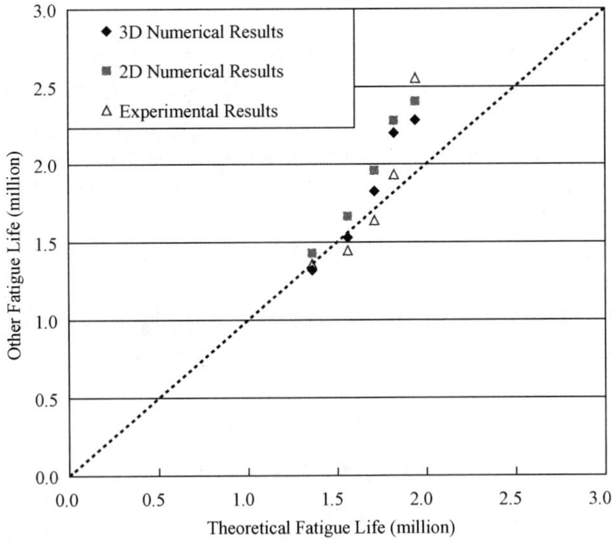

Figure 5 Relationships among the numerical, theoretical and experimental fatigue lives

CONCLUSIONS

This paper presented an analytical model for predicting fatigue life of steel connections under combined loads. The theoretical models are developed on the basis of Paris Law. The stress intensity factors for the cracks emanating from a circular hole in a plate subjected to biaxial stress is a modification of the formulae given by Tada et al. [9]. To verify its accuracy, the model was applied in predicting the fatigue life of specimens which had been tested. The good agreement between the theoretical and experimental results disclosed that the analytical method is able to predict the fatigue life with acceptable engineering accuracy.

ACKNOWLEDGEMENT

This research is being undertaken within the CSIRO Wealth from Oceans Flagship Cluster on Subsea Pipelines with funding from the CSIRO Flagship Collaboration Fund and Monash University.

REFERENCES

[1] Busby, R.F., "Underwater inspection/testing/monitoring of offshore structures", *Ocean Engineering*, 1979, 6 (4), pp. 355-491.

[2] Langley, R.S., "Random dynamic analysis of towed pipelines", *Ocean Engineering*, 1989, 16(1), pp. 85-98.

[3] Macdonald, K.A. and Maddox, S.J., "New guidance for fatigue design of pipeline girth welds", *Engineering Failure Analysis*, 2003, 10(2), pp. 177-197

[4] Xiao, Z.G. and Zhao, X.L., "Stress analysis of free spanning subsea pipelines with finite element method", *Proceedings of 10th International Symposium on Structural Engineering for Young Experts*, Changsha, China, 2008.

[5] Xiao, Z.G. and Zhao, X.L., "Prediction of natural frequency of free spanning subsea pipelines", *International Journal of Steel Structures*, 2010, 10(1), pp. 81-89.

[6] Liu, H. B. and Zhao, X. L., "Fatigue behaviours of steel connections under combined actions", *Journal of Advances in Structural Engineering*, 2011.

[7] Liu, H.B., Zhao, X.L. and Xiao Z.G., "Fatigue testing of subsea pipeline steel connections under combined actions", *Proceedings of 21st Australian Conference on the Mechanics of Structures and Materials*, Melbourne, 2010, pp.6 49-655.

[8] Australia Standards 1391, "Methods of tensile testing of metals", Standards Australia, Sydney, 1991.

[9] Tada, H., Paris, P.C. and Irwin, G.R., "The stress analysis of cracks handbook", *The American Society of Mechanical Engineers*, New York, 2000, p. 289.

[10] Albrecht P. and Yamada K., "Rapid calculation of stress intensity factors", *Journal of the Structural Division, ASCE*, 1977, 103(ST2), pp. 377-389.

[11] British Standards 7910, "Guide to methods for assessing the acceptability of flaws in metallic structures", *British Standard Institute*, 2005, pp. 57-58.

SECOND-ORDER ANALYSIS FOR LONG SPAN STEEL STRUCTURE PROTECTING A HERITAGE BUILDING

Y.P. Liu, S.W. Liu, Z.H. Zhou and * S.L. Chan

The Hong Kong Polytechnic University, Department of Civil and Structural Engineering Hong Kong, China
* Email: ceslchan2@gmail.com

KEYWORDS

Second-order analysis, steel structures, structural design.

INTRODUCTION

Second-order analysis has been widely referred in codes and claimed by structural analysis software. There are actually many types of second-order analysis with the simplest one that considers only the P-Δ effect and other more advanced ones that include the $P-\Delta-\delta$ effect with careful consideration of initial imperfections as well as plastic hinges. In Hong Kong, which has one of the most highest world's density in high-rise steel composite buildings with height greater than 150 m, this method of design is widely used and the success is partly due to the use of one-element-per-member model which saves time and efforts in design and improves safety as well. This paper is concerned with the protection of a heritage building. A long span steel building of clear span 30 m and clear height 14 m is designed to house the heritage building using the second-order analysis to Eurocode - 3 [1] without the assumption of effective (or buckling) length. The computer method employs the reliable curved element with curvature set as member initial imperfection so that the member P-δ effect could be directly considered in the analysis [2, 3]. It further utilizes the critical mode as initial imperfection mode for the global frame $P-\Delta$ imperfection in its checking of sway stability. The advantages of the proposed design method based on second-order analysis include its ignorance of the uncertain assumption of effective length for the vertical columns which are subject to heavy axial forces from the loads of a restaurant at the top of the steel building. As the design method uses only the basic material properties of the material as well as their imperfections, it could be applied to the analysis of structures of different ages allowing for deterioration of material properties under various scenarios like seismic [4] and materials by simply adopting the properties of materials adopted for the structures. Consequently, the proposed new method could become a unified design method for design for structures under complex loads and scenarios and this new codified method is replacing the old effective length method.

GENERAL

Heritage buildings are with significant architectural, cultural and economical values to societies, and its protection has gained wide acceptance by the public. The first protection method is to strengthen the heritage buildings directly, which might inevitably affect the original architecture. The second method, which is commonly used for the heritage buildings in low-seismic zone, like Hong Kong or southern China, is to build a larger framed structure to cover the entire historical building so that it can prevent them from further deterioration caused by natural hazards or damage by tourists. This paper is mainly focused on the second method and a second-order design method is proposed for the design of a new structure to protect a heritage building.

One of the main concerns for structural design is stability check. Since the concept of elastic Euler's buckling load introduced a century ago, the linear analysis and the effective length design method was developed and used for practical design, code development and research reference in the past several decades. The design practiced by most engineers is still based on this concept of undeformed geometry for equilibrium which is obviously incorrect as no deformation leads to no displacement, no strain and no stress which cannot balance the eternal forces. In actual practice, the method leads to the need of using separate codes for design of members made of different materials, for example, Eurocode-2 [5] is used for reinforced concrete members, Eurocode-3 [1] for steel members and Eurocode-4 [6] for composite members and no code for design of special structural forms like pre-tensioned steel trusses or bamboo scaffolds. This brings much inconvenience and sometimes confusion to the design engineer. Chan *et al.* [7] propose a practical numerical approach for structures made of various common building materials with the consideration of geometrical instability and material nonlinearity that all structural members can be designed in a unified way with the section capacity check carried out for all members to insure safety in stability, strength and ductility.

Although the description for the stability check in steel, concrete and composite codes may be different, the requirement for consideration of second-order effects such as $P-\Delta$ and $P-\delta$ effects is conceptually and numerically similar. It is noted that the $P-\delta$ effect and imperfections are still commonly ignored in most previous structural analysis and design software and therefore the tedious member buckling strength design by code is still needed and the software cannot be used for a proper second-order analysis. In this paper, by using the Pointwise-Equilibrium-Polynomial (PEP) element [8] and allowing for initial imperfection in a robust nonlinear incremental-iterative procedure, both the $P-\delta$ effect of individual members and the $P-\Delta$ effect and imperfections of the structural system can be simulated. Thus, a unified design method is developed for the present project and no tedious member design checks to various codes are required.

The proposed method has been extended to inelastic analysis by using the refined plastic hinge approach for performance-based seismic design, advanced analysis and progressive analysis. In order to capture the gradual yielding under the interaction of axial and bending effects, the first and full yield surfaces for a section (which might contain steel, reinforcement and concrete materials) are necessary for the analysis. Herein, the sectional fibre approach can be adopted to calculate the stress resultants of the concrete, the reinforcement and the structural steel. Both the strength reduction and stiffness deterioration can be represented in the proposed method. However, in this paper, the plastic hinge approach for second-order plastic analysis of steel member is mainly discussed, while the application of the method to structures of other materials is presented elsewhere [7].

In this paper, the basic element formulation for considering the geometric nonlinearity with initial imperfections and plastic hinge formulation is discussed. Finally, two examples are presented for demonstration of the validity, accuracy and advantages of the proposed practical second-order analysis method for practical design.

BASIC ASSUMPIONS AND DEFINITIONS

In the formulation of the beam-column element, the following widely used assumptions for practical design are adopted: (1) The Euler-Bernoulli hypothesis is valid and warping is neglected; (2) strains are small but the deflections can be large; (3) plane section normal to the centroid axis before deformation remains plane after deformation and normal to the axis; (4) the concept of lumped plasticity is employed, i. e., the yielding of material is assumed to be concentrated at the both ends of beam-column element; and (5) loads are conservative and shear distortions are negligible.

GEOMETRIC NONLINEARITY

The $P-\Delta$ and $P-\delta$ effects are two principal parameters needed to be considered in the second-order or advanced analysis. In the topic of formulating a design element capturing the $P-\delta$ effect of a member, Chan and Zhou [8] developed several elements with different features to simplify the analysis procedure and make the advanced analysis practicable. In this paper, the Pointwise Equilibrating Polynomial (PEP) element proposed by Chan and Zhou [8] is adopted. The PEP element is capable of modelling the member initial curvature which is mandatory for buckling design in various codes. The basic force-displacement relations in an element are illustrated in Figure 1 and more details about its formulation can be referred to the original papers.

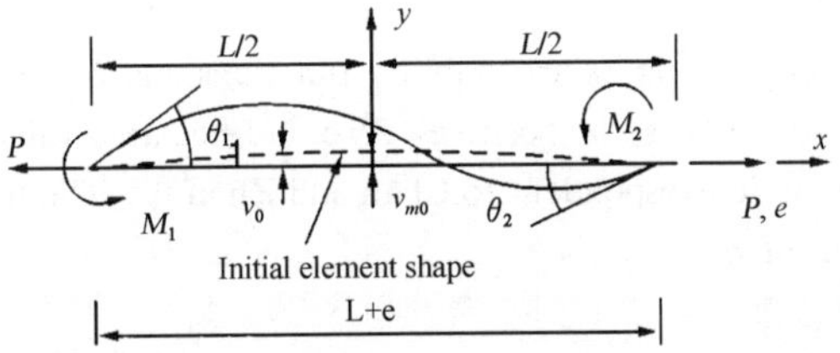

Figure 1 Basic forces versus displacements relations in an elastic element

The equilibrium equation is given by

$$EI\ddot{v} = P(v+v_0)+\frac{M_1+M_2}{L}\left(\frac{L}{2}+x\right)-M_1 \tag{1}$$

in which E is the Young's modulus of elasticity, I the second moment of area, L the member length, v the lateral displacement due to applied loads expressed in equation 2 below, v_0 the initial member deflection, P the axial force, and M_1 and M_2 the nodal end moments. A super dot represents a differentiation with respect to the distance x along an element.

$$v = \sum^{i=0,5} a_i x^i \tag{2}$$

$$v = \frac{M_1}{P}\left[\frac{\sin(\phi - kx)}{\sin\phi} - \frac{L-x}{L}\right] - \frac{M_2}{P}\left[\frac{\sin kx}{\sin\phi} - \frac{x}{L}\right] + \frac{q}{1-q} v_{m0}\sin\frac{\pi x}{L} \tag{3}$$

v_{m0} is the amplitude of initial displacement.

By superimposing the deflection to the initial imperfection, we have the final offset of the element centroidal axis from the axis joining the two ends of the element as,

$$v = v_1 + v_0 = \frac{M_1}{P}\left[\frac{\sin(\phi - kx)}{\sin\phi} - \frac{L-x}{L}\right] - \frac{M_2}{P}\left[\frac{\sin kx}{\sin\phi} - \frac{x}{L}\right] + \frac{1}{1-q} v_{m0} \sin\frac{\pi x}{L} \tag{4}$$

in which,

$$k = \sqrt{\frac{P}{EI}}; \tag{5}$$

$$\phi = kL; \tag{6}$$

$$q = \frac{P}{P_{cr}} = \frac{PL^2}{\pi^2 EI} \tag{7}$$

where P_{cr} is the buckling axial force parameter given by $P_{cr} = \frac{\pi^2 EI}{L^2}$.

By differentiating Eq. (3) with respect to x, and expressing the rotations at two ends as the nodal rotations as $dv/dx\mid_{x=0} = \theta_1$ and $dv/dx\mid_{x=L} = \theta_2$, we have

$$M_1 = \frac{EI}{L}\left[c_1\theta_1 + c_2\theta_2 + c_0\left(\frac{v_{m0}}{L}\right)\right] \tag{8}$$

$$M_2 = \frac{EI}{L}\left[c_2\theta_1 + c_1\theta_2 - c_0\left(\frac{v_{m0}}{L}\right)\right] \tag{9}$$

Axial strain can be expressed in terms of the nodal shortening, u and the bowing due to initial imperfection and deflection as

$$\varepsilon = \frac{u}{L} + \frac{1}{2}\int_L \left[\frac{dv_0}{dx}\right]^2 - \frac{1}{2}\int_L \left[\frac{dv}{dx}\right]^2 dx \tag{10}$$

$$P = EA\varepsilon = EA\left[\frac{u}{L} - b_1(\theta_1+\theta_2)^2 - b_2(\theta_1-\theta_2)^2 - b_{vs}\frac{v_{m0}}{L}(\theta_1-\theta_2) - b_{vv}\left(\frac{v_{m0}}{L}\right)^2\right] \tag{11}$$

In Eqs. (8), (9) and (11), c_1, c_2 and c_0 are stability functions and b_1, b_2, b_{vs} and b_{vv} are curvature functions. They are derived for the case of positive, zero and negative values of axial force parameter, q. in which, c_1, c_2, b_1, b_2 are correspondent to Chan and Zhou [8]. The terms c_0, b_{vs} and b_{vv} are new and can be expressed in terms of q as

- For compression, $q > 0$

$$c_0 = -\frac{\pi q \phi \sin\phi}{(1-q)(1-\cos\phi)} \tag{12}$$

$$b_{vs} = \frac{\phi[2\sin\phi - (1-q)(\phi - \sin\phi)]}{2\pi(1-q)^2(1-\cos\phi)} \tag{13}$$

$$b_{vv} = \frac{q\phi[(1-q)(\phi-\sin\phi) - 4\sin\phi]}{2(1-q)^3(1-\cos\phi)} + \frac{\pi^2 q(2-q)}{4(1-q)^2} \tag{14}$$

in which $\phi^2 = \frac{PL^2}{EI} = \pi^2 q$.

- For no axial force, $q = 0$

$$c_0 = 0, \tag{15}$$

$$b_{vs} = 2/\pi \tag{16}$$

$$b_{vv} = 0 \tag{17}$$

- For tension, $q < 0$

$$c_0 = -\frac{\pi q \psi \sinh\psi}{(1-q)(\cosh\psi - 1)} \tag{18}$$

$$b_{vs} = \frac{\psi[2\sinh\psi - (1-q)(\psi - \sinh\psi)]}{2\pi(1-q)^2(1-\cosh\psi)} \tag{19}$$

$$b_{vv} = \frac{q\psi[(1-q)(\psi-\sinh\psi)-4\sinh\psi]}{2(1-q)^3(\cosh\psi-1)} + \frac{\pi^2 q(2-q)}{4(1-q)^2} \tag{20}$$

in which $\psi^2 = -\dfrac{PL^2}{EI} = -\pi^2 q$

The variation of the parameters c_0, b_{vs} and b_{vv} with respect to the axial force, q is found to be considerable when q is large. b_{vs} and b_{vv}, can be expressed in terms of q, c_1, c_2 and c_0 as follows,

$$b_{vs} = \frac{c_1 - c_2}{\pi(1-q)^2} + \frac{c_2 c_0}{2(c_1+c_2)(c_1-c_2)} \tag{21}$$

$$b_{vv} = \frac{\pi^2 q(2-q)}{4(1-q)^2} + \frac{2c_0}{\pi(1-q)^2} + \frac{c_2 c_0^2}{2(c_1+c_2)(c_1-c_2)^2} \tag{22}$$

Furthermore, the axial force parameter q can be written in place of the explicit expression in Eq. (11) as

$$q = \frac{\lambda^2}{\pi^2}\left(\frac{u}{L} - c_b - c_{b0}\right) \tag{23}$$

in which λ = slenderness ratio given by $\lambda = \dfrac{L}{\sqrt{I/A}}$.

The coefficients c_b and c_{b0} are for the effect of deformed and initial curvatures on the axial force parameter q, given by the following expressions

$$c_b = b_1(\theta_1+\theta_2)^2 + b_2(\theta_1-\theta_2)^2 \tag{23}$$

$$c_{b0} = b_{vs}\frac{v_{m0}}{L}(\theta_1-\theta_2) + b_{vv}\left(\frac{v_{m0}}{L}\right)^2 \tag{24}$$

Repeating the procedure for the other principal axes, the general stiffness equation for an element in 3D space can be written as,

$$M_{1n} = \frac{EI_n}{L}\left[c_{1n}\theta_{1n} + c_{2n}\theta_{2n} + c_{0n}\left(\frac{v_{m0n}}{L}\right)\right] \tag{25}$$

$$M_{2n} = \frac{EI_n}{L}\left[c_{2n}\theta_{1n} + c_{1n}\theta_{2n} - c_{0n}\left(\frac{v_{m0n}}{L}\right)\right] \tag{26}$$

$$P = EA\left[\frac{u}{L} - \sum_{n=y,z}\left(b_{1n}(\theta_{1n}+\theta_{2n})^2 + b_{2n}(\theta_{1n}-\theta_{2n})^2 + b_{vsn}\frac{v_{m0n}}{L}(\theta_{1n}-\theta_{2n}) + b_{vvn}\left(\frac{v_{m0n}}{L}\right)^2\right)\right] \tag{27}$$

where the torsional degree of freedom is taken as Chan and Zhou [8]

$$M_t = \left(\frac{GJ + Pr_0^2}{L}\right)\phi \tag{28}$$

in which M_{1n} and M_{2n} are conjugate moments to θ_{1n} and θ_{2n}, M_t is the torsional moment, P is the axial force with sign positive for compression, A is the cross sectional area, I_n is the second moment of area about the n-axis, E is the Young's modulus of elasticity, GJ is the torsional constant, r_0 is the polar radius of gyration given by $r_0 = \sqrt{\int(y^2+z^2)\mathrm{d}A/A}$. c_{1n}, c_{2n} and c_{0n} are the stability functions associated with bending moments and n-axis, b_{1n}, b_{2n}, b_{vsn} and b_{vvn} are the bowing functions for n-axis.

By expressing the axial force parameter q from Eq. (27), we have

$$q = \frac{\lambda^2}{\pi^2}\left[\frac{u}{L} - \sum_{n=y,z}\left(b_{1n}(\theta_{1n}+\theta_{2n})^2 + b_{2n}(\theta_{1n}-\theta_{2n})^2 + b_{vsn}\frac{v_{m0n}}{L}(\theta_{1n}-\theta_{2n}) + b_{vvn}\left(\frac{v_{m0n}}{L}\right)^2\right)\right] \tag{29}$$

The secant stiffness matrix, which relates the equilibrium equations between forces, moments, displacements and rotations, can be obtained by the energy principle. For incremental-iterative nonlinear procedure, the tangent stiffness matrix which relates the incremental forces, moments to rotations and displacements is needed and can be formed by the second variation of the total potential energy functional.

The remarkable advantage of proposed advanced analysis by PEP element is its automatic computation of primary linear and secondary non-linear stresses such that the assumption of K-factor or effective length factor is avoided. The influence on member stiffness in the presence of axial load is also allowed for in the stress computation and analysis. The first global eigenvalue buckling mode shape is used to determine the direction of local member imperfection and the global frame imperfection due to out-of-plumbness which is normally taken in the range from 1/1 000 to 1/200 of the building height H. The member initial imperfection can be obtained from codes such as Table 5.1 in Eurocode-3 [1] and Table 6.1 in HKSC [9] and the imperfections of HKSC [9] are adopted herein as they are more consistent with the buckling curves. The tangent stiffness matrix could be obtained by the second variation of the total potential energy.

MATERIAL NONLINEARITY

For inelastic analysis, it is necessary to assume a yield function to monitor the gradual plastification of a section. This refined plastic hinge approach introduced by Chan and Chui [10] is revised and adopted in this study. The nodal rotations of a deformed PEP element with pseudo-springs at the end nodes are shown in Figure 2.

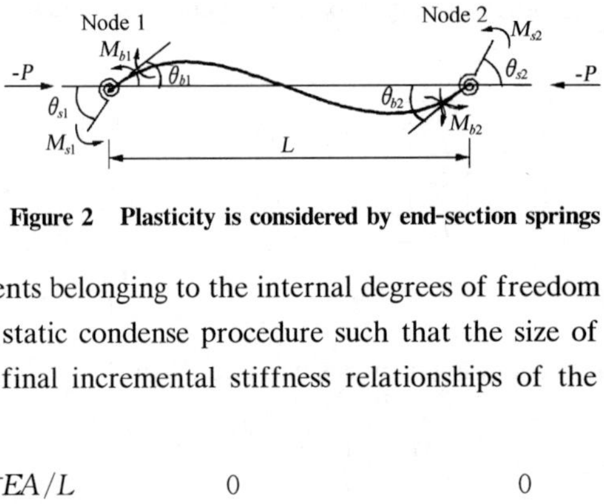

Figure 2 Plasticity is considered by end-section springs

The zero-length spring elements belonging to the internal degrees of freedom of beam-column element can be eliminated by a standard static condense procedure such that the size of the element stiffness matrix will not be increased. The final incremental stiffness relationships of the hybrid element can also be formulated.

$$\begin{Bmatrix} \Delta P \\ \Delta_e M_1 \\ \Delta_e M_2 \end{Bmatrix} = \begin{bmatrix} EA/L & 0 & 0 \\ 0 & S_1 - S_1^2(K_{22}+S_2)/\beta & S_1 S_2 K_{12}/\beta \\ 0 & S_1 S_2 K_{21}/\beta & S_1 - S_2^2(K_{11}+S_1)/\beta \end{bmatrix} \begin{Bmatrix} \Delta L \\ \Delta_e\theta_1 \\ \Delta_e\theta_2 \end{Bmatrix} \tag{30}$$

in which S_1 and S_2 are the stiffness of the end springs, K_{ij} are the flexural stiffness of the PEP element considering the presence of axial force, ΔP is the axial force increment, $\Delta_e M_1$ and $\Delta_e M_2$ are the incremental nodal moments at the junctions between the spring and the global node and between the beam and the spring, ΔL is the axial deformation increment, $\Delta_e\theta_1$ and $\Delta_e\theta_2$ are the incremental nodal rotations corresponding to these moments, and

$$\beta = \begin{vmatrix} K_{11}+S_1 & K_{12} \\ K_{21} & K_{22}+S_2 \end{vmatrix} = (K_{11}+S_1)(K_{22}+S_2) - K_{12}K_{21} > 0 \tag{31}$$

The section spring stiffness, S, can be calculated by the following equation,

$$S=\frac{6EI}{L}\left(\left|\frac{M_{pr}-M}{M-M_{er}}\right|+\rho\right) \tag{32}$$

where EI is the flexural constant, L is the member length and M_{er} and M_{pr} are respectively the first and full yield moments reduced due to the presence of axial force and the ρ represents a strain-hardening parameter. From this equation, the section stiffness varies from infinity to a small strain-hardening value which represents three sectional stages, i. e., elastic, partially plastic and fully plastic with strain-hardening stages.

EXAMPLE: A long span steel frame for housing the heritage building

Two examples are chosen to demonstrate the application of the second-order analysis for buckling design and analysis of steel frames. The first one is the deep toggle frame where the member is under a high axial load and the second example is about the elastic and plastic second-order analysis of a long-span portal frame housing a heritage building which is not allowed to have contact with the protected portal frame.

In this example, a long span steel building of clear span 30 m and clear height 14 m is designed to house the heritage building using the proposed second-order analysis to Eurocode-3 [1]. The structural layout is shown in Figure 3 and all members are steel grade S275. There is a restaurant at the lower level of the truss supported by the steel columns for tourists and the live load is taken as 5 kPa. A downward wind of 2 kPa from the top is considered in the load case studied due to the windy location of structure. The applied live and wind loads are shown Figure 4.

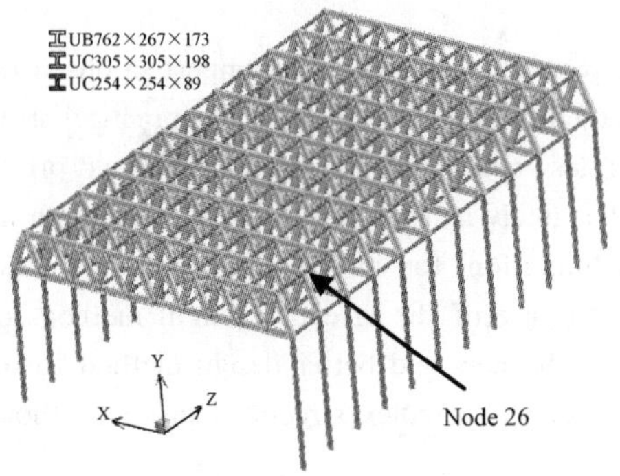

Figure 3 Structural layout

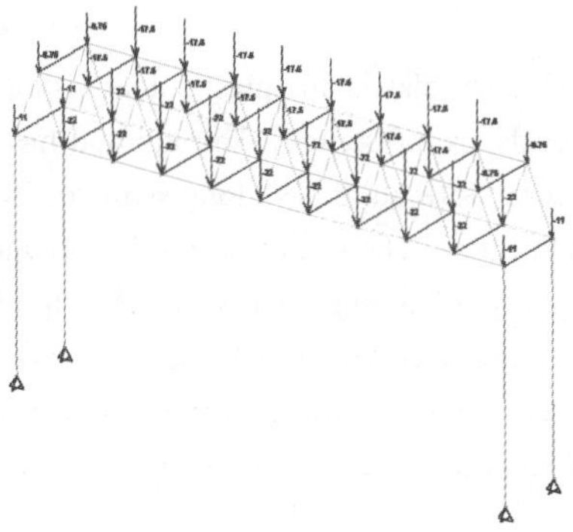

Figure 4 Design loads (one typical bay)

The traditional second-order analysis without consideration of initial imperfection does not adopt the most critical buckling mode as the initial geometry of the structure and the design is therefore not addressing to the critical scenario. Further, the $P-\delta$ moment is also not considered properly and completely in the conventional second-order analysis since the member curvature of the conventional cubic element does not very with external load and it is unable to capture the buckling behaviour of the member.

In this paper, the first eigen-mode shape (see Figure 5) is used to determine the global frame and local member initial imperfection and therefore the second-order effects are captured. As a result, only section capacity check is required and the member design by effective length method is avoided. The structure is first designed by the second-order elastic analysis in conjunction with the first plastic hinge concept. Secondly, the second-order plastic analysis is used to check the ultimate strength reservation. The load-deflection curves obtained from the proposed second-order elastic and plastic analysis are shown in Figure 6. The second-order elastic design load factor is obtained as 1.57 compared with the maximum load factor 1.74 obtained by second-order plastic analysis. In the former second-order elastic analysis, the design load is assumed to be the minimum load causing the formation of the first plastic hinge and it therefore ignores strength reserve after the first plastic hinge. On the other hand, the second-order plastic analysis continues the analysis until the maximum load in the load-deflection plot is obtained and therefore it represents the true collapse load of the structure under the design assumptions.

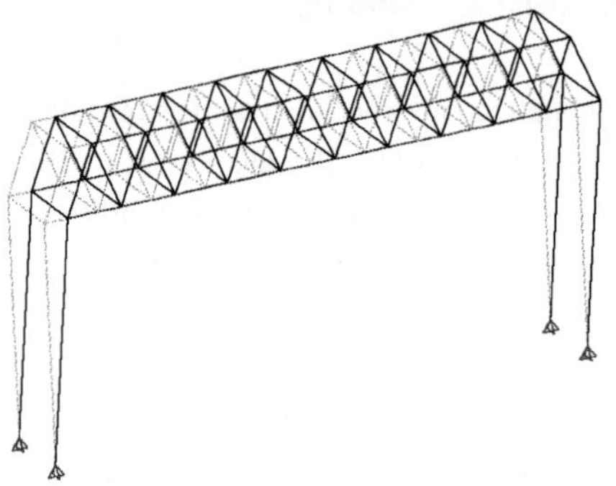

Figure 5 Imperfection using 1st eigen-mode (one typical bay)

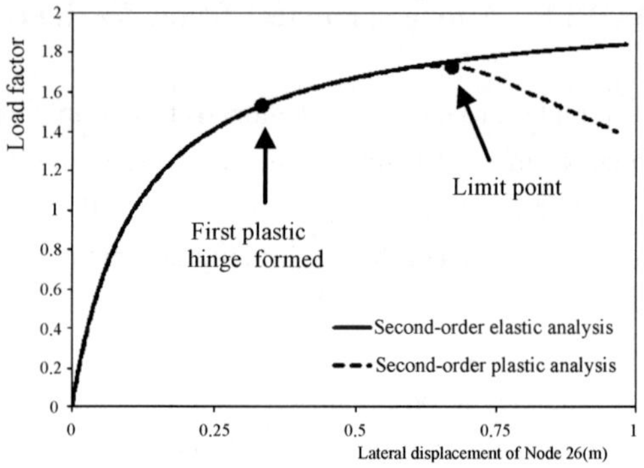

Figure 6 Results from second-order elastic and plastic analysis

CONCLUSIONS

The second-order elastic and plastic analysis allowing second-order effects and imperfections at frame and member levels are presented. The method has been used in design of a number of practical structures as steel and composite buildings, long-span roofs, scaffolds, space frames, conventional and pre-tensioned trusses and others. The era of design by simulation has likely arrived and the effective length method is becoming an obsolete approach which has too many limitations for contemporary structural design. For this reason, many codes downgrade in recent years the use of the effective length method for simple stocky structures that competent engineers to-date use the new and better design method for design of structures of simple forms composed of a few members to complex structures made of thousands of members.

ACKNOWLEDGEMENT

The authors acknowledge the financial support by the Research Grant Council of the Hong Kong Special Administrative Region Government under the projects "Collapse Analysis of Steel Tower Cranes and Tower Structures (PolyU 5 119/10E)" and "Simulation-based Second-order and Advanced Analysis for Strength, Stability and Ductility Design of Steel Structures (PolyU 5 120/09E)".

REFERENCES

[1] CEN, Eurocode No. 3, *Design of Steel Structures. Part 1.1: General Rules and Rules for Buildings. ENV 1993-1-1: 2005*, The European Committee for Standardization, Brussels, 2005.

[2] Chan, SL and Zhou, ZH, "Pointwise equilibrating polynomial element for nonlinear-analysis of frames", *Journal of Structural Engineering-ASCE*, Vol.120, pp. 1703-1717, 1994.

[3] Chan, SL and Zhou, ZH, "2nd-order elastic analysis of frames using single imperfect element per member", *Journal of Structural Engineering-ASCE*, Vol.121, pp. 939-945, 1995.

[4] Liu, SW, Liu, YP and Chan, SL, "Pushover analysis by one element per member for performance-based seismic design", *International Journal of Structural Stability and Dynamics*, Vol. 10, pp. 111-1126, 2010.

[5] CEN, Eurocode No. 2, *Design of Concrete Structures. Part 1.1: General Rules and Rules for Buildings. ENV 1992-1-1:2004*. The European Committee for Standardization, Brussels, 2004.

[6] CEN, Eurocode No. 4, *Eurocode 4: Design of Composite Steel and Concrete Structures. ENV 1994-1-1:2004*. The European Committee for Standardization, Brussels, 2005.

[7] Chan, SL, Liu, SW and Liu, YP, "Advanced analysis of hybrid frame structures by refined plastic-hinge approach", in: *Proceedings of 4th International Conference on Steel & Composite Structures*, 21-23 July 2010, Sydney, Australia, 2010.

[8] Chan, SL and Zhou, ZH, "Non-linear integrated design and analysis of skeletal structures by 1 element per member", *Engineering Structures*, Vol. 22, pp. 246-257, 2000.

[9] *Code of Practice For The Structural Use of Steel* 2011, Buildings Department, Hong Kong SAR Government, 2011.

[10] Chan, S.L. and Chui, P.P.T., *Non-linear Static and Cyclic Analysis of Semi-rigid Steel Frames*, Elsevier Science, 2000, pp.336.

IJSSD Symposium on Progress in Structural Stability and Dynamics
14-16 April 2012, Nanjing, China

BUCKLING BEHAVIOUR OF CONTINUOUS BEAMS AND FRAMES SUBJECTED TO PATCH LOADING

C. Basaglia and * D. Camotim

Department of Civil Engineering and Architecture, ICIST, Instituto Superior Técnico,
Technical University of Lisbon, Av. Rovisco Pais, 1049-001 Lisboa, Portugal
* Email: dcamotim@civil.ist.utl.pt

KEYWORDS

Thin-walled steel beams, thin-walled steel frames, generalised beam theory (GBT), buckling analysis, effect of the load point of application, localised buckling, patch loading.

ABSTRACT

This work illustrates the application and potential of a recently developed GBT (Generalised Beam Theory) finite element formulation to perform buckling analyses of thin-walled continuous beams and frames acted by transverse loads applied at various member cross-section points. After briefly addressing this GBT formulation, numerical results concerning the localised, local and global buckling behaviour of (i) two-span I-beams and (ii) "L-frames" with I-section members, all subjected to transverse point loads acting at the top or bottom flanges (*i.e.*, patch loading) are presented and discussed. For validation, the GBT-based results are compared with values yielded by ANSYS shell finite element analyses.

INTRODUCTION

The members of a slender steel structure used in the construction industry are often subjected to loads applied at various cross-section locations—*e.g.*, beams acted by concentrated transverse loads (patch loading). Even so, it seems fair to say that virtually all the information available on how the locations of the points of load application influence the member stability deals with global buckling, almost always beam lateral-torsional buckling—aside from classical results included in books or manuals (*e.g.*, [1]), the works of Wang and Kitiporchai [2], Mohri *et al.* [3] and Andrade *et al.* [4] deserve mention. As far as the local, distortional and localised buckling phenomena are concerned, the amount of research on the effects stemming from the transverse load position is still rather scarce. In this context, it is worth mentioning the works of (i) Gonçalves [5], who used an approximate one-dimensional model to study the distortional buckling behaviour of hat-section cantilevers acted by a tip point load, (ii) Samanta and Kumar [6], who used shell finite element models to investigate the lateral-torsional-distortional buckling of singly symmetric I-section single-span beams acted by transverse loads applied at their top and bottom flanges, and (iii) Silva *et al.* [7], who developed a beam finite element formulation based on Generalised Beam Theory (GBT) to analyse rigorously (iii_1) problems that involve localised buckling (*e.g.*, web crippling) and (iii_2) the effects due to the point of load application in the local, distortional and global buckling behaviour of isolated beams.

The authors have recently developed and numerically implemented GBT-based beam finite elements that make it possible to analyse the local, distortional and global buckling behaviour of continuous beams and frames [8, 9]— however, such finite elements are only valid for the case of transverse loads acting along the longitudinal axis passing through the cross-section shear centres.

The objective of this paper is (ⅰ) to briefly present the main concepts involved in developing a GBT formulation intended to analyse the localised, local, distortional and global buckling behaviour of thin-walled continuous beams and frames subjected to transverse loads applied at various member cross-section points (away from their shear centres), and (ⅱ) to illustrate its application, by investigating the buckling behaviour of (ii_1) two-span I-beams and (ii_2) "L-frames" with I-section members, all of them subjected to transverse point loads acting at the top or bottom flanges. The accuracy of the GBT-based results is assessed through the comparison with "exact" values yielded by rigorous shell finite element analyses carried out in the code ANSYS [10].

INCORPORATION OF LOAD APPLICATION EFFECTS IN THE GBT BUCKLING ANALYSIS

In a GBT formulation, the cross-section displacement field is expressed as a combination of deformation modes, leading to a very convenient and most unique form of expressing the member equilibrium equations— this is achieved by performing a cross-section analysis, which allows for a much better understanding of the member structural behaviour [11]. Figure 1 shows the dimensions and GBT discretisations of the I-sections dealt with in this work—Young's modulus $E = 210$ GPa and

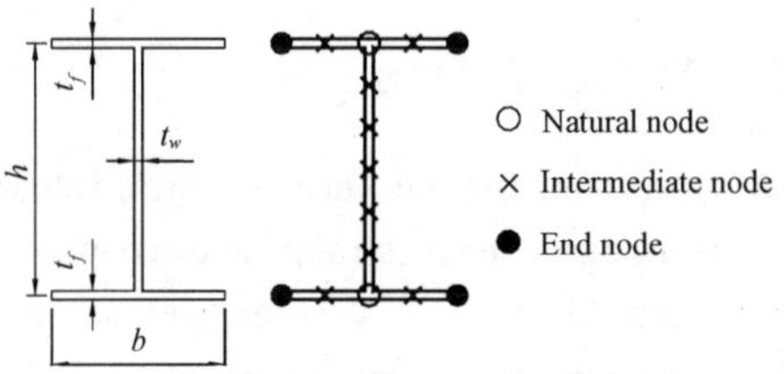

Designation	web (mm)	flange (mm)	thickness (mm)
I300×150	300	150	5.0
I300×200	300	200	5.0

Figure 1 Dimensions of I-section and discretisation adopted in the GBT analyses

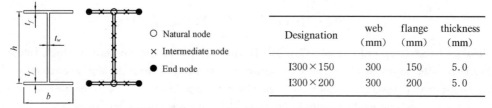

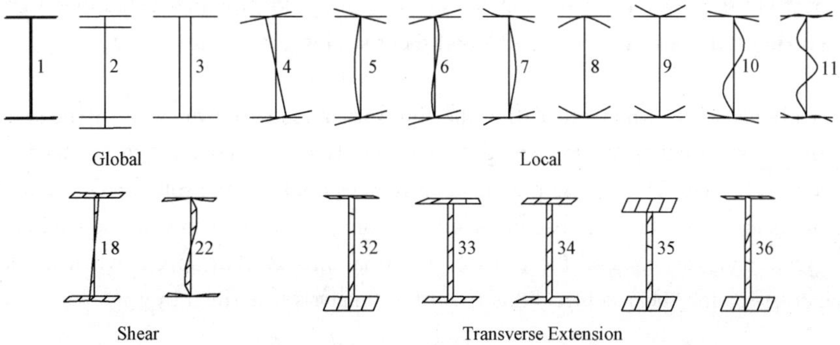

Figure 2 Main features of the most relevant I-section deformation modes

Poisson's ratio $\nu = 0.3$ are always assumed. Figure 2 shows the main features of the deformation modes that are more relevant to the buckling analyses presented and discussed in this paper: 11 conventional, 2 shear and 5 transverse extension deformation modes.

The *conventional* modes, based on the assumption of null membrane shear strains and transverse extensions, constitute the core of GBT and can still be subdivided into (i) *global* modes (cross-section in-plane rigid-body motions: axial extension, major/minor axis bending and torsion), (ii) *distortional* modes and (iii) *local* modes—the last two involve cross-section in-plane deformation (distortion and/or transverse wall bending). The *shear* modes concern the *non-linear* variation of the warping displacements along the cross-section mid-line (the cross-section experiences no in-plane deformation). The *transverse extension* modes involve only in-plane displacements and obviously account for the cross-section deformation due to wall transverse extensions.

After performing the cross-section analysis, *i. e.*, determining the deformation mode shapes and evaluating the corresponding mechanical properties, one readily establishes and solves the member/frame buckling eigenvalue problem. Indeed, the usual application of the principle of virtual work leads to

$$\delta V = \delta U + \delta \Pi_{\sigma-x} + \delta \Pi_{\sigma-s} + \delta \Pi_{\tau} = \int_{\Omega} \sigma_{ij} \delta \varepsilon_{ij} \, d\Omega + \int_{\Omega} \sigma^0_{xx} \delta \varepsilon_{xx} \, d\Omega + \int_{\Omega} \sigma^0_{ss} \delta \varepsilon_{ss} \, d\Omega + \int_{\Omega} \tau^0_{xs} \delta \gamma_{xs} \, d\Omega = 0, \tag{1}$$

where (i) Ω is the structural system volume (n plates), (ii) δU is the first variation of the strain energy, given by the tensor product between the internal stresses σ_{ij} ($\sigma_{xs} \equiv \tau_{xs}$) and strain variations $\delta\varepsilon_{ij}$ ($\delta\varepsilon_{xs} \equiv \delta\gamma_{xs}$) associated with the buckling action/mode, and (iii) $\delta\Pi_{\sigma-x}$, $\delta\Pi_{\sigma-s}$ and $\delta\Pi_{\tau}$ are the works done by the pre-buckling longitudinal normal (σ^0_{xx}), transverse normal (σ^0_{ss}) and shear (τ^0_{xs}) stresses. Then, the terms potential energy δU and $\delta\Pi_i$ read

$$\delta U = \int_L (C_{ik}\phi_{k,xx}\delta\phi_{i,xx} + D^1_{ik}\phi_{k,x}\delta\phi_{i,x} + D^2_{ik}\phi_k\delta\phi_{i,xx} + D^2_{ki}\phi_{k,xx}\delta\phi_i + B_{ik}\phi_k\delta\phi_i)\,dx \tag{2}$$

$$\delta\Pi_{\sigma-x} = \lambda\int_L (X^{\sigma-x}_{jik}\phi^0_{j,xx}\phi_{k,x}\delta\phi_{i,x})\,dx \qquad \delta\Pi_{\sigma-s} = \lambda\int_L (X^{\sigma-s}_{jik}\phi^0_j\phi_k\delta\phi_i)\,dx$$

$$\delta\Pi_{\tau} = \lambda\int_L (X^{\tau}_{jik}\phi^0_{j,x}\phi_{k,x}\delta\phi_i + X^{\tau}_{jki}\phi^0_{j,x}\phi_i\delta\phi_{k,x})\,dx, \tag{3}$$

where (i) the second-order tensors C_{ik}, B_{ik}, D^1_{ik} and D^2_{ik} account for the linear stiffness values that are associated with longitudinal extensions, transverse extensions, shear strains and coupling between longitudinal and transverse extensions (Poisson effects), respectively, (ii) λ is the load parameter, (iii) ϕ^0_j are the pre-buckling mode amplitude functions and (iv) the third-order tensors $X^{\sigma-x}_{jik}$, $X^{\sigma-s}_{jik}$ and X^{τ}_{jik}, included in the geometric stiffness terms, concern the works done by the longitudinal normal, transverse normal and shear stresses over the variations of the non-linear longitudinal extensions ($\varepsilon^{NL}_{xx} = (v^2_{,x} + w^2_{,x})/2$, transverse extensions ($\varepsilon^{NL}_{ss} = w^2_{,s}/2$) and shear strains ($\gamma^{NL}_{ss} = w_{,x}w_{,s}$).

It should be mentioned that, when calculating the third-order tensor (geometric stiffness) components, the inclusion of the pre-buckling stresses and deformations effects is accomplished by means of the modal amplitude functions ϕ^0_j (see (3)). These pre-buckling stresses (i) are the solution of the member/frame first-order analysis under a reference loading profile (loading profile multiplying the load parameter λ in buckling analyses), and (ii) include the transverse normal stresses that appear when the loads are not applied at the cross-section shear centre—this first-order analysis is defined by

$$C_{ij}\phi^0_{j,xxxx} - D_{ij}\phi^0_{j,xx} + B_{ij}\phi^0_j = q_i, \tag{4}$$

where q_i is an external applied load vector.

In order to incorporate the frame joint behaviour into the frame analysis, one must (i) "transform" the modal degrees of freedom into *nodal* ones (generalised displacements of the point where the joint is deemed located), by means of a "*joint element*" concept, and (ii) impose *joint compatibility conditions* to enforce compatibility between the end section displacements and rotations caused by warping (due to torsion and/or distortion) and wall transverse bending (Camotim *et al.* [9]).

The solution of the non-linear systems (1) and (4) can be obtained by means of a GBT-based beam finite element formulation analogous to that developed and implemented by Basaglia *et al.* [11]—this finite element approximates the modal amplitude functions $\phi_k(x)$ by linear combinations of (i) Lagrange cubic polynomial primitives (axial extension and shear modes) and (ii) Hermite cubic polynomials (transverse extension and remaining conventional modes).

NUMERICAL RESULTS

In order to illustrate the application and capabilities of the proposed GBT finite element formulation, two sets of numerical results are presented and discussed in this section. They concern the local, global and localised buckling behaviour of (i) two-span I-beams and (iii) I-section "L-frames"—all subjected to transverse loads applied at the top and bottom flanges. The GBT-based results are compared with values yielded by ANSYS [10] shell finite element analyses—members and frames discretised into fine meshes of SHELL 181 elements.

Two Span I-beams

One investigates the buckling behaviour of two-span symmetric beams with (i) overall length $L = 400$ cm (2×200 cm) or $L = 800$ cm (2×400 cm) and (ii) I300 × 150 cross-sections. They are acted by two identical mid-span transverse point loads, applied at either the top or bottom flange-web corner. Concerning the support conditions, (i) the end sections are locally/globally pinned and can warp freely, and (ii) all in-plane cross-section displacements are fully restrained by the intermediate support.

While Table 1 compares the GBT and ANSYS beam critical loads corresponding to the two lengths and two loadings, Figs. 3(a) and 3(c) ($L = 400$ cm) and Figs. 4(a)-(b) ($L = 800$ cm) show the (i) ANSYS 3D views and (ii) GBT modal amplitude functions $\phi_k(x)$ concerning the buckling mode shapes of the four beam-loading combinations. As for Figure 3(b), it provides 3D representations of the $L = 400$ cm beam buckling mode shapes yielded by the GBT analyses. After comparing these various sets of beam buckling results, the following comments are appropriate:

(i) The ANSYS and GBT critical buckling loads are virtually identical (all differences below 3%)—the GBT values are obtained with 28 beam finite elements that include (i_1) 8 conventional (**2−7+10+11**), 2 shear (**18+22**) and 5 transverse extension (**32-36**) modes in the first-order analyses, and (i_2) 7 conventional (**3−7+10+11**) modes in the buckling analyses. Moreover, the two buckling mode representations are remarkably similar. Indeed, the comparison between those shown in Figs. 3(a)-(b), concerning the $L = 400$ cm beams, provides ample evidence of this striking resemblance (the GBT views are 3D representations of buckling modes yielded by *beam* finite element analyses).

(ii) The coincidence between the GBT and ANSYS critical load values stems from the accuracy of the proposed GBT formulation in capturing the pre-buckling web transverse normal stress distributions. This assessment can by confirmed by looking at Figs. 4(a)-(b), which provide these ANSYS and GBT normal stress values in the mid-span cross-section webs of the L = 400 cm beams loaded at both the top and bottom flanges.

TABLE 1 GBT AND ANSYS TWO-SPAN BEAM CRITICAL LOAD VALUES (kN)

Length (cm)	400		800	
Load application	*Top*		*Bottom*	
GBT	250.96	644.37	65.65	278.13
ANSYS	244.69	652.56	64.18	276.39

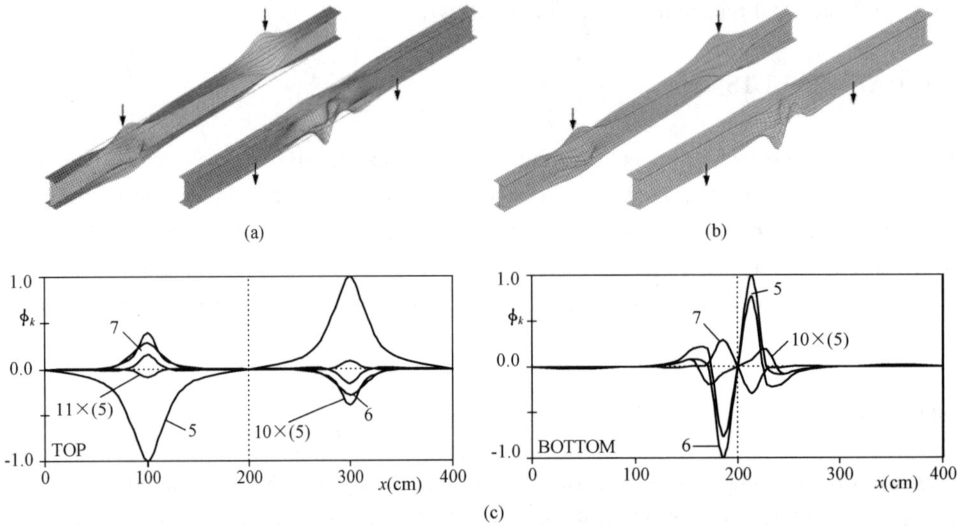

Figure 3 L=400 cm two-span beam with top and bottom loading: critical buckling mode shapes yielded by (a) ANSYS and (b) GBT-based formulation, and (c) modal amplitude functions

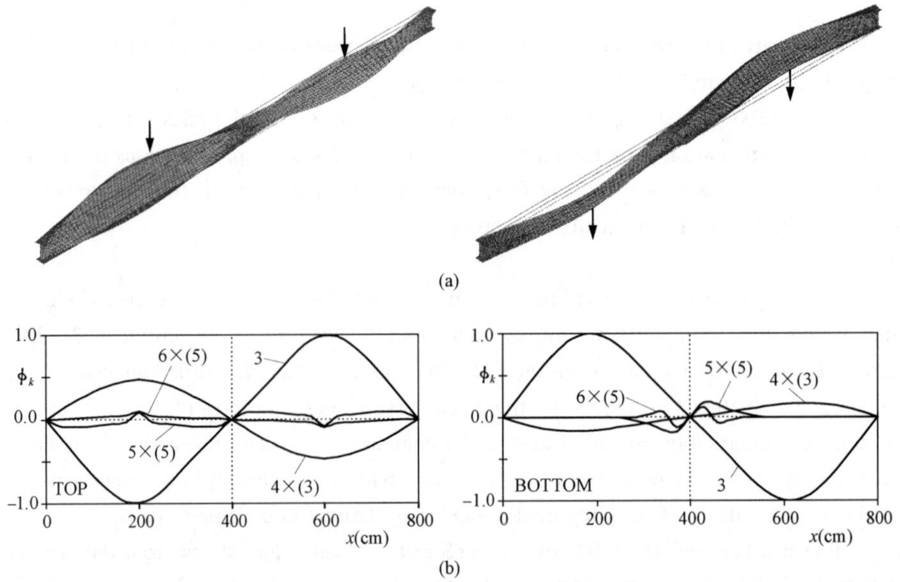

Figure 4 L=800 cm two-span beam with top and bottom loading: (a) critical buckling mode shapes yielded by ANSYS and (b) modal amplitude functions

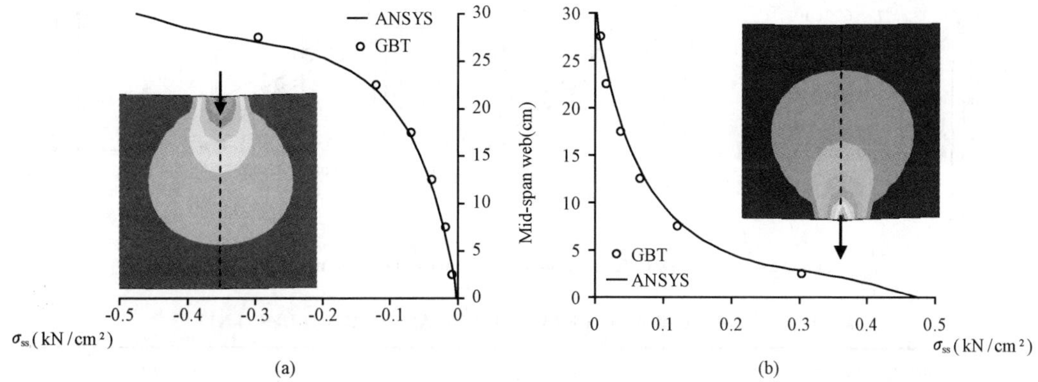

Figure 5 Pre-buckling transverse normal stresses in the mid-span cross-section webs (ANSYS and GBT) and mid-span regions (ANSYS) of the L=400 cm beam—(a) top and (b) bottom loading

(iii) The local deformation modes (**5, 6, 7, 10, 11**) have visible contributions to the $L = 400$ cm beam critical buckling modes. While the instability of the beam loaded at the top flange is triggered by localised web buckling occurring in the close vicinity of the loaded cross-section, the buckling pattern of the beam loaded at the bottom flange is governed by the local deformation of the web and compressed flange taking place in the neighbourhood of the intermediate support (beam central region).

(iv) Although the $L = 800$ cm beam critical buckling is clearly global (lateral-torsional—dominance of deformation modes **3, 4**), there are also minor participations from local deformation modes (**5, 6**), perceptible in the vicinity of the loaded cross-sections (top flange loading) or near the intermediate support (bottom flange loading)—see the modal amplitude functions in Figure 4(b). Note also that such local deformations are barely visible in the ANSYS 3D views displayed in Figure 4(a) and would certainly remain unnoticed without the information provided by the GBT analyses.

"L-shaped" Frames

The "L-shaped" frame exhibiting the geometry, support conditions and loadings shown in Figure 6 is analysed in this section—the loadings consist of a vertical point load P applied at the beam mid-span top or bottom flange. The frame is formed by equal-length ($L_c = L_b = 600$ cm) orthogonal members (column and beam) with I300 × 200 cross-sections and connected by a box-stiffened joint (web continuity). Concerning the support conditions, (i) the column is always laterally unrestrained and has a fully fixed end support and (ii) the beam has a "fork-type" end support (locally/globally pinned and free-to-warp end section) and is laterally restrained by two intermediate point supports, located at the web-flange corners of the $X_b = 100$ cm cross-section—*localised displacement restraints* (see the right hand side of Figure 6).

Table 2 compares the ANSYS and GBT critical buckling loads of all the frames analysed. As for Figs. 7 (a) − (b), they provide two representations of the frame buckling mode shapes: (i) Ansys 3D views and (ii) GBT modal amplitude functions. In this case, the GBT analyses involved 32 beam finite elements (8 in the column and 24 in the beam), which include (i) 7 conventional (**1-7**), 2 shear (**18+22**) and 5 transverse extension (**32-36**) modes in the first-order analyses, and (ii) 7 conventional (**1-7**) modes in the buckling analyses. This amounts to totals of 960 (first-order analyses) and 448 (buckling analyses)

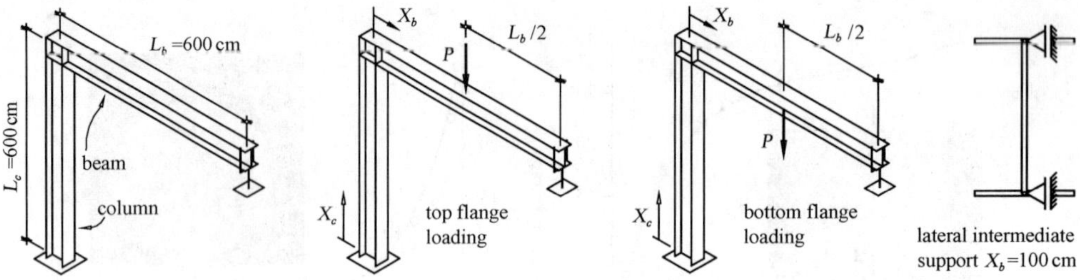

Figure 6 "L-shaped" frame geometry, support condition and loading

TABLE 2 GBT AND ANSYS FRAME CRITICAL LOAD VALUES (kN)

Load application	*Top*	*Bottom*
GBT	86.24	112.57
ANSYS	85.26	115.34

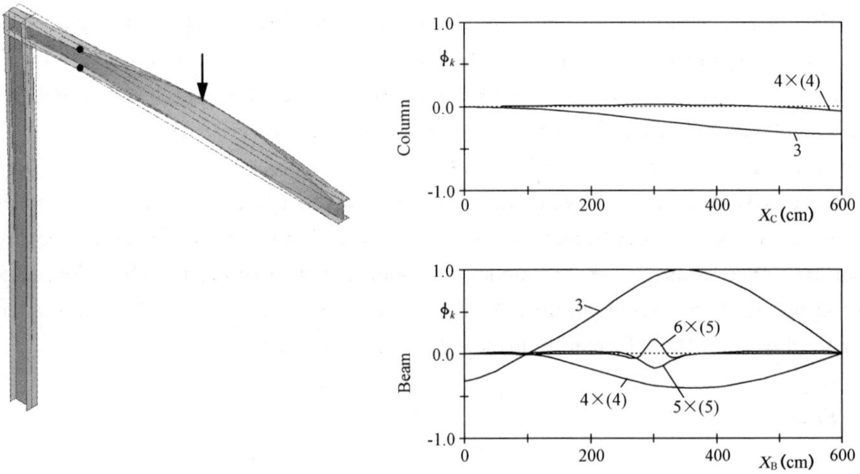

(a)

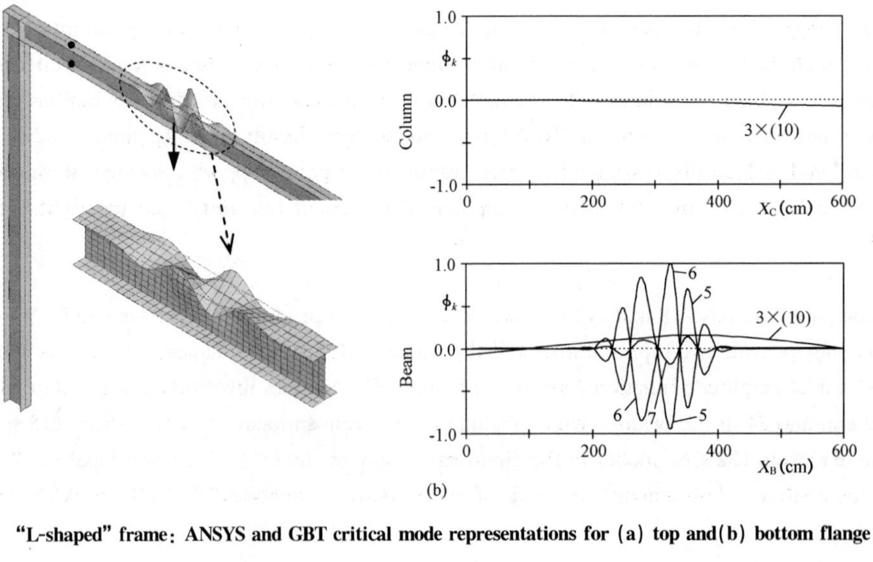

(b)

Figure 7 "L-shaped" frame: ANSYS and GBT critical mode representations for (a) top and (b) bottom flange loading

degrees of freedom—it is worth stressing that the ANSYS shell finite element analyses involve more than 11 700 degrees of freedom. The following conclusions can be drawn from the analysis of the frame buckling results presented in Table 2 and Figs. 7(a)-(b):

(i) The GBT and ANSYS critical loads are again extremely close—indeed, the values presented in Table 2 show that all differences are below 2.4%. Note also that there is very close agreement between the two critical buckling mode representations, even if the GBT modal amplitude functions provide more in-depth insight on the frame buckling mechanics.

(ii) The critical buckling mode switches from (ii_1) predominantly lateral-torsional, for top flange loading, to (ii_2) mostly local (modes **5**, **6**, **7**), with the deformation occurring in the beam mid-span area, for bottom flange loading—in both cases, the frame buckling is triggered by the beam instability.

(iii) The influence of the load point of application on the frame critical buckling load is significant—the load concerning bottom flange loading exceeds its top flange loading counterpart by about 30%.

CONCLUSION

Initially, a very brief overview of the most relevant concepts and procedures involved in the derivation and numerical implementation of a GBT beam finite element formulation was presented—this finite element makes it possible to analyse the buckling behaviour of thin-walled continuous beams and frames acted by transverse loads applied at various member cross-section points (away from the shear centre). Then, in order to illustrate the application and capabilities of this novel computational tool, a set of numerical results were presented and discussed. They concerned the local, global and localised buckling behaviour of (i) two-span I-section beams and (ii) "L-frames" with I-section members, all subjected to patch loading—point loads acting on the top or bottom flange. Most GBT-based results (critical loads, critical buckling mode shapes and transverse normal stresses) were compared with numerical values yielded by ANSYS shell finite element analyses—a virtually perfect agreement was found in all cases. Taking full advantage of the GBT modal nature, it was possible (i) to discuss the results in great structural detail and also (ii) to unveil and/or shed some new light on a number of interesting and scarcely known behavioural aspects, namely the relative importance of the localised, local and global deformations.

ACKNOWLEDGEMENTS

Financial support provided by *Fundação para a Ciência e Tecnologia* (FCT—Portugal), through project PTDC/ECM/108146/2008, is gratefully acknowledged. The first author also acknowledges FCT for his postdoctoral scholarship, SFRH/BPD/62904/2009.

REFERENCES

[1] Trahair, N.S., *Flexural-Torsional Buckling of Structures*, E & FN Spon (Chapman & Hall), London, 1993, pp. 360.

[2] Wang, C.M. and Kitipornchai, S., "Buckling capacities of monosymmetric I-beams", *Journal of Structural Engineering*, ASCE, 1986, 112(11), pp. 2372-2391.

[3] Mohri, F., Brouki, A. and Roth, J.C., "Theoretical and numerical stability analyses of unrestrained, mono-symmetric thin-walled beams", *Journal of Constructional Steel Research*, 2003, 59(1), pp. 63-90.

[4] Andrade, A., Camotim, D. and Providência e Costa, P., "On the evaluation of elastic critical moments in

doubly and singly symmetric I-section cantilevers", *Journal of Constructional Steel Research*, 2007, 63(7), pp. 894-908.

[5] Gonçalves, R., "*Analysis of Thin-Walled Beams with Deformable Cross-Sections: New Formulations and Applications*", PhD thesis in Civil Engineering, IST, Technical University of Lisbon, 2007. (Portuguese)

[6] Samanta, A. and Kumar, A., "Distortional buckling in monosymmetric I-beams", *Thin-Walled Structures*, 2006, 44(1), pp. 51-56.

[7] Silva, N. F., Camotim, D. and Silvestre, N., "Generalized Beam Theory formulation capable of capturing localized web buckling and load application effects", *Proceedings of 19th Specialty Conference on Cold-Formed Steel Structures*, R. LaBoube, W.-W. Yu (Eds.), 2008, pp. 33-59.

[8] Camotim, D., Silvestre, N., Basaglia, C. and Bebiano, R., "GBT-based buckling analysis of thin-walled members with non-standard support conditions", *Thin-Walled Structures*, 2008, 46(7-9), pp. 800-815.

[9] Camotim, D., Basaglia, C. and Silvestre, N., "GBT buckling analysis of thin-walled steel frames: a state-of-the-art report", *Thin-Walled Structures*, 2010, 48(10-11), pp. 726-743.

[10] SAS (Swanson Analysis Systems Inc.), *ANSYS Reference Manual* (version 12), 2009.

[11] Basaglia, C., Camotim, D. and Silvestre, N., "Local, Distortional and Global Post-Buckling Behaviour of Thin-Walled Steel Frames Using Generalised Beam Theory", *Proceedings of Tenth International Conference on Computational Structures Technology* (CST 2010—Valencia, 14-17/9), B. H. V Topping *et al*. (Eds), Civil-Comp Press, 2010. (full paper in CD-ROM Proceedings)

FREE VIBRATION AND BUCKLING CHARACTERISTICS OF COMPOSITE PANELS HAVING ANISOTROPIC DAMAGE IN A SINGLE LAYER

* P. K. Datta and S. Biswas

Department of Aerospace Engineering, Indian Institute of Technology, Kharagpur, India—721302
* Email: pkdatta@aero.iitkgp.ernet.in

KEYWORDS

Anisotropic damage, composite structure, curved panels, vibration, buckling, curvature effects

ABSTRACT

The present study deals with the effects of single layer damage on the free vibration and the buckling characteristics of doubly curved composite panels. First order shear deformation theory is used to model the panels, considering the effects of rotary inertia and shear deformation according to Sanders' first approximation for doubly curved shells. An element based anisotropic damage formulation is used to model damaged composite panels using Finite Element Method. The orthogonality of the damaged model is established and subsequently, suitable damage parameters are chosen for the present study. Cross-ply and angle-ply cylindrical, spherical and hyper-paraboloid curved laminates as well as flat plates have been studied. The results obtained have been compared with equivalent undamaged panels as well as with panels having damage throughout the thickness. It has been observed that the damage model shows a strong orthogonality and damage in the fiber direction affects free-vibration and buckling characteristics more than damage in any other orthogonal direction. Curvature greatly affects the free vibration and buckling characteristics of damaged panels. A damage located on a layer stressed due to curvature greatly influences free vibration and buckling characteristics than damage located in an unstressed or less stressed layer. It has also been observed that the fundamental natural frequency and buckling load of a single layer damaged panel always lie between the corresponding undamaged case and a case where damage exists throughout the thickness.

INTRODUCTION

Damage in composite structures may arise from various factors like growth of micro-cracks, accidents, projectile impacts and chemical corrosion. Sometimes a reduction in stiffness is also caused by fatigue or wear and tear due to extensive usage. Voids in the matrix, arising out of faulty manufacturing process, also cause reduction in the load bearing capacity of a laminate. While a damage due to projectile impact or collision manifests itself in all the layers of a laminate, that due to surface wear and tear, fiber breakage, chemical corrosion and local delamination may be thought of as local damage existing only in some particular layer.

Specific problems can be solved by "hard-wiring" the damage as a part of the geometry of the structure [1, 2]. However, a parametric model of damage is more versatile in describing a range of problems [3-5]. In general continuum damage mechanics introduces a continuous parameter that can be related to the density of defects. This method is helpful in describing the deterioration of the material before the initiation of micro-cracks.

In the present study, an anisotropic damage formulation is used to parametrically model damage. The effect of damage in a single layer on the free vibration and buckling characteristics of panels is studied and is compared with that of panels having damage throughout the thickness.

MATHEMATICAL FORMULATION

Anisotropic Damage

In a two dimensional structure, viz. a thin plate, anisotropic damage is parametrically incorporated into the formulation by considering the parameter Γ_i. This parameter is essentially a representation of reduction in effective area and is given by

$$\Gamma_i = \frac{A_i - A_i^*}{A_i} \tag{1}$$

where A_i is the effective area (with unit normal) after damage and $i \in \{1, 2, 3\}$ are the three orthogonal directions. For a thin plate, only Γ_1 and Γ_2 has to be considered. Γ_1 represents the damage in the direction of the fiber while Γ_2 refers to an orthogonal damage [4].

Assuming that the internal forces acting on any damaged section are same as the ones before damage, the following relationship between the damaged stress tensor $[\sigma_{ij}^*]$ and the undamaged stress tensor $[\sigma_{ij}]$ can be established

$$\begin{Bmatrix} \sigma_{11}^* \\ \sigma_{22}^* \\ \sigma_{12}^* \\ \sigma_{21}^* \end{Bmatrix} = \begin{bmatrix} \frac{1}{1-\Gamma_1} & 0 & 0 \\ 0 & \frac{1}{1-\Gamma_2} & 0 \\ 0 & 0 & \frac{1}{1-\Gamma_1} \\ 0 & 0 & \frac{1}{1-\Gamma_2} \end{bmatrix} \begin{Bmatrix} \sigma_{11} \\ \sigma_{22} \\ \sigma_{12} \end{Bmatrix} \tag{2}$$

It may be noted that the damaged stress tensor is not symmetric. This equation can be abbreviated as $\{\sigma^*\} = [\Psi]\{\sigma\}$, where $\{\Psi\}$ is a transformation matrix and can be used to relate a damaged stress-strain matrix with an undamaged one, $[D^*]^{-1} = [\Psi]^T[D]^{-1}[\Psi]$. The damaged stress-strain matrix for a two-dimensional laminate can be written as

$$[D^*] = \begin{bmatrix} f_1 E_1 & f_{12} E_2 \nu_{12} & 0 \\ f_{21} E_1 \nu_{21} & f_2 E_2 & 0 \\ 0 & 0 & 2\frac{f_1 f_2}{f_1 + f_2} G \end{bmatrix} \tag{3}$$

The factors, $f_1 = \frac{(1-\Gamma_1)^2}{1-\nu_{12}\nu_{21}}$, $f_2 = \frac{(1-\Gamma_2)^2}{1-\nu_{12}\nu_{21}}$ and $f_{12} = \frac{(1-\Gamma_1)(1-\Gamma_2)}{1-\nu_{12}\nu_{21}}$ are defined as stiffness reduction factors.

For a damaged region, the stress-strain relation can be written as $\{\sigma^*\} = [D^*]\{\varepsilon\}$. This relation can then be transformed into the general coordinate system as used in general undamaged cases.

Problem Definition

To solve the present problem, an eight-noded, 5 degrees of freedom per node, finite element has been used. The panels studied here are thin, square panels having curvature ratio, $b/R_x = 0.1$. The thickness ratio b/h is 100. The ratio of R_x/R_y is 1, 0 and 1 for spherical, cylindrical and hyper-paraboloid panels respectively.

Each layer has the following material properties:

$E_{11} = 40E_{22}$; $G_{12} = G_{13} = 0.6E_{22}$; $G_{23} = 0.5E_{22}$; $\nu_{12} = 0.25$

The damaged area is a square patch having a size of 4% of the total panel area and located at the center of the panel.

RESULTS AND DISCUSSION

Table 1 compares the free vibration natural frequencies of undamaged panels with those reported by Bert and Chen [6]. It can be observed that the present formulation agrees well with the results available in the literature.

TABLE 1 NON-DIMENSIONAL FREE VIBRATION FREQUENCIES FOR AN UNDAMAGED, SIMPLY SUPPORTED, ANGLE PLY, COMPOSITE PLATES. FIBER ORIENTATION: ($\theta/-\theta/\theta/-\theta$). b/h =10. THE RESULTS ARE COMPARED TO THOSE REPORTED BY BERT AND CHEN [6].

θ	5	15	30	45
Present	14.742 3	15.606 4	17.634 7	18.463 7
Bert and Chen	14.74	15.61	17.63	18.46

Orthogonality of Damage Parameter

Since the model of damage being investigated is anisotropic in nature, it is essential to understand the nature of orthotropy exhibited by this formulation. It may be inferred that the contribution of damage in the direction of fiber, Γ_1 may not be equal to that in the orthogonal direction to the fiber, Γ_2. In order to understand this, one of these orthogonal parameters must be fixed while the other must be varied to obtain the static and the dynamic characteristics of the structure.

Figures 1(a) and (b) plot the change in the fundamental natural frequency and the buckling load of a typical, centrally damaged, (0/90/0) cross-ply cylindrical panel with variation of one of the damage parameters while the other is fixed. It can be observed that damage in the direction of fiber, i.e. Γ_1, shows a steeper deterioration of both natural frequency as well as buckling characteristics when compared to the influence of Γ_2. The influence of damage parameter orthogonal to the fiber direction Γ_2 is negligible when compared to the undamaged values. It may be noted that a similar behaviour is observed in all panels irrespective of the curvature properties or ply layouts. Owing to this, for all further computational purposes, the value of Γ_2 is fixed at 0.1, while the intensity of damage is represented by the damage ratio, Γ_1/Γ_2. A mild damage maybe represented with a damage ratio $0.0<\Gamma_1/\Gamma_2<3.0$, while a heavy damage may be denoted as $7.0<\Gamma_1/\Gamma_2<9.0$.

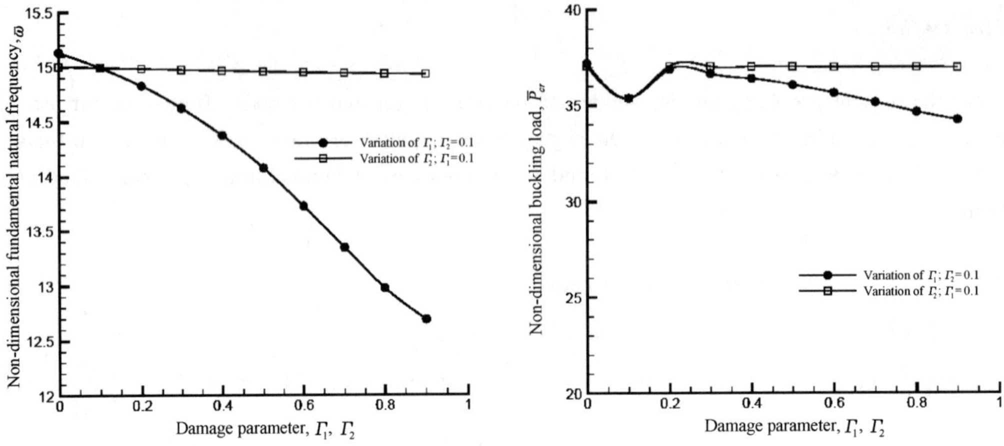

Figure 1 Variation of (a) fundamental natural frequency and (b) buckling load of a centrally damaged, square, simply supported (0/90/0) cross-ply cylindrical panel with individual damage parameters Γ_1 and Γ_2 while other is kept fixed at 0.1.

Damage in a Single Layer

Tables 2 and 3 tabulate the variation of fundamental natural frequency for simply supported, cross-ply and angle-ply panels respectively for the occurrence of damage in individual layer. It can be observed that fundamental frequency is greatly dependent on the location of the damaged layer across thickness and for various curvatures.

For all the cases of ply layout, the plates exhibit symmetry across the thickness. It can be observed that damage in interior layer(s) causes less reduction in its fundamental frequency as compared to plates whose surface layers are damaged.

TABLE 2 VARIATION OF NON-DIMENSIONAL FUNDAMENTAL NATURAL FREQUENCY WITH DAMAGE LAYER FOR THREE LAYER AND FIVE LAYER SIMPLY SUPPORTED CROSS-PLY PANELS. DAMAGE RATIO, $\Gamma_1/\Gamma_2=9.0$.

Panel	ud	Layer 1	Layer 2	Layer 3	Layer 4	Layer 5	through
				(0/90/0)			
Plate	18.829 9	17.201 2	18.772 9	17.201 2	–	–	15.4872
Cyl.	27.458 5	27.212 1	27.191 0	26.212 7	–	–	27.059 3
Sph.	54.329 0	51.485 6	54.252 2	53.541 0			43.416 2
Hyp.-Par.	53.932 4	53.915 9	53.649 7	49.435 3			42.975 6
				(0/90/0/90/0)			
Plate	18.843 9	17.926 4	18.582 8	18.834 3	18.582 8	17.926 4	15.670 8
Cyl.	39.508 3	39.406 5	38.038 6	39.459 4	38.858 0	39.058 8	33.829 7
Sph.	59.573 4	57.721 8	58.568 2	59.359 3	59.139 5	59.296 4	47.942 3
Hyp.-Par.	59.203 6	59.212 0	57.869 8	58.990 4	58.624 4	56.667 1	47.251 2

Note: "ud" refers to undamaged panels and "through" refers to panels with damage throughout the thickness.

TABLE 3 AS IN TABLE 2, BUT FOR THREE LAYER ANGLE-PLY PANELS.

Panel	ud	Layer 1	Layer 2	Layer 3	through
		(15/−15/15)			
Plate	18.829 9	17.201 2	18.772 9	17.201 2	15.487 2
Cyl.	27.458 5	27.212 1	27.191 0	26.212 7	27.059 3
Sph.	54.329 0	51.485 6	54.252 2	53.541 0	43.416 2
Hyp.-Par.	53.932 4	53.915 9	53.649 7	49.435 3	42.975 6
		(30/−30/30)			
Plate	18.843 9	17.926 4	18.582 8	18.834 3	15.670 8
Cyl.	39.508 3	39.406 5	38.038 6	39.459 4	33.829 7
Sph.	59.573 4	57.721 8	58.568 2	59.359 3	47.942 3
Hyp.-Par.	59.203 6	59.212 0	57.869 8	58.990 4	47.251 2
		(45/−45/45)			
Plate	18.843 9	17.926 4	18.582 8	18.834 3	15.670 8
Cyl.	39.508 3	39.406 5	38.038 6	39.459 4	33.829 7
Sph.	59.573 4	57.721 8	58.568 2	59.359 3	47.942 3
Hyp.-Par.	59.203 6	59.212 0	57.869 8	58.990 4	47.251 2

For any curved panel, the asymmetry in results across thickness is expected and is observed for all cases of cross-ply and angle-ply laminates. In the case of the spherical panels, the curvature is positive in both the axes. In such a scenario, damage in the top layer shows greater reduction in fundamental frequency than damage in the bottom layer. As one of the curvatures is reduced to zero, as in the case of cylindrical panels, this trait is in the process of reversal and the influence of damage in the bottom layer increases. When the curvature is further decreased to obtain a negative curvature in one of the axes, i. e. hyper-paraboloid panels, the behaviour is reversed. A larger drop in fundamental frequency is obtained for a damaged bottom layer as compared to a similar panel with its upper layer damaged.

In all the cases it can be observed that the fundamental natural frequency for a panel with an individual damaged layer lies between an undamaged, similar panel and a similar panel with damage in all the layers. It can also be observed that curvature greatly improves the free vibration characteristics in general.

TABLE 4 VARIATION OF NON-DIMENSIONAL BUCKLING LOAD WITH DAMAGE LAYER FOR THREE LAYER C-F-S-S CROSS-PLY PANELS. DAMAGE RATIO, $\Gamma_1/\Gamma_2=1.0$.

Panel	ud	Layer 1	Layer 2	Layer 3	through
		(0/90/0)			
Plate	13.244 2	13.121 4	13.200 1	13.121 4	13.246 4
Cyl.	48.599 5	48.059 7	48.100 1	48.044 2	48.502 4
Sph.	48.388 8	47.853 4	47.880 9	47.816 5	48.279 5
Hyp.-Par.	49.873 7	49.342 2	49.398 6	49.353 2	49.752 8
		(15/−15/15)			
Plate	12.394 6	12.301 7	12.351 6	12.301 7	12.436 2
Cyl.	23.873 2	22.911 9	23.047 5	22.972 0	23.050 2
Sph.	23.873 2	23.644 4	23.777 1	23.652 0	23.724 7
Hyp.-Par.	25.835 9	25.545 9	25.758 7	25.631 5	25.582 5
		(30/−30/30)			
Plate	12.087 5	12.044 8	12.041 7	12.044 8	12.142 8

continued

Panel	ud	Layer 1	Layer 2	Layer 3	through
Cyl.	20.987 7	20.891 7	20.918 4	20.927 5	20.931 0
Sph.	21.517 0	21.423 9	21.448 4	21.447 8	21.471 9
Hyp.-Par.	22.764 8	22.656 8	22.701 5	22.692 1	22.638 5
		(45/-45/45)			
Plate	10.356 4	10.330 0	10.325 7	10.330 0	10.369 9
Cyl.	18.222 7	18.187 9	18.190 7	18.201 2	18.179 7
Sph.	18.317 1	18.284 0	18.287 1	18.294 1	18.290 6
Hyp.-Par.	18.854 2	18.823 6	18.823 5	18.831 5	18.797 8

Table 4 tabulates the non-dimensional buckling load for mildly damaged C-F-S-S① panels. The regularity obtained in simply supported cases are not observed owing to asymmetrical boundary conditions. In most of the cases, the panels show marginally better vibration and buckling characteristics if the damage is located in the middle layer as compared to damage in the surface layers. The plates show symmetrical values across the thickness, as expected.

As with thenatural frequencies, it can be observed that the buckling loads for a panel with an individual damaged layer lies between an undamaged panel and one with damage in all the layers. Also, the introduction of curvature increases the buckling loads as compared to flat plates.

CONCLUSIONS

An anisotropic damage model has been used to solve the free vibration and buckling characteristics of damaged composite panels. Using limiting cases the formulation has been verified with results available in the literature. The orthogonality of damage parameter on fundamental natural frequency and buckling load as well as the influence of the location of damage across the thickness of the panel can be clearly established in the results presented. The observations can be summarised as follows:

1. The anisotropic damage formulation has a strong orthogonality. It has been observed that damage in the fiber direction affects free vibration and buckling characteristics of panels more than damage in any other direction. An increase in the intensity of damage in the fiber direction causes a greater deviation in free vibration and buckling characteristics, while one in the perpendicular direction of the fibre hardly causes any significant change.

2. The introduction of curvatures in the panel have increased fundamental natural frequency and buckling load in comparison to the corresponding flat plates.

3. For damage occurring in a particular layer of the laminate, reduction in the free vibration and buckling characteristics is greatly influenced by the curvature of the laminate and asymmetry in these characteristics is observed in curved laminates due to asymmetric stress concentration across the thickness.

4. The reduced characteristics of a single layer damaged laminate always lie between a corresponding undamaged case and one that is damaged throughout the thickness.

---

① Clamped and free edges are opposite to each other. In-plane compressive loading is applied on the free edge.

REFERENCES

[1] Martha, L. F., Wawrzynek, P. A., and Ingraffea, A. R., "Arbitrary crack representation using solid modeling", *Engineering with Computers*, 1993, 9(2), pp. 63-82.

[2] Belytschko, T. and Black, T., "Elastic crack growth in finite elements with minimal remeshing", *International Journal for Numerical Methods in Engineering*, 1999, 45(5), pp. 601-620.

[3] Ladeveze, P. and LeDantec, E., "Damage modelling of the elementary ply for laminated composites", *Composites Science and Technology*, 1992, 43(3), 257-267.

[4] Valliappan, S., Murti, V., and Wohua, Z., "Finite element analysis of anisotropic damage mechanics problems", *Engineering Fracture Mechanics*, 1990, 35(6), pp. 1061-1071.

[5] Talreja, R., "A continuum mechanics characterization of damage in composite materials", *Proceedings of the Royal Society of London. Series A, Mathematical and Physical Sciences*, 1985, 399(1817), pp. 195-216.

[6] Bert, C.W. and Chen, T.L.C., "Effect of shear deformation on vibration of antisymmetric angle-ply laminated rectangular plates", *International Journal of Solids and Structures*, 1978, 14(6), pp. 465-473.

DYNAMIC STABILITY OF PIEZOELECTRIC BRAIDED COMPOSITE PLATES

* S. Kitipornchai[1], J. Yang[2], T. Yan[1], Y. Xiang[3]

[1] Department of Civil and Architectural Engineering, City University of Hong Kong, Kowloon, Hong Kong, China
[2] School of Aerospace, Mechanical and Manufacturing Engineering, RMIT University, PO Box 71, Bundoora, Victoria, 3083 Australia
[3] School of Engineering, University of Western Sydney, Penrith South, NSW 1797, Australia
* Email: bcskit@cityu.edu.hk

KEYWORDS

Braided composite materials, piezoelectric materials, rectangular plate, dynamic stability, higher-order shear deformation plate theory.

ABSTRACT

Three-dimensional braided composites are a class of textile composite materials with a fully integrated, continuous spatial fiber network that eliminates the interface problem by the inter-lacing of the tows in the thickness direction. These braided composites have attracted considerable attention in recent years due to their unique advantages such as low fabrication cost, improved flexure strength, and higher impact resistance. This paper presents a dynamic stability analysis of a simply supported 3D braided composite laminated plate with surface-bonded piezoelectric layers and subjected to electrical and periodic in-plane mechanical loads. A fiber inclination model is used to predict the effective stiffness matrices of the braided composite laminates. Theoretical formulations are based on higher-order shear deformation plate theory and include piezoelectric effects. Double Fourier series is employed to convert the dynamic governing equations into a linear system of Mathieu-Hill equations from which the boundary points on the unstable regions are determined by Bolotin's method. Numerical illustrations are given in both tabular and graphical forms, showing the influence of fiber volume fraction, braiding angle, inclination angle, static load level, applied voltage, and aspect ratio on the dynamic stability of the braided composite plate.

INTRODUCTION

Composite plates are often subjected to time-varying periodic in-plane forces which may cause dynamic instability through parametric resonance. It is therefore of prime importance to understand the dynamic stability characteristics of laminated plates for engineering design and assessment of structural failure. The integration of a composite structure with smart materials can potentially enhance and control the performance of the structure. This has triggered intense research interests in composite structures bonded or embedded with piezoelectric sensors/actuators. However, studies on the dynamic stability of composite laminated plates with piezoelectric effects are limited. Birman [1] discussed the active control of dynamic stability of arbitrarily laminated rectangular plates reinforced by stiffeners which include piezoelectric and composite layers using a dynamic electric field with a frequency equal to that of the in-plane load. Wang et al. [2] studied the dynamic stability of negative-velocity feedback control of piezoelectric composite plates using a finite element model and Lyapunov's energy functional based governing equations of motion. Kim and Kim [3] considered a plate actively damped by piezoelectric layer and presented finite element results for the dynamic stability of such a structure by employing the first-order shear deformation plate theory. Similar work was also done by Bhattacharya et al. [4] who analyzed the parametric instability of laminated plates integrated with active piezoelectric layers used for feedback damping and most recently, by Pradyumna and Gupta [5].

This paper investigates the dynamic stability of 3D braided composite laminated plates with surface-bonded piezoelectric layers within the framework of higher-order shear deformation plate theory by using a fiber inclination model, double Fourier series solution and Bolotin's method. The effects of fiber volume fraction, braiding angle, inclination angle, static load level, applied voltage, and aspect ratio on the dynamic stability are discussed.

THEORETICAL FORMULATIONS

Figure 1 shows a laminated composite plate with a braided composite substrate and two piezoelectric layers bonded to the upper and lower surfaces of the substrate. The length, width and total thickness of the plate are a, b and h. The thickness of the substrate is h_c, while the thickness of each piezoelectric layer is h_p. The braided composite is produced by the three-dimensional four-step 1×1 technique. Its preform is fabricated in the x-direction, and the total numbers of row and column yarn carriers on the machine bed for the rectangular preform are M and N, respectively.

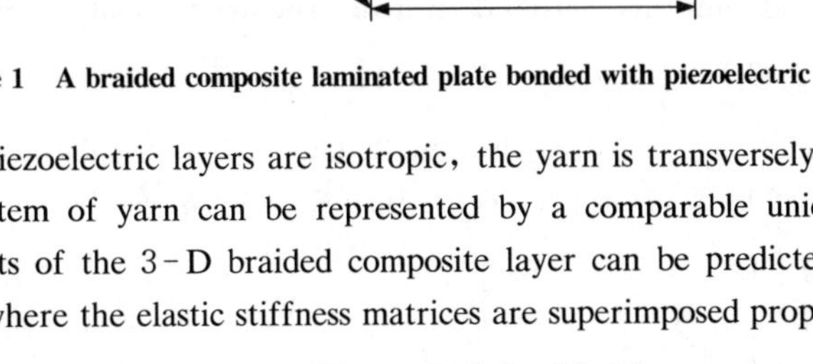

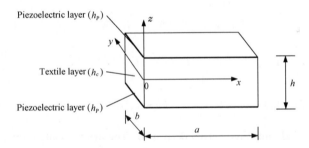

Figure 1 A braided composite laminated plate bonded with piezoelectric layers

It is assumed that the piezoelectric layers are isotropic, the yarn is transversely isotropic, the matrix is isotropic, and each system of yarn can be represented by a comparable unidirectional lamina. The effective elastic constants of the 3-D braided composite layer can be predicted by a fiber inclination laminated plate model where the elastic stiffness matrices are superimposed proportionately

$$[\overline{Q}_{ij}]_k = V_f[T]_k[C_f][T]_k^T + V_m[T]_k[C_m][T]_k^T \quad (k = 1 \sim 6) \tag{1}$$

where the yarn volume fraction V_f and the matrix volume fraction V_m are related by $V_f + V_m = 1$, $[C_f]$ and $[C_m]$ are stiffness matrices for the yarn and the matrix, and $[T]_k$ is the transformation matrix (Sun and Qiao [6]) dependent on the interior braiding angle γ, surface braiding angle θ, inclination angle ϕ, and braiding angle α as shown in Figure 2. For a braided composite plate fabricated by four-step 1×1 method, the inclination angle ϕ is 45°.

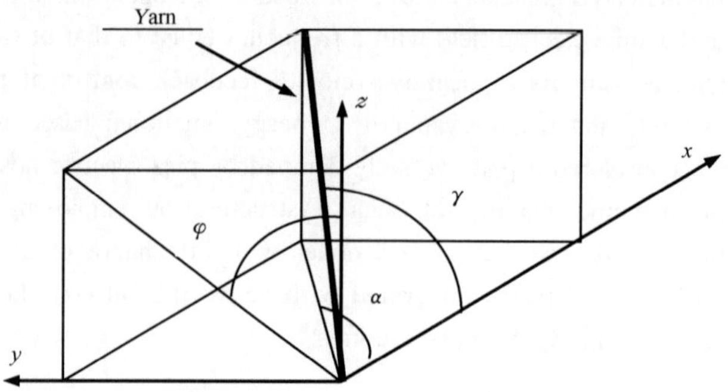

Figure 2 A yarn direction in the (x, y, z) coordinate system

Let t be time, (U, V, W) the displacements along x-, y- and z-axes, Ψ_x and Ψ_y the mid-plane rotations. The plate is under a constant electric field E_z and a periodic uniform in-plane load per length p_x

$$p_x(t) = p_s + p_d \cos \theta t \tag{2}$$

where p_s and p_d are the amplitudes of static and dynamic components, respectively, and θ is the excitation frequency. Based on Reddy's higher order shear deformation plate theory [7] and introducing Airy stress function relating to stress resultants by $N_x = F_{,yy}$, $N_y = F_{,xx}$ and $N_{xy} = -F_{,xy}$, the equations of motion for the plate can be derived as

$$\begin{aligned} L_{11}(W) - L_{12}(\Psi_x) - L_{13}(\Psi_y) + L_{14}(F) + p_x \frac{\partial^2 W}{\partial x^2} + p_y \frac{\partial^2 W}{\partial y^2} - L_{15}(N^P) - L_{16}(M^P) \\ = L_{17}(\ddot{W}) + I_8 \left(\frac{\partial \ddot{\Psi}_x}{\partial x} + \frac{\partial \ddot{\Psi}_y}{\partial y} \right) \end{aligned} \tag{3}$$

$$L_{21}(F) + L_{22}(\Psi_x) + L_{23}(\Psi_y) - L_{24}(W) - L_{25}(N^P) = 0 \tag{4}$$

$$L_{31}(W) + L_{32}(\Psi_x) - L_{33}(\Psi_y) + L_{34}(F) - L_{35}(N^P) - L_{36}(S^P) = I_9 \frac{\partial \ddot{W}}{\partial x} + I_{10} \ddot{\Psi}_x \tag{5}$$

$$L_{41}(W) - L_{42}(\Psi_x) + L_{43}(\Psi_y) + L_{44}(F) - L_{45}(N^P) - L_{46}(S^P) = I_9 \frac{\partial \ddot{W}}{\partial y} + I_{10} \ddot{\Psi}_y \tag{6}$$

where a superposed dot indicates time partial derivative. The inertia-related terms are

$$I_8 = (I_2 \bar{I}_2 / I_1) - \bar{I}_3 - c_1 \bar{I}_5, \quad I_9 = c_1(\bar{I}_5 - \bar{I}_2 I_4 / I_1), \quad I_{10} = \bar{I}_2 \bar{I}_2 / I_1 - \bar{I}_3 \tag{7}$$

where $c_1 = 4/3h^2$, I_j and $\bar{I}_j$ $(j = 1 \sim 5)$ are given by Reddy [7], $L_{ij}()$ are defined by Yang et al. [8]. The forces and moments caused by the constant electric field $E_z = V_k / h_k$ can be calculated by

$$\begin{bmatrix} N_x^P & M_x^P & P_x^P \\ N_y^P & M_y^P & P_y^P \\ N_{xy}^P & M_{xy}^P & P_{xy}^P \end{bmatrix} = \sum_k \int_{-h_{k-1}}^{h_k} \begin{bmatrix} B_x \\ B_y \\ B_{xy} \end{bmatrix}_k (1, z, z^3) \frac{V_k}{h_k} \mathrm{d}z, \quad \begin{bmatrix} S_x^P \\ S_y^P \\ S_{xy}^P \end{bmatrix} = \begin{bmatrix} M_x^P \\ M_y^P \\ M_{xy}^P \end{bmatrix} - \frac{4}{3h^2} \begin{bmatrix} P_x^P \\ P_y^P \\ P_{xy}^P \end{bmatrix} \tag{8}$$

$$\begin{bmatrix} B_x \\ B_y \\ B_{xy} \end{bmatrix} = - \begin{bmatrix} Q_{11} & Q_{12} & Q_{16} \\ Q_{12} & Q_{22} & Q_{26} \\ Q_{16} & Q_{26} & Q_{66} \end{bmatrix} \begin{bmatrix} 1 & 0 \\ 0 & 1 \\ 0 & 0 \end{bmatrix} \begin{bmatrix} d_{31} \\ d_{32} \end{bmatrix} \tag{9}$$

where V_k is the applied voltage across the kth piezoelectric layer, h_k is the thickness of the kth ply, d_{31} and d_{32} are piezoelectric strain constants. Note that the terms containing the forces N^P, bending *moments* M^P, and higher-order moments S^P will vanish when the electric field varies in the z-direction only. For a laminated plate simply supported on all edges, movable in the x-axis but immovable in the y-axis, the boundary conditions require

$$W = M_x = P_x = \Psi_y = 0, \quad \int_0^b N_x \mathrm{d}y + p_1 b = 0 \quad (\text{at } x = 0,\ a) \tag{10a}$$

$$W = M_y = P_y = \Psi_x = 0, \quad V = 0 \quad (\text{at } y = 0,\ b) \tag{10b}$$

The immovability condition $V = 0$ is fulfilled on the average sense as

$$\int_0^a \int_0^b \frac{\partial V}{\partial y} \mathrm{d}x\mathrm{d}y = \int_0^a \int_0^b \left[A_{22}^* \frac{\partial^2 F}{\partial x^2} + A_{12}^* \frac{\partial^2 F^2}{\partial y^2} - (A_{12}^* N_x^P + A_{22}^* N_y^P) \right] \mathrm{d}x\mathrm{d}y = 0 \tag{11}$$

NUMERICAL SOLUTION

Double Fourier series solutions are used in the present analysis

$$W(x,\ y,\ t) = \sum_{m=1}^{\infty} \sum_{n=1}^{\infty} w_{mn}(t) \sin \frac{m\pi x}{a} \sin \frac{n\pi y}{b} \tag{12a}$$

$$\Psi_x(x,\ y,\ t) = \sum_{m=1}^{\infty} \sum_{n=1}^{\infty} \psi_{xmn}(t) \cos \frac{m\pi x}{a} \sin \frac{n\pi y}{b} \tag{12b}$$

$$\Psi_y(x,\ y,\ t) = \sum_{m=1}^{\infty} \sum_{n=1}^{\infty} \psi_{ymn}(t) \sin \frac{m\pi x}{a} \cos \frac{n\pi y}{b} \tag{12c}$$

$$F(x,\ y,\ t) = -\frac{1}{2}(p_y x^2 + p_x y^2) + \sum_{m=1}^{\infty} \sum_{n=1}^{\infty} f_{mn}(t) \cos \frac{m\pi x}{a} \cos \frac{n\pi y}{b} \tag{12d}$$

Re-write the periodic in-plane load as

$$p_x(t) = p_{cr}(\mu_{1s} + \mu_{1d} \cos \theta t) \tag{13}$$

where p_{cr} is the critical buckling load, $\mu_{1s} = p_s / p_{cr}$ and $\mu_{1d} = p_d / p_{cr}$ are the static and dynamic load parameters. Substituting Eqs (12) and (13) into Eqs (3 - 6) and boundary conditions (10), and performing Galerkin integral, a linear dynamic matrix system is obtained as

$$\boldsymbol{M}\ddot{\delta} + (\boldsymbol{K}_0 + \Delta V \boldsymbol{K}_P + \mu_{1s} p_{cr} \boldsymbol{K}_G)\boldsymbol{\delta} + (\mu_{1d} p_{cr} \boldsymbol{K}_G \cos \theta t)\boldsymbol{\delta} = \boldsymbol{0} \tag{14}$$

where $\boldsymbol{M}$ is the "mass matrix", $\boldsymbol{K}_0$ is the "stiffness matrix", $\boldsymbol{K}_P$ is the matrix showing the piezoelectric effect, $\boldsymbol{K}_G$ is the "geometric stiffness matrix", δ is composed of unknowns, and ΔV is defined as $\Delta V = (|V_U| + |V_L|)/2$ where V_U and V_L are external voltages at the upper and lower piezoelectric layers.

Eq. (14) is a Mathieu-Hill type equation describing the dynamic stability behavior of the laminated plate under a time-dependent periodic in-plane force. The Bolotin method is employed to locate the points on the boundaries of unstable regions (Yang et al. [8]).

NUMERICAL RESULTS AND DISCUSSIONS

A parametric study is carried out to investigate the dynamic stability of three-dimensional braided laminated composite plates with surface-bonded piezoelectric layers. The side length of the square laminated plate is $a = b = 48$ mm, the thicknesses of the braided composite plate and each piezoelectric layer are $h_c = 6.21$ mm and $h_p = 1.0$ mm, respectively. The elastic constants of the carbon fiber/epoxy braided composite layer ($M \times N = 5 \times 42$) are $E_{f11} = 230$ GPa, $E_{f22} = 40$ GPa, $G_{f12} = 24$ GPa, G_{f23}

= 14.3 GPa, ν_{f12} = 0.25, ρ_f = 1 750 kg/m^3 for carbon fiber, and E_m = 3.5 GPa, ν_m = 0.35, ρ_m = 1 200 kg/m^3 for epoxy. The elastic constants of the piezoelectric layers are E_P = 63.0 GPa, ν_P = 0.3, ρ_P = 7 600 kg/m^3, d_{31} = d_{32} = 2.54e − 10 m/V.

Figs 3-5 give the principal unstable regions (PURs) of the laminated plates with different fiber volume fraction V_f, braiding angle α, applied voltage V_U and V_L, aspect ratio β, and the static load level μ_{1s}, where the dimensionless excitation frequency $\theta^* = \theta a^2\sqrt{12\rho_f/E_{f11}h_c^2}$ is plotted versus the dynamic load parameter ϕ_d for a fixed value of static load parameter μ_{1s}. Each PUR is bounded by two curves in the $\theta^*-\phi_d$ plane at which the excitation frequency is twice the fundamental frequency of the plate

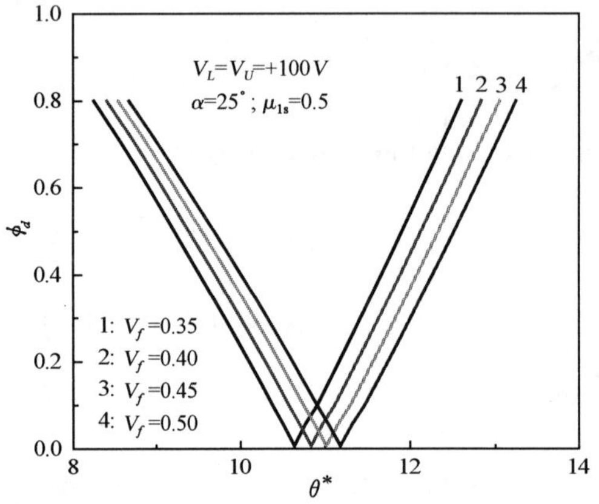

Figure 3 PUR of piezoelectric braided composite square plates: effect of fiber volume fractions.

Figure 3 discusses the effect of fiber volume fraction V_f on the PURs for braided composite plates under a positive applied voltage. The excitation frequencies and the size of PURs increase with an increase in fiber volume fraction. A higher fiber volume fraction implies a higher volumetric percentage of carbon fiber whose elastic modulus is much larger than epoxy and consequently, leads to a plate with improved bending stiffness.

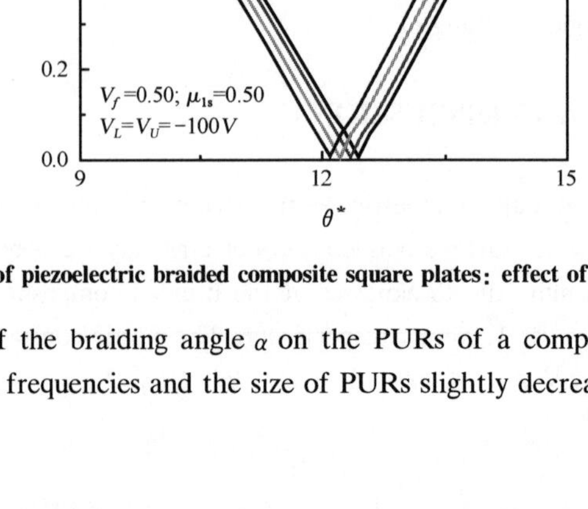

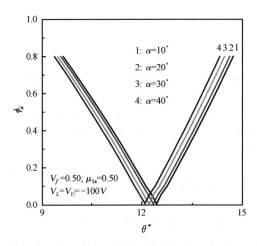

Figure 4 PUR of piezoelectric braided composite square plates: effect of braiding angles.

Figure 4 shows the effect of the braiding angle α on the PURs of a composite plate under a negative voltage. Both the excitation frequencies and the size of PURs slightly decrease as α increases.

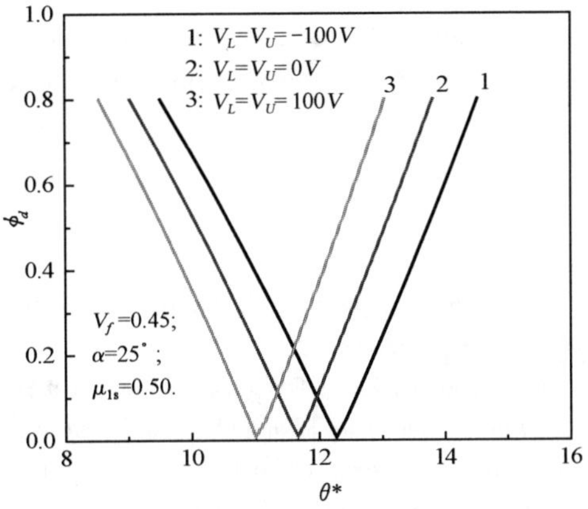

Figure 5 PUR of piezoelectric braided composite square plates: effect of applied voltage

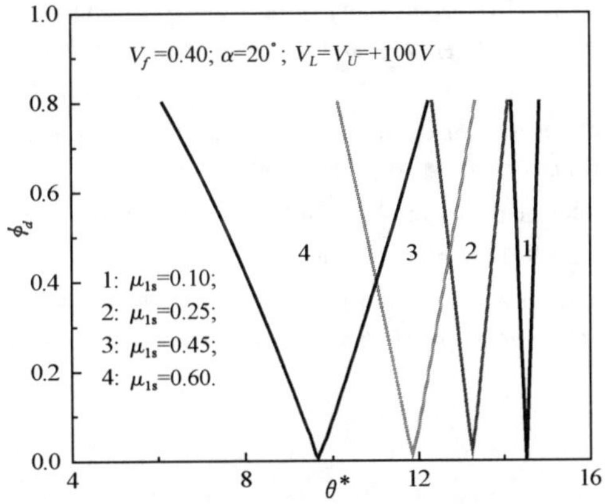

Figure 6 PUR of piezoelectric braided composite square plates: effect of static compressive load

The PURs of the plates under different applied voltages are compared in Figure 5. As expected, the negative voltage raises the excitation frequency and slightly widens the PUR of the laminated plate.

Figure 6 presents the results for the braided composite plate under an electro-mechanical load with different static load levels. Uniform compressive edge forces (μ_{1s} = 0.10~0.60) are considered. As a greater static compression is exerted on the plate edge, the PUR not only becomes much wider but also moves closer to the coordinate origin.

CONCLUSIONS

Dynamic stability of a simply supported braided composite plate bonded with piezoelectric actuators under a combined action of electrical and periodic in-plane loads is investigated. It is found that the principal unsteady region of the plate is sensitive to the applied voltage and static load level but less sensitive to the braiding angle and the fiber volume fraction. The size of the PUR is remarkably increased with an increase in static load level but is slightly widened by the application of negative voltage.

ACKNOWLEDGEMENT

The work described in this paper was fully supported by a grant from the Research Grants Council of the Hong Kong Special Administrative Region, China [Project no. 9041668 (CityU 114911)]. The authors are grateful for this support.

REFERENCES

[1] Birman, V., "Active control of composite plates using piezoelectric stiffeners", *International Journal of Mechanical Sciences*, 1993, 35(5), pp. 387-396.

[2] Wang, S. Y., Quek, S. T. and Ang, K. K., "Dynamic stability analysis of finite element modeling of piezoelectric composite plates", *International Journal of Solids and Structures*, 2004, 41(3-4), pp. 745-764.

[3] Kim, H. W. and Kim, J. H., "Effect of piezoelectric damping layers on the dynamic stability of plate under thrust", *Journal of Sound and Vibration*, 2005, 284(3-5), pp. 597-612.

[4] Bhattacharya, P., Rose, M. and Homann, S., "Effects of piezo-actuated damping on parametrically excited laminated composite plates", *Journal of Reinforced Plastics and Composites*, 2008, 25(8), pp. 801 - 813.

[5] Pradyumna, S. and Gupta, A., "Dynamic stability of laminated composite plates with piezoelectric layers subjected to periodic in-plane load", *International Journal of Structural Stability and Dynamics*, 2011, 11(2), pp. 297 - 311.

[6] Sun, H.Y. and Qiao, X., "Prediction of the mechanical properties of three-dimensionally braided composites", *Composites Science and Technology*, 57(6), pp. 623-629.

[7] Reddy, J.N., "A refined nonlinear theory of plates with transverse shear deformation", *International Journal of Solids and Structures*, 1984, 20(9-10), pp. 818-896.

[8]Yang, J., Liew, K.M. and Kitipornchai, S., "Dynamic stability of laminated FGM plates based on higher-order shear deformation theory", *Computational Mechanics*, 2004, 33(4), pp. 305-315.

REDUCING HYDROELASTIC RESPONSE OF VERY LARGE FLOATING STRUCTURE USING FLEXIBLE LINE CONNECTOR AND GILL CELLS

C. M. Wang, * R. P. Gao and C. G. Koh

Department of Civil and Environmental Engineering, National University of Singapore,
Kent Ridge, Singapore, 119260
* Email: g0801482@nus.edu.sg

KEYWORDS

Hydroelasticity, very large floating structures, flexible line connector, gill cells, optimization.

ABSTRACT

This paper investigates the effects of flexible line connector and gill cells on reducing the hydroelastic response of pontoon-type, very large floating structure (VLFS) under wave action. Gill cells are compartments in VLFS with holes or slits at their bottom surfaces to allow water to enter or leave freely. They are modelled by eliminating the associated hydrostatic buoyancy forces. In the hydroelastic analysis, the water is assumed to be an ideal fluid and its motion is irrotational so that a velocity potential exists. The VLFS is modelled as an isotropic plate according to the Mindlin plate theory. In order to decouple the fluid-structure interaction problem, the modal expansion method is adopted for the hydroelastic analysis which is carried out in the frequency domain. The boundary element method is used to solve the Laplace equation for the velocity potential, whereas the finite element method is employed for solving the equations of motion of the floating plate. It is found that by appropriately positioning the flexible line connector and gill cells, the hydroelastic response and stress resultants of the VLFS can be significantly reduced.

INTRODUCTION

Very large floating structures (VLFSs) technology allows the creation of artificial land from the sea with minimal effect on marine eco-system, water quality, tidal and natural current flows (Wang *et al*. [1]). This kind of structures have been proposed for various applications such as floating airports, bridges, breakwaters, piers and docks, fuel storage facilities, emergency bases, performance stages, recreation parks, mobile offshore military bases, and even for habitation (Watanabe *et al*. [2]; Wang *et al*. [1]). Usually, moored by dolphin-frame guide systems that restrain the horizontal movement of the floating structures, VLFS are free to move in the vertical direction (heave motion) according to the tidal variations, varying pay-loads and wave actions. Owing to their large surface areas and relatively small depths, VLFS deforms elastically under wave action. Therefore, it is important to reduce the hydroelastic responses of the floating structures for applications that demands stringent serviceability requirements. Researchers have proposed various ways to reduce the deflection due to wave action. For instance, conventional methods such as the bottom-founded type or floating-type breakwaters are used to attenuate the wave forces impacting on the VLFS. Besides that, innovative antimotion devices such as the submerged plate antimotion devices, oscillating water column (OWC) breakwaters, air-cushion, and the articulated attachments have been proposed. It was reported that the OWC approach effectively reduces the hydroelastic response of the main structure. However, box-shaped devices, submerged plates and OWC breakwaters have a relatively large draft requirement and hence induce larger drift forces. Researchers have also investigated different types of mechanical joints and shapes of VLFS that would minimize the hydroelastic response. The mechanism and performance of these mitigation devices and methods are extensively reviewed by Wang *et al.* [3].

Riyansyah *et al.* [4] suggested the use of semi-rigid joints in reducing the hydroelastic response of a VLFS modeled by interconnected beams, based on the innovative idea of auxiliary or sacrificial floating structures (Furukawa *et al.* [5], Khabakhpasheva and Korobkin [6], Kim *et al.* [7]) attached to the main VLFS body. This idea is further extended and investigated by Wang *et al.* [8] in reducing the hydroelastic response of a VLFS modeled by the Mindlin plate theory. It is found that hinge line connector when appropriately located can significantly reduce the hydroelastic response of the VLFS as well as the stress resultants (Gao *et al.* [9]). It is also pointed out that line hinge connectors could lead to larger stress resultants in certain portion along the length of the plate, especially in the front portion where more wave energy is received.

In order to overcome the drawback of using line hinge connector in reducing the hydroelastic response, we proposed herein the combined use of a hinge line connector and gill cells to achieve a greater reduction in the hydroelastic response of the VLFS under wave action. Gill cells are compartments in VLFS with holes or slits at their bottom surfaces to allow water to enter or leave freely. At these gill cell locations, the buoyancy forces are eliminated and this allows uneven buoyancy forces acting at the bottom hull of the VLFS to somewhat counterbalance the wave loading (Wang *et al.* [10], Pham *et al.* [11]).

MATHEMATICAL FORMULATION

Plate-water Model

Figure 1 shows the schematic diagram of the coupled plate-water problem. The VLFS has a length L, width B, height h and it is assumed to be of zero draft. The VLFS is modeled as a floating plate and assumed to be perfectly flat with free edges. The water is assumed to be an ideal fluid (inviscid and incompressible) and its flow is irrotational. The water domain is denoted by Ω. The symbols S_{HB}, S_F and S_{SB} represent the plate domain, the free water surface and the seabed, respectively. The water motion and plate deflection are assumed to be in a steady state harmonic motion with a circular frequency ω. The free and undisturbed water surface is at $z = 0$ while the seabed is found at $z = -H$. An incident wave ϕ_I of period T and wave height $2A$ enters the computational domain at a wave angle θ with wavelength-to-structure length ratio $\alpha = \lambda / L$. The deflection w of the plate is measured from the free and undisturbed water surface. The line connection is located at βL from the fore end of the floating structure and the gill cells are indicated by the filled grids, as shown in Figure 1.

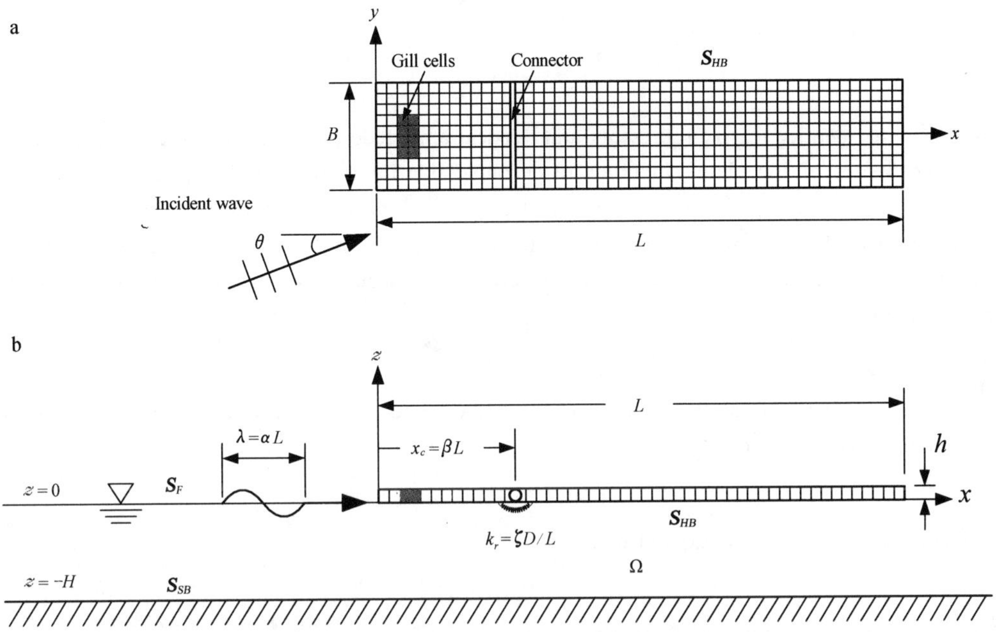

Figure 1 Schematic diagram of coupled plate-water problem (a) plan view (b) side view

Formulation of Equation of Motion

The Mindlin plate theory is used to model the VLFS. The plate material is assumed to be isotropic and obeys Hooke's law. For simplicity, the zero draft assumption is made. The equation of motion of the plate structure is established using the finite element method and takes the form of

$$(-\omega^2[M]+[K])\{w\} = \sum_e \int \{N\} p dS_{HB} \tag{1}$$

where $[M]$ and $[K]$ are mass and stiffness matrices. $\{w\}$ is displacement vector that includes vertical displacements and rotations. ω the circular frequency of motion, $\{N\}$ the shape function vector and p the wave pressure comprising hydrostatic and hydrodynamic pressure. The subscript e denotes numerical summation on elements.

Water is assumed to be an ideal fluid (inviscid and incompressible) and has an irrotational flow so that a velocity potential exists. Thus the single frequency velocity potential ϕ of water must satisfy the Laplace's equation

$$\nabla^2\phi = 0 \tag{2}$$

Based on the linear potential theory, the velocity potential ϕ may be separated into three parts

$$\{\phi\} = \{\phi_I\} + \{\phi_S\} + \{\phi_R\} \tag{3}$$

where $\{\phi_I\}$, $\{\phi_S\}$ and $\{\phi_R\}$ are incident, scattering and radiation potentials, respectively. $\{\phi_I\}$ can be expressed in an analytical expression whereas $\{\phi_S\}$ and $\{\phi_R\}$ are to be determined using the boundary element method (Wang *et al.* [8]). At the fluid-structure interface, the following boundary conditions are imposed for $\{\phi_S\}$ and $\{\phi_R\}$.

$$\frac{\partial\{\phi_S\}}{\partial z}=-\frac{\partial\{\phi_I\}}{\partial z} \tag{4}$$

$$\frac{\partial\{\phi_R\}}{\partial z}=-i\omega[I]\{w_z\} \tag{5}$$

where $[I]$ is the identity matrix represents unit vertical amplitude on each node and $\{w_z\}$ the vertical displacement vector of the plate. In order to decouple this interaction problem into a hydrodynamic problem in terms of the velocity potential and a fluid-plate vibration problem in terms of the generalized displacement, we adopt the modal expansion method as proposed by Newman [12]. In this method, the displacement of the plate $\{w\}$ is expanded by N series of products of the modal function $[\psi_w]$ and the complex amplitudes $\{\zeta_w\}$

$$\{w\}=[\psi_w]\{\zeta_w\} \tag{6}$$

In particular, the vertical displacement vector of the plate $\{w_z\}$ is expanded by modal function corresponding to deflection of plate as

$$\{w_z\}=[\psi_w^z]\{\zeta_w\} \tag{7}$$

The radiation potential $\{\phi_R\}$ is expanded as

$$\{\phi_R\}=[\Phi_R]\{\zeta_\phi\} \tag{8}$$

Therefore, the following relations is obtained

$$\frac{\partial[\Phi_R]}{\partial z}=-i\omega[I][\psi_w] \tag{9}$$

where $\Phi_R(i, j)$ indicates the value of radiation potential on ith node for unit vertical motion on jth node, and $\{\zeta_\phi\}$ is the corresponding complex amplitudes which is assumed to be the same as $\{\zeta_w\}$ (Newman [12]). $[\Phi_R]$ can be obtained as $[\tilde{\Phi}_R]\{\psi_w\}$ from the boundary element method (Wang *et al.* [8]). The pressure on plate element can be interpolated by nodal pressures as

$$p=\{N\}^T\{p\}_e \tag{10}$$

The nodal water pressure $\{p\}_e$ can be evaluated from the linearized Bernoulli equation

$$\{p\}_e=-\rho g\{w_z\}_e+i\omega\rho\{\phi\}_e \tag{11}$$

where ρ is the fluid density. The final combined governing equation for fluid-structure interaction is derived by combining Eqns. (1), (3), (8), (10) and (11) as

$$(-\omega^2[M]-\omega^2[M_w]-i\omega[C_w]+[K]+[K_w])\{w\}=\{f\} \tag{12}$$

By substituting Eqn. (6) into Eqn. (12), one obtains

$$\{\psi_w\}^T(-\omega^2[M]-\omega^2[M_w]-i\omega[C_w]+[K]+[K_w])\{\psi_w\}\{\zeta_w\}=\{\psi_w\}^T\{f\}=\{f_D\} \tag{13}$$

where $[M_w]$, $[C_w]$ and $[K_w]$ are added mass, radiation damping and hydrostatic stiffness matrices, respectively. $\{f\}$ is the wave force vector. They have the following forms

$$[M_w]=\sum_e\int(-\rho/\omega)\{N\}\{N\}^T\mathrm{Im}[\tilde{\Phi}_R]_e\mathrm{d}S_{HB} \tag{14}$$

$$[C_w]=\sum_e\int\rho\{N\}\{N\}^T\mathrm{Re}[\tilde{\Phi}_R]_e\mathrm{d}S_{HB} \tag{15}$$

$$[K_w]=\sum_e\int k\{N\}\{N\}^T\mathrm{d}S_{HB} \tag{16}$$

$$\{f\}=\sum_e\int i\omega\rho\{N\}\{N\}^T\mathrm{d}S_{HB}(\{\phi_I\}+\{\phi_S\}) \tag{17}$$

Modelling of Flexible Line Connector

If ith node and jth node are connected with elastic rotational springs, the components of stiffness matrix in Eqn. (13) are modified as follow

$$
\begin{aligned}
K(i_3, i_3)_{new} &= K(i_3, i_3) + k_r, \\
K(i_3, j_3)_{new} &= K(i_3, j_3) - k_r, \\
K(j_3, j_3)_{new} &= K(j_3, j_3) + k_r, \\
K(j_3, i_3)_{new} &= K(j_3, i_3) - k_r,
\end{aligned} \tag{18}
$$

where $k_r = \zeta D/L$ is spring constants of rotational spring about y-axis, and $D = Eh^3/[12(1-\nu^2)]$ the flexural rigidity of the plate. i_3 and j_3 are the degrees of freedom for rotational motion about y-axis of ith and jth node, respectively. In the case of a hinge line connection, $k_r = 0$ and the bending moment about y-axis is zero at the connection.

Modelling of Gill Cells

Gill cells are compartments in the VLFS without buoyancy because they have holes or slits at their bottom floors that allow water to flow in and out freely (and hence they are called gill cells). Therefore, the gill cells are modelled by eliminating the buoyancy force in the corresponding region. The constant k in Eqn. (16) that models the buoyancy force is then given by

$$k = \rho g \text{ in the region without gill cells} \tag{19}$$

$$k = 0 \text{ in the region with gill cells} \tag{20}$$

Based on the foregoing equations, an FEM program is developed in MATLAB (2008) for the hydroelastic analysis of the VLFS. Upon solving the coupled fluid-structure equation (13), we obtain the complex amplitudes $\{\zeta_w\}$ and then we back-substitute the amplitudes into Eqn. (6) to obtain the deflection and rotations of the plate $\{w\}$, hence the stress resultants of the plate.

RESULTS AND DISCUSSIONS

The VLFS considered by Sim and Choi [13] is used as an example for this study. The length, width and height of the floating plate are 300, 60 and 2 m, respectively. The following material properties of the plate are assumed: Poisson's ratio $\nu = 0.13$, Young's modulus $E = 1.19 \times 10^{10}$ N / m^2, and the weight density $\rho_p = 256.25$ kg / m^3. Water density $\rho = 1\ 025$ kg / m^3 and a water depth $H = 20$ m. In the numerical example, two wavelength cases ($\lambda = 30$, and 60 m) are considered and the rectangular VLFS is subjected to a head sea wave condition (i.e. incident wave angle $\theta = 0°$).

Effect of Gill Cells

First, we introduce gill cells into the continuous VLFS and investigate the corresponding hydroelastic responses and stress resultants. As the locations of gill cells are discontinuous, genetic algorithms (GA) are adopted as an optimization tool to optimize the layout of the gill cells. The objective is to minimize the deflection parameter defined by

$$\Psi = \frac{1}{AL^2}\int_0^L \int_{-B/2}^{B/2} |w| \, dy dx \tag{21}$$

The optimal layouts of the gill cells are shown in Figure 2. The hydroelastic response of continuous VLFS, VLFS with gill cells in front, and VLFS with proposed gill cells layout are compared in Figure 3. It can be seen that in order to reduce the hydroelastic response of the VLFS, it is better to place gill cells at the regions where larger deflections occur. This characteristic will be utilized in the following section to develop a hybrid system combing a flexible connector and gill cells to achieve greater reduction of hydroelastic response as well as the stress resultants in the VLFS.

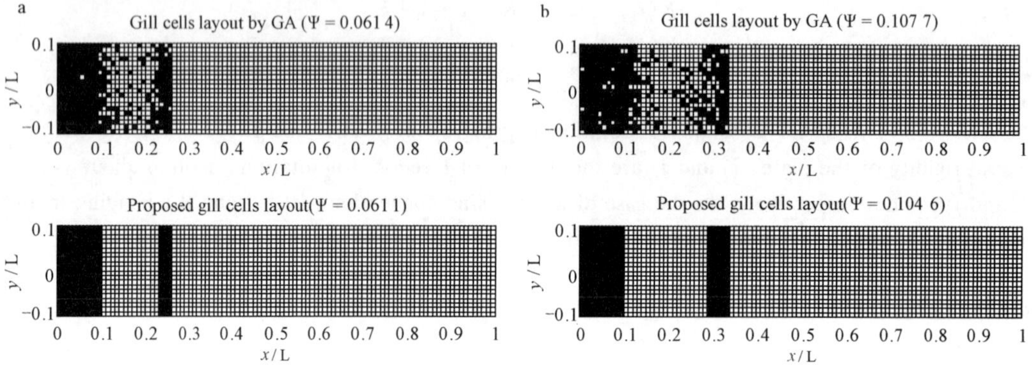

Figure 2 Gill cells layouts for (a) α=0.1 and (b) α=0.2

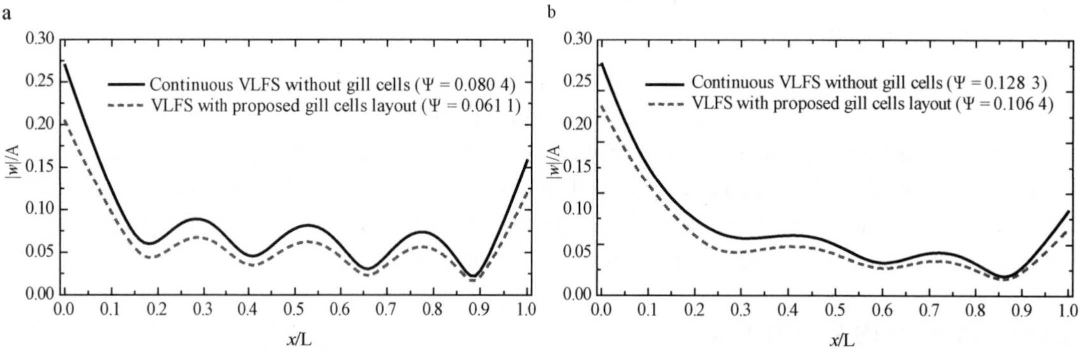

Figure 3 Normalized responses of continuous VLFS with gill cells for (a) α=0.1 and (b) α=0.2

Effect of Hybrid Reduction System

The effectiveness of hinge line connectors in reducing hydroelastic response of VLFS has been demonstrated in an earlier paper by Gao *et al.* [9]. However, it is noted that line hinge connectors could lead to larger stress resultants in certain portion along the length of the plate, especially in the front portion where more wave energy is received. In order to resolve this problem, we propose a hybrid system that combines the use of line hinge connector and gill cells to further reduce the hydroelastic response and stress resultants of the VLFS. The gill cells can be easily applied to any region in the VLFS and hence reduce the responses at these regions.

Figure 4 shows the normalized deflection, nondimensionalized bending moments ($\overline{M}_{xx} = M_{xx} L / D$) and shear forces ($\overline{Q}_x = Q_x L^2 / D$) along the longitudinal centreline of the plate for two wavelength cases (i.e. α = 0.1 and α = 0.2) with a head sea wave condition. It can be seen that the hydroelastic response of the

VLFS is further reduced when combing the use of hinge connector and gill cells, especially in the front end where gill cells are placed. Moreover, the corresponding stress resultants are further reduced. More importantly, the peak values of both bending moments and shear forces (generally in the front part) which have been increased by using only a hinge connector, are also reduced (i. e. $\alpha = 0.1$ case) or maintained (i.e. $\alpha = 0.2$ case). Thus, it can be concluded that this hybrid system is effective in reducing hydroelastic response and stress resultants of a VLFS.

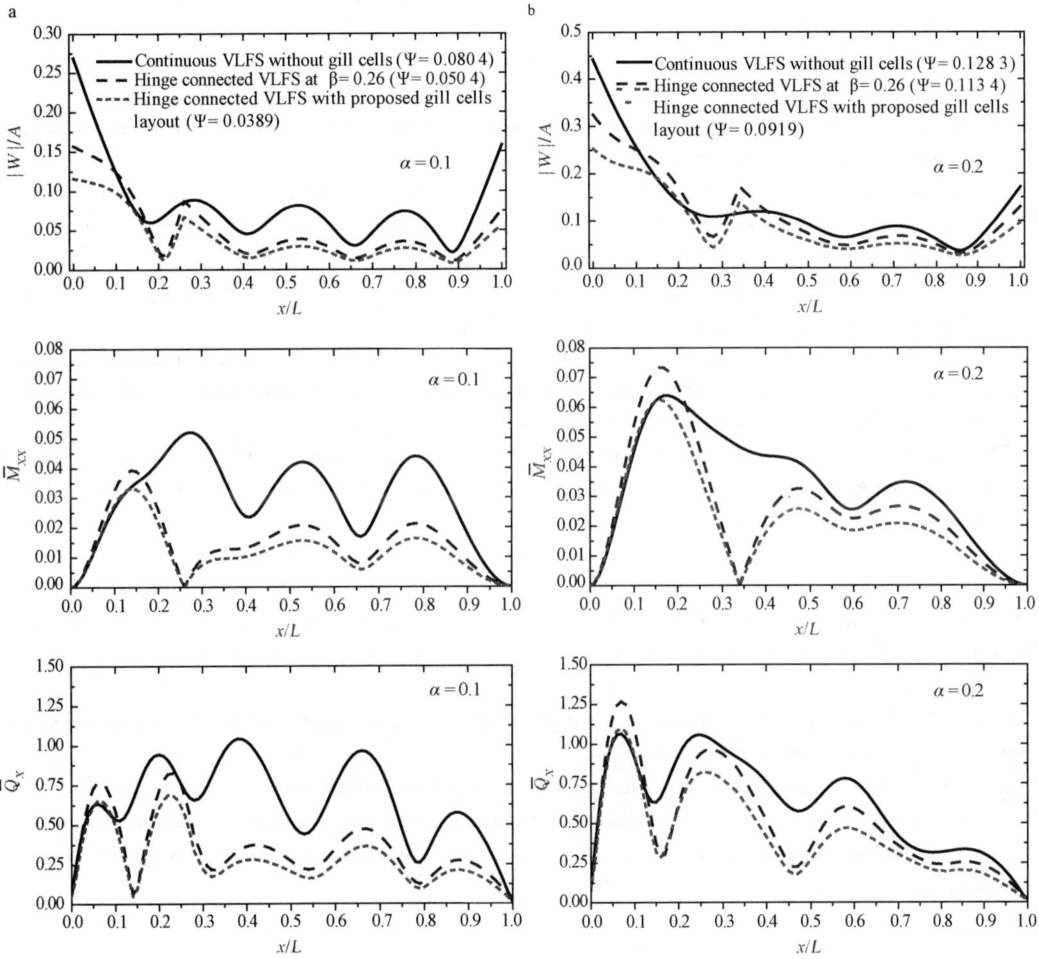

Figure 4 Normalized maximum deflection, nondimensionalized bending moments and shear forces along the centreline of VLFS for (a) α=0.1 and (b) α=0.2 subjected to head sea wave condition θ= 0°.

CONCLUDING REMARKS

We have investigated a hybrid system that uses a hinge line connector and gill cells on the hydroelastic response of a longish rectangular VLFS under wave actions. It is found that with the combined presence of a hinge line connector and appropriately positioned gill cells, a greater reduction in both hydroelastic response and stress resultants of VLFS can be achieved. For a longish rectangular VLFS under a head sea, it is found that gill cells should be placed in the front part of the VLFS for maximum reduction of hydroelastic response and stress resultants. Work is in progress to study this hybrid reduction system comprising multiple flexible connectors and gill cells.

REFERENCES

[1] Wang, C.M., Watanabe, E. and Utsunomiya, T., *Very Large Floating Structures*, Taylor & Francis, 2008, pp. 236

[2] Watanabe, E.,Utsunomiya, T. and Wang, C.M., "Hydroelastic analysis of pontoon-type VLFS: a literature survey", *Engineering Structures*, 2004, 26(2), pp.245-256.

[3] Wang, C.M., Tay, Z.Y., Takagi, K. and Utsunomiya, T., "Literature review of methods for mitigating hydroelastic response of VLFS under wave action", *Applied Mechanics Reviews*, 2010, 63(3), pp.1-18.

[4] Riyansyah, M., Wang, C.M. and Choo, Y.S., "Connection design for two-floating beam system for minimum hydroelastic response", *Marine Structures*, 2010, 23(1), pp.67-87.

[5] Furukawa, T., Yamada, Y., Furuta, H. and Tachibana, E., "Experimental study on vibration control of unit-linked floating structures", *Proceedings of First International Conference on Advances Structural Engineering and Mechanics*, Seoul, South Korea, 1999, pp. 833-838.

[6] Khabakhpasheva, T.I. and Korobkin, A.A., "Hydroelastic behaviour of compound floating plate in waves", *Journal of Engineering Mathematics*, 2002, 44(1), pp.21-40.

[7] Kim, B.W., Kyoung, J.H., Hong, S.Y. and Cho, S.K., "Investigation of the effect of stiffness distribution and structure shape on hydroelastic response of very large floating structures", *Proceedings of 15th International Offshore and Polar Engineering Conference*, Seoul, South Korea, 2005, pp.229-238.

[8] Wang, C.M., Tay, Z.Y., Gao, R.P. and Koh, C.G.,"Hydroelastic response of VLFS with a hinge or semi-rigid line connection", *Proceedings of the ASME 2010 29th International Conference on Ocean, Offshore and Arctic Engineering*, Shanghai, China, 2010, pp.1-8.

[9] Gao, R.P.,Tay, Z.Y., Wang, C.M. and Koh, C.G., "Hydroelastic response of very large floating structures with a flexible line connection", *Ocean Engineering*, 2011, 38(17-18), pp.1957-1966.

[10] Wang, C.M., Wu, T.Y., Choo, Y.S., Ang, K.K., Toh, A.C., Mao, W.Y. and Hee, A.M., "Minimizing differential deflection in a pontoon-type very large floating structure via gill cells", *Marine Structures*, 2006, 19 (1), pp.70-82.

[11] Pham, D.C., Wang, C.M., "Optimal layout of gill cells for very large floating structures", *Journal of Structural Engineering*, 2010, 136(7), pp.907-916.

[12] Newman, J.N., "Waveeffects on deformable bodies", *Applied Ocean Research*, 1994, 16(1), pp.47-59.

[13] Sim, I.H. and Choi, H.S.,1998, "An analysis of the hydroelastic behavior of large floating structures in oblique waves", *Proceedings of Second International Conference on Hydroelasticity in Marine Technology*, Fukuoka, Japan, pp.195-199.

ASSESSMENT OF SHELL AND MEMBRANE MODELS FOR PREDICTING WRINKLING PHENOMENON IN ANNULAR GRAPHENE UNDER IN-PLANE SHEAR

* Z. Zhang[1], W.H. Duan[2], C.M. Wang[1,3]

[1]Department of Civil and Environmental Engineering, National University of Singapore, Singapore, Kent Ridge, Singapore 119260
[2]Department of Civil Engineering, Monash University, Australia
[3]Engineering Science Programme, National University of Singapore, Kent Ridge, Singapore 119260
* Email: ceezzhen@nus.edu.sg

KEYWORDS

Wrinkles, shear, single layer-graphene sheet, finite element method (FEM), shell, membrane.

ABSTRACT

In this paper, we study the wrinkling phenomenon in annular graphenes due to inplane shear by using molecular dynamics simulations. The in-plane shear is applied to the inner edge of the annular graphene while its outer edge is fixed. The wrinkles are found to be closely concentrated near the inner edge and take the form of highly curved wave crest at the inner edge and gradually flatten out near the outer edge. Also in this paper, the shell model and the membrane model proposed by previous researchers for graphene will be assessed on their ability to reproduce the wrinkling phenomenon in such loaded annular graphene sheet. It will be shown herein that the shell model and the membrane model are unable to predict accurately the number of wrinkling waves, the wave amplitude and wavelength that are obtained from molecular dynamics simulation, but they are useful in providing upper and lower bounds for these aforementioned wrinkling parameters.

INTRODUCTION

Graphene, a carbon atom bonded honeycomb lattice, has been discovered to have wonderful mechanical, electronic as well as optical properties[1-3]. Together with the convenient and increasingly cheap technology for large-scale production, graphene has been regarded as a revolutionary material for many electronic devices[4] and nano-scale sensors[5]. Recently, it is reported that wrinkles in the graphenes can change their electronic properties as well as the gauge field[6, 7]. These wrinkles may originate from the geometry of substrates and external strains. Nevertheless, controllable wrinkle patterns may be tuned and its tailored properties be exploited for suitable applications. In this paper, the wrinkling phenomenon in an annular graphene sheet under an inplane shear applied to the inner edge is investigated by using molecular dynamics simulation. The observed wrinkles indicate the interplay of stretching and bending. The wrinkles take the form of a highly curved wave crest at the inner edge and gradually flatten out near the outer edge. Also in this paper, continuum mechanics models based on membrane and shell theories are assessed for their abilities to capture the observed wrinkling phenomenon (such as the number of waves, the wave amplitude, and wavelength) in the annular graphene sheet under inplane shear acting on its inner edge.

WRINKLES IN ANNULAR GRAPENE

Wrinkles in the single layer annular graphene are formed by rotating the internal edge while keeping the outer edge fixed. Molecular dynamics (MD) simulations are performed on the annular graphene sheet. For the atomistic modeling, the COMPASS force field is used for the atom to atom interaction. While holding the outer edge, the inner edge is rotated by a small angular displacement and then both edges are held to allow for geometrical relaxation by the conjugate gradient method. The rotation angle is monitored to ensure that wrinkles are well developed and no bond breaks during this loading process. The MD model of the annular graphene considered is shown in Figure 1. The inner and outer edges are fixed by constraining the movement of 5 layers of atoms at each edge.

As shown in Figure 1, wrinkles are generated by rotating the inner edge. These wrinkles are narrowly concentrated at the inner edge and tend to be flattened out when approaching the outer edge. This wrinkling pattern intuitively indicates that bending is dominant around the inner edge and a transition from bending to stretching is expected in the graphene domain away from the inner edge. However, it is noted that the use of MD simulation is prohibitively expensive, especially with a large number of atoms in the model. Continuum mechanics models such as the shell model and the membrane model therefore becomes more attractive for modeling the graphene sheet. But are these continuum mechanics models capable of accurately capturing the observed wrinkling phenomenon in such a loaded annular graphene sheet? This is the question that will be addressed in this paper.

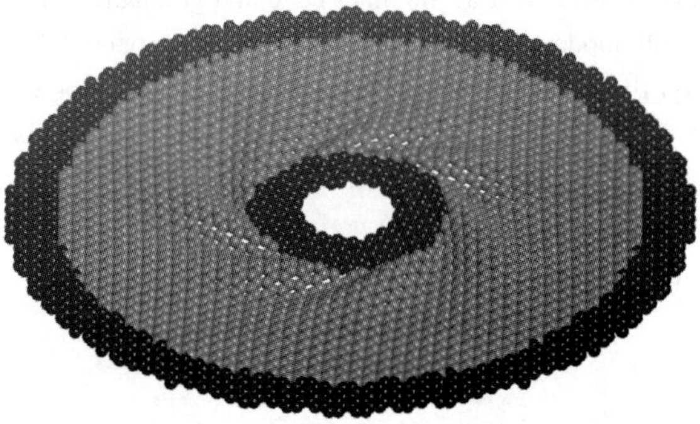

Figure 1 Wrinkles due to in-plane rotation in the inner boundary

NUMERICAL MODELS BASED ON CONTINUUM MECHANICS THEORY: SHELLS AND MEMBRANES

In order to investigate these continuum mechanics models, the wrinkling process in MD simulation is reproduced by implementing a quasi-static procedurein ABAQUS/Explicit (ver. 6.10). For the purpose of illustration, we consider an annular graphene sheet with an inner radius of 1.5 nm and the outer radius 4.5 nm, and the atomistic model involves 3 220 atoms. Investigations on different model sizes will be presented in the following sections. For the shell and membrane models, we use the commonly adopted Young's modulus E of 5.5 TPa, Poisson's ratio υ of 0.19 and a thickness t of 0.066 nm[8]. The ABAQUS shell element used is S4Rwhich is a 4-node element finite membrane strains and arbitrarily large rotations and therefore the element is suitable for large-strain analysis. The ABAQUS membrane element adopted is M3D4 which is a three-dimensional general membrane element with full integration and having 4 nodes per element.

The present finite element models consider the non-smooth boundary due to the constrained carbon bonds in MD as shown in Figure 2. Therefore the mesh strategy is to use the 4-node elements in majority while using 3-node element, for example S3R and M3D3, when necessary. To get reliable FEM results from ABAQUS, a mesh study is conducted using different mesh designs involving 1 454, 2 420, and 28 336 elements for S4R and 1 454, 20 916, 42 829, and 78 386 elements for M3D4. The wrinkled shapes furnished by ABAQUS are shown in Figure 2. Table 1 shows the pertinent results based on an 18° rotation of the inner edge. With respect to the wave number and the maximum wave amplitude, the mesh design of 1 454 S4R elements gives converged results. Based on the considered mesh designs, convergence was not achieved with the membrane element M3D4. We choose the mesh design of 78 386 for M3D4 and regard it as the final result given by the membrane model. Based on Richardson's extrapolation[9], the result for this mesh design gives a convergence error of 6.2%.

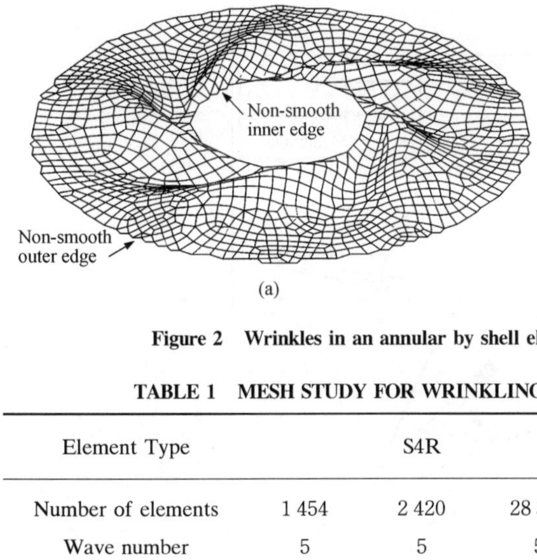

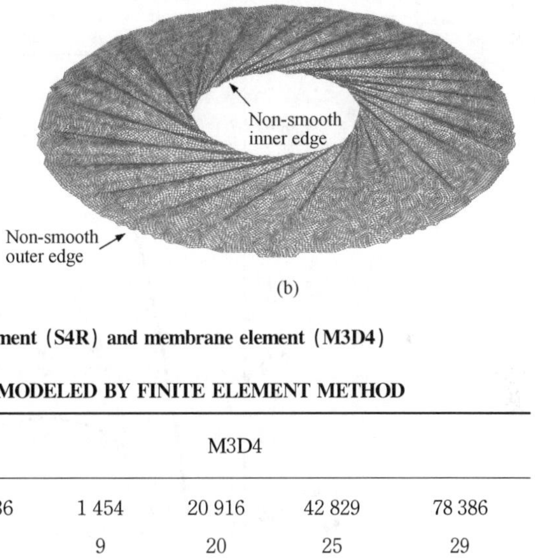

Figure 2 Wrinkles in an annular by shell element (S4R) and membrane element (M3D4)

TABLE 1 MESH STUDY FOR WRINKLING MODELED BY FINITE ELEMENT METHOD

Element Type	S4R			M3D4			
Number of elements	1 454	2 420	28 336	1 454	20 916	42 829	78 386
Wave number	5	5	5	9	20	25	29
Maximum wave amplitude (nm)	0.306	0.301	0.300	0.201	0.096	0.082	0.069

RESULTS AND DISCUSSION

This section discusses the wrinkling phenomenon results predicted by the continuum mechanics models as well as how these results compare with those observed from MD simulations. The wrinkling behavior is studied for the graphene sheet whose inner edge is rotated from 10° to 18°. During this rotation process, the wrinkling features that we focus on are the wave number, the wave amplitude and the wavelength. As an example, Figures 3 and 4 presents the tangential (ξ-ζ plane) and radial profiles (η-ζ plane) of a wrinkle at the location of maximum peak of an annular graphene of inner to outer radius ratio of 1 : 3. The figures show a typical snapshot of the wrinkle due to a rotation within the whole process.

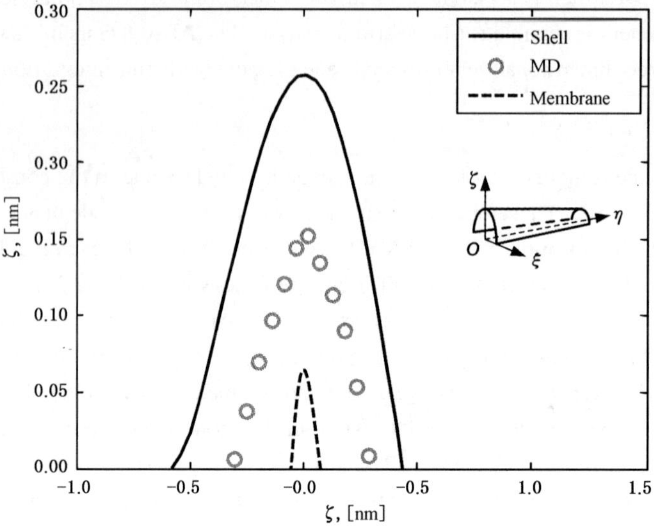

Figure 3 Tangential profile of the maximum wrinkle, θ=15°

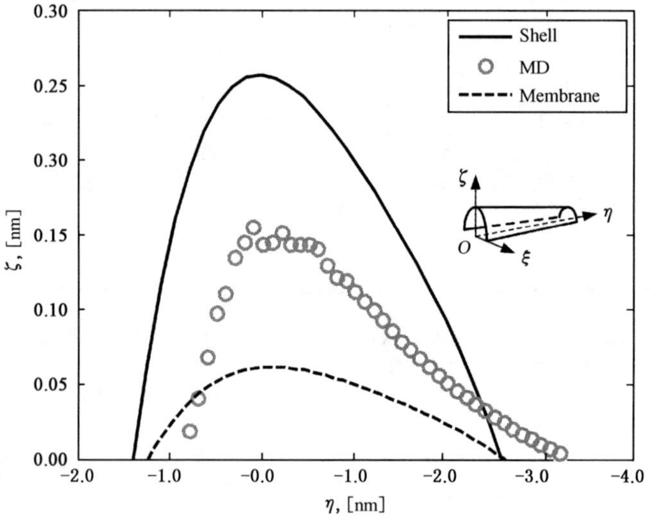

Figure 4 Radial profile of the maximum wrinkle, θ=15°

As it can be seen from Figure 3, the shell theory gives a maximum wave amplitude of 0.27 nm whereas the membrane theory gives a maximum wave amplitude of 0.06 nm. The MD simulation furnishes a wave amplitude of 0.16 nm. Thus, the shell element overestimates the maximum wave amplitude by approximately 69% whereas the membrane element underestimates the wave amplitude by about 63%. With regard to the half wavelength, MD simulation gives a value of 0.59 nm while the shell model and the membrane model furnish a half wavelength of 1.03 nm and 0.14 nm, respectively. It is clear that the shell and membrane models provide prediction bounds of the wrinkle dimensions. Further studies should be carried out to investigate the effect of Young's modulus and Poisson's ratio of the shell and membrane models on the wave amplitude and wavelength.

Next, we explore the wave profile in the radial direction. Figure 4 shows the shape of the maximum wrinkle in the η-ζ plane (see the inset of Figure 4). As η varies from-2 nm to 4 nm, one moves from the inner edge to the outer edge of the annular graphene. It is seen that the shape of the wrinkle in the radial direction resembles a tilted sine wave. The projected length of radial extension in η axis furnished by the shell model is larger than that predicted by a membrane model. The projected length for all models is about 4 nm for the annular graphene with an inner radius to outer radius ratio of $r_i/r_0=\frac{1}{3}$. The wrinkle shape furnished by MD simulations reveals that the annular graphene is highly bent at the region close to the inner edge, while stretching dominates the interplay between stretching and bending for the region far from the inner edge. Although the tilted wrinkled crest predicted by the membrane theory captures this transition, the absence of bending stiffness in membrane leads to the mismatch of the tilted angle of and crest amplitude when compared to the MD results.

It is found that the prediction of wave number, wave amplitude, and wave length may be well bounded by the shell and membrane theories. To further study the validity of these observations, annular graphene of different sizes were analysed for inner edges rotations from 10° to 18°. These models cover the inner radius to outer radius from 1/3 to 1/5, involving 3220, 5376, and 8072 atoms respectively. The results are given in TABLE 2. It can be seen that the wave amplitude of MD simulation falls between the results predicted by shell and membrane models. With increasing rotations in the inner boundary, the increase in the wave amplitude demonstrates that the behavior of graphene is closer to that of the shell model. This indicates the role of flexural stiffness in a wrinkled graphene. As a comparison, membrane theory predicts the wave amplitude with marginal increment. Nevertheless, if one wishes to accurately capture the wrinkling behavior in MD simulations, the traditional thick shell and membrane theories may not be adequate. It is therefore necessary to develop a better model to quantify the wrinkling characteristics.

TABLE 2 SIZE EFFECT ON THE BOUNDS PREDICTED BY MEMBRANE AND SHELL THEORIES

Model size	Rotations (degrees)	Wave number			Wave amplitude (nm)		
		Membrane	MD	Shell	Membrane	MD	Shell
$\frac{r_i}{r_0}=\frac{1}{3}$	10.5	29	7	5	0.053	0.109	0.232
	11.5	29	7	5	0.055	0.123	0.239
	15	29	7	5	0.061	0.161	0.272
	18	29	7	5	0.066	0.187	0.296
$\frac{r_i}{r_0}=\frac{1}{4}$	10	25	7	5	0.065	0.108	0.264
	13	25	7	5	0.075	0.151	0.285
	16.5	25	7	5	0.081	0.182	0.311
	18.5	25	7	5	0.085	0.199	0.333
$\frac{r_i}{r_0}=\frac{1}{5}$	11	24	7	5	0.066	0.121	0.272
	13	24	7	5	0.074	0.159	0.295
	16.5	24	7	5	0.084	0.172	0.335
	18.5	24	7	5	0.087	0.189	0.354

CONCLUSION

Controllable wrinkles in graphenes are useful in the design of tunable nano-scale devices. This study presents a way to introduce wrinkles in annular graphene sheets, by imposing an inplane shear action on the inner edge. By using MD simulations, one can study the formation of the wrinkles in the annular graphene sheet. If one wishes to capture the wrinkling behavior in the annular graphene sheet using continuum mechanics models based on shell and membrane theories, one needs to tweak the material properties and the thickness because it was shown herein that the commonly adopted material properties and thickness lead to the shell model over estimating the wrinkle wave shape while the membrane model underestimating the wrinkle wave shape. It is however worth noting that the shell and membrane models may be used to bound the wrinkle amplitude, wave length, and wave number furnished by MD simulations. Study is now underway to develop a better structural model to predict the wrinkling phenomenon observed in the annular graphene under an inplane shear action applied at the inner edge. Such a structural model may take the form of a grillage system.

REFERENCES

[1] Bunch, J. S., van der Zande, A. M., Verbridge, S. S., Frank, I. W., Tanenbaum, D. M., Parpia, J. M., Craighead, H. G. and McEuen, P. L., "Electromechanical resonators from graphene sheets", *Science*, 2007, 315(5811), pp. 490-493.

[2] Castro Neto, A. H., Guinea, F., Peres, N. M. R., Novoselov, K. S. and Geim, A. K., "The electronic properties of graphene", *Reviews of Modern Physics*, 2009, 81(1), pp. 109-162.

[3] Wang, X., Zhi, L. J. and Mullen, K., "Transparent, conductive graphene electrodes for dye-sensitized solar cells", *Nano Letters*, 2008, 8(1), pp. 323-327.

[4] Lin, Y. M., Valdes-Garcia, A., Han, S. J., Farmer, D. B., Meric, I., Sun, Y. N., Wu, Y. Q., Dimitrakopoulos, C., Grill, A., Avouris, P. and Jenkins, K. A., "Wafer-scale graphene integrated circuit", *Science*, 2011, 332(6035), pp. 1294-1297.

[5] Jensen, K., Kim, K. and Zettl, A., "An atomic-resolution nanomechanical mass sensor", *Nature Nanotechnology*, 2008, 3(9), pp. 533-537.

[6] Guinea, F., Horovitz, B. and Le Doussal, P., "Gauge field induced by ripples in graphene", *Physical Review B*, 2008, 77(20), Art. No.: 205421.

[7] Guinea, F., Katsnelson, M. I. and Vozmediano, M. A. H., "Midgap states and charge inhomogeneities in corrugated graphene", *Physical Review B*, 2008, 77(7), Art. No.: 075422.

[8] Yakobson, B. I., Brabec, C. J. and Bernholc, J., "Nanomechanics of carbon tubes: Instabilities beyond linear response", *Physical Review Letters*, 1996, 76(14), pp. 2511-2514.

[9] Richardson, L. F., "The approximate arithmetical solution by finite differences of physical problems involving differential equations, with an application to the stresses in a masonry dam", *Philosophical Transactions of the Royal Society A: Mathematical, Physical and Engineering Sciences*, 1911, 210, pp. 307-357.

NONDESTRUCTIVE METHOD FOR PREDICTING BUCKLING LOADS OF ELASTIC SPHERICAL SHELLS

S.N. Amiri and * H.A. Rasheed

Department of Civil Engineering, Kansas State University 2118 Fiedler Hall, Manhattan, KS 66506-5000, USA

* Email: hayder@ksu.edu

KEYWORDS

Nondestructive method, buckling load, elastic spherical shells.

ABSTRACT

In this paper, the Southwell's nondestructive method for columns is analytically extended to spherical shells subjected to a uniform external pressure acting radially. Subsequently, finite element simulation and experimental work proved that the theory is also applicable to spherical shells with an arbitrary axisymmetrical loading. The results show that the technique provides a useful estimate of the elastic buckling load provided care is taken in interpreting of the results. The usefulness of the method lies in its generality, simplicity and in the fact that, it is non-destructive. Moreover, it does not make any assumption regarding the number of buckling waves or the exact localization of buckling.

INTRODUCTION

Determination of the buckling loads of shells either experimentally or theoretically is very important. Since in most cases, the true behavior of the shell is not known or very difficult to obtain, the best thing is to make some assumptions and then to verify these assumptions by means of model tests. Southwell method for determining the minimum buckling load is a nondestructive test for pin-ended, initially imperfect struts. In 1932, Southwell presented his analysis for the special case of a pin-ended strut of constant flexural rigidity[1]. This paper investigates a nondestructive method for finding the critical buckling load in spherical shells. For this purpose, Southwell's nondestructive method for columns is extended to spherical shells subjected to a uniform external pressure acting radially, and then by means of experiments, it is shown that the theory is applicable to spherical shells with an arbitrary axisymmetrical loading.

THE SOUTHWELL METHOD APPLIED TO SHELLS

If a spherical shell of radius a and thickness t_0 is subjected to a uniform external pressure p, it may retain its spherical form, and undergo a uniform compression whose magnitude is:

$$\sigma = \frac{pa}{2t_0} \tag{1}$$

So, for values of pressure increasing from zero, the shell will at first deform in a rather uniform manner. This process persists until the external pressure p reaches a certain critical value p_{cr} called the initial buckling pressure. At this value of pressure, the shell no longer deforms in a uniform manner but snaps into another non-adjacent equilibrium configuration. The pressure to which the shell snaps is called the final buckling pressure. In order to calculate this critical pressure, the buckled surface is assumed to be symmetrical with respect to the diameter of the sphere[2].

Equations of Equilibrium of a Spherical Shell

Assuming the symmetrical deformation in the buckling process, one of the displacement components vanishes, and the others are only functions of angle θ. Therefore, $u = f_1(\theta)$, $v = 0$ and $w = f_2(\theta)$ [3], where u, v and w are the meridian, circumferential and radial displacements.

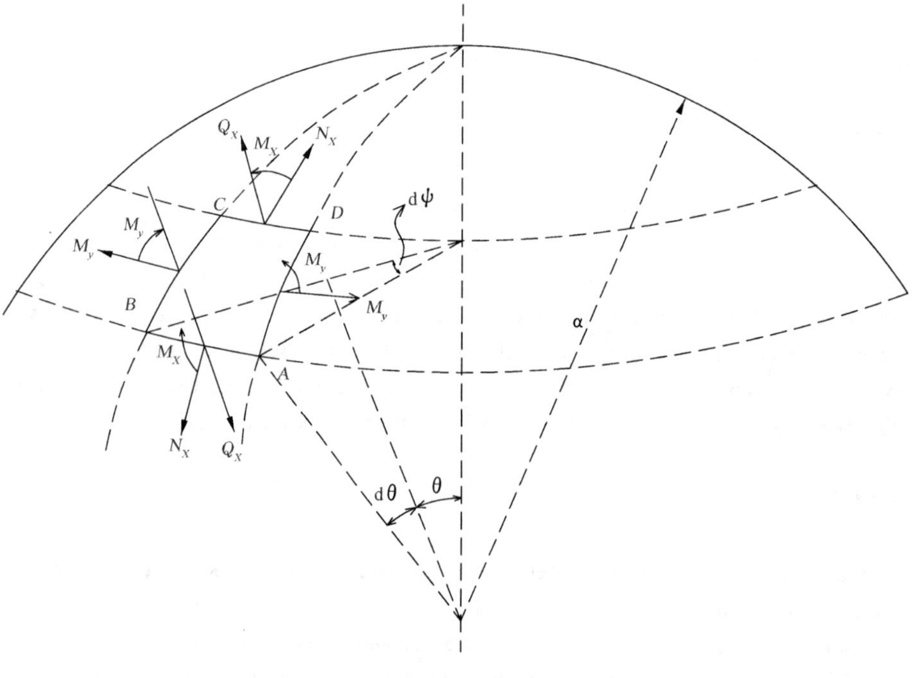

Figure 1 Spherical shell element and corresponding forces

In the case of symmetrical deformation, there are only three equations to be considered as the projections of forces on the x, and z axes and moments of forces with respect the $y-$ axes. Therefore after simplifications, the three equations of equilibrium become[4];

$$\frac{dN_x}{d\theta} + (N_x - N_y)\cot\theta - Q_x = 0 \tag{2}$$

$$\frac{dQ_x}{d\theta} + Q_x\cot\theta + N_x + N_y + pa = 0 \tag{3}$$

$$\frac{dM_x}{d\theta} + (M_x - M_y)\cot\theta - Q_x a = 0 \tag{4}$$

Derivation of Formula Based on the Southwell Method

In the derivation of the formula, for the classical theory of shell buckling, it is assumed that the displacements u and w may be expressed as, $\psi = \frac{du}{d\theta} = \sum_{n=0}^{\infty} A_n P_n$ and $w = \sum_{n=0}^{\infty} B_n P_n$ where P_n is the Legendre functions of the orders n and A_n and B_n are the real constants. Furthermore, because of manufacturing imperfections which are unavoidable, it is assumed that these imperfections may be expressed as; $\psi_0 = \sum_{n=0}^{\infty} A'_n P_n$ and $w_0 = \sum_{n=0}^{\infty} B'_n P_n$. Moreover, for the sake of simplicity, it is assumed that the manufacturing imperfections of ψ_0 is equal to zero. Thus, it is introduced only with the direction w. When the compressive load q is applied to shell, each point of the middle surface undergoes elastic displacements u and w, and its normal distance from the reference sphere then becomes $w + w_0$. It is assumed of course, that w_0 is of the order of an elastic deformation, and then the element of the shell looks like the deformed elements, which are used to establish the differential equations of the buckling problem. Thus, the following set of differential equations can be derived[5]:

$$H(\psi + w) + \alpha H(w) - (1-\upsilon)(\psi + w) - \phi(\psi + w + w_0) = 0 \tag{5}$$

$$\begin{aligned} &\alpha HH(\psi + w) - (1+\upsilon)H(\psi) - (3+\upsilon)\alpha H(w) + 2(1+\upsilon)(\psi + w) \\ &\quad + \phi[-H(\psi) + H(w + w_0) + 2(\psi + w + w_0)] = 0 \end{aligned} \tag{6}$$

in which α, ϕ, and H are defined as

$$\alpha = \frac{D(1-\upsilon^2)}{a^2 E t_0} = \frac{t_0^{\,2}}{12a^2},\ \phi = \frac{pa(1-\upsilon^2)}{2Et_0},\ H = \frac{d^2()}{d\theta^2} + \cot\theta \frac{d()}{d\theta} + 2()$$

By following the procedure that is used for the classical buckling theory of spherical shells[3] the following set of algebraic equations can be obtained:

$$A_n(\lambda_n + 1 + \upsilon + \phi) + B_n(\alpha\lambda_n + 1 + \upsilon + \phi) = -B'_n\phi \tag{7}$$

$$\begin{aligned} &A_n(\alpha\lambda_n^2 + \lambda_n + 2 + \upsilon\lambda_n + 2\upsilon + \phi\lambda_n + 2\phi) \\ &+ B_n(\alpha\lambda_n^2 + 3\alpha\lambda_n + \upsilon\alpha\lambda_n + 2 + 2\upsilon - \phi\lambda_n + 2\phi) = B'_n\phi(\lambda_n - 2) \end{aligned} \tag{8}$$

Thus, the problem is reduced to solving the foregoing set of equations. Eliminating A_n from Eqns. (7) and (8), one obtains

$$\begin{aligned} &[\alpha(\alpha-1)\lambda_n^3 + (\phi\alpha - 2\alpha + \phi)\lambda_n^2 + (\upsilon\alpha + \upsilon^2 + 2\phi\upsilon + \phi + 2\phi^2 - 1 - \alpha - 3\upsilon\alpha - \upsilon^2\alpha + \upsilon\phi - \alpha\phi \\ &- \upsilon\alpha\phi)\lambda_n]B_n = -B'_n\phi[(\alpha+1)\lambda_n^2 + (2\upsilon + 2\phi)\lambda_n] \end{aligned} \tag{9}$$

Therefore, the coefficient B_n becomes;

$$B_n = -\frac{B_n{}'\phi\lambda_n[(\alpha+1)\lambda_n + 2(\upsilon+\phi)]}{[-\alpha\lambda_n^2(1-\alpha) + (\phi - 2\alpha + \phi\alpha)\lambda_n + \upsilon(3\phi - 2\alpha) + \phi + 2\phi^2 + \upsilon^2 - \upsilon^2\alpha - \alpha - \alpha\phi - \upsilon\alpha\phi - 1]\lambda_n} \tag{10}$$

After canceling λ_n, and neglecting the small quantities as α, ϕ and their products in comparison with unity;

$$B_n \cong -\frac{B'_n\phi[\lambda_n + 2(\upsilon+\phi)]}{-\alpha\lambda_n^2 + (\phi - 2\alpha + \phi\alpha)\lambda_n + \upsilon^2 - 1} \tag{11}$$

Returning to the definition of the displacement w, one may write the equation, $w = \sum_{n=0}^{\infty} B_n P_n$. By substituting the values of the Legendre Polynomials in their places, one obtains:

$$w=[B_0+0.25B_2+\frac{9}{64}B_4+\cdots\cdots]+[B_1+\frac{3}{8}B_3+\cdots\cdots]\cos\theta+[B_2+\cdots\cdots]\cos 2\theta+\cdots\cdots \quad (12)$$

Also according to the definition of; $\lambda_n=n(n+1)-2$ for which the minimum value is $\lambda=-2$. Therefore, the minimum value is obtained for n equals to minus one and zero. Since n must be an integer, it is chosen as zero, which yields, $\lambda=-2$ and corresponds to B_0 which is a function of λ_n and gets smaller when λ_n becomes greater. Thus, it is possible to neglect all the terms and simply write $w\cong B_0$ since the terms which contains $\cos\theta, \cos 2\theta\ldots$ are much more smaller so, buckling is usually expected at the places where θ is large. Accordingly, it is possible to write

$$w\cong\frac{B'_0\phi[-2+2(\upsilon+\phi)]}{4\alpha+(\phi-2\alpha+\phi\alpha)2+1-\upsilon^2} \quad (13)$$

or rearranging the terms,

$$w\cong\frac{2B'_0\phi[\upsilon+\phi-1]}{2\phi(1+\alpha)+1-\upsilon^2}. \quad (14)$$

By neglecting the small terms since α and ϕ are small in comparison with unity so, it can be written as

$$w\cong\frac{B'_0(\upsilon-1)}{1+\dfrac{1-\upsilon^2}{2\phi}}. \quad (15)$$

Writing ϕ in detail, $\phi=\frac{p(1-\upsilon^2)}{2Em}$ or $\frac{2Em}{1-\upsilon^2}=\frac{p}{\phi}$ in which m is the ratio of the thickness to the radius of the sphere. The classical critical load for a spherical shell as found before is[3]; $p_{cr}=\frac{2Em^2}{\sqrt{3(1-\upsilon^2)}}$ or $p_{cr}^2=\frac{4E^2m^4}{3(1-\upsilon^2)}$. Therefore, $\frac{2Em}{(1-\upsilon^2)}=3\frac{p_{cr}^2}{2Em^3}$. By equating the two relations, one obtains $\frac{p}{\phi}=\frac{3p_{cr}^2}{2Em^3}$ or $\frac{1}{\phi}=\frac{3}{2}\times\frac{p_{cr}^2}{Em^3}\times\frac{1}{p}$. By substituting it back in $w=\frac{B'_0(\upsilon-1)}{1+\frac{3(1-\upsilon^2)p_{cr}^2}{4Em^3p}}$ and performing the cross multiplication, one gets

$$w+\frac{3(1-\upsilon^2)}{4Em^3}\times p_{cr}^2\times\frac{w}{p}=B'_0(\upsilon-1) \quad (16)$$

which is the equation of a straight line if the x axis is taken as w and the y axis as $\frac{w}{p}$. Thus the inverse slope of this line gives the critical load with a minus sign. Therefore obtaining the slope of this line experimentally, $p_{cr}^2=\frac{-4Em^3}{3(1-\upsilon^2)S}$ where S denotes the slope of the line. Thus, the Southwell procedure is applicable for uniformly compressed spherical shells.

NONLINEAR FINITE ELEMENT ANALYSIS (FEA)

All FEA for this investigation was performed using the general purpose program ABAQUS Version 6.7. Eight-node shell element was used to model hemispherical shells. This element is a general purpose quadratic shell element. The material of the shells is assumed as homogeneous, isotropic, incompressible and elastic. In order to check for the accuracy attainable by this method, a number of spherical shells with different kinds of boundary condition and loading were solved (Figs. 2 to 5).

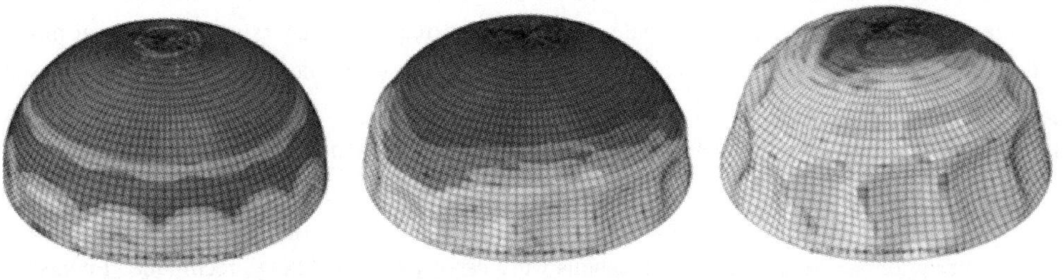

Figure 2　Deformation of hemispherical shell with hinge support under radially uniform pressure

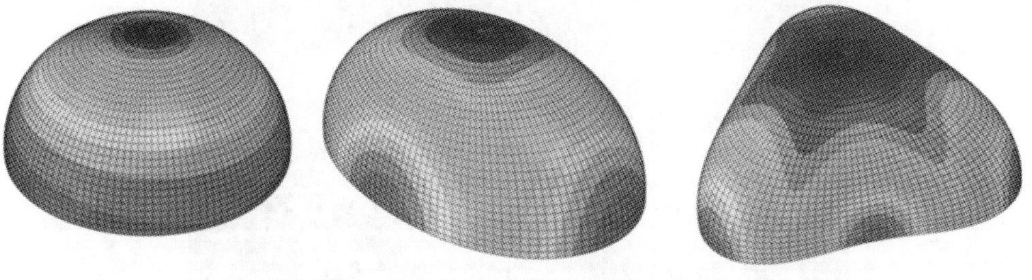

Figure 3　Different deformation modes in hemispherical shell with roller support under radially uniform pressure

Figure 4　Deformation of the hemispherical shell with hinge support under ring load at $\frac{R}{2}$

Figure 5　Buckling of the hemispherical shell with hinge support under gravity loading

EXPERIMENTAL PROGRAM

A total of six thin walled polyethylene hemispherical shells were constructed and tested under uniform suction pressure. The base diameters of these shells were 15 cm and 10 cm and their wall thickness were 0.05 cm yielding $\frac{R}{t}$ ratios of 150 and 100 respectively. It is evident that the construction of these shells through machining would have been difficult and for the following reasons, the shells were made of solid polyethylene plastic which posses good tensile, flexural, and impact strengths and its flexural modulus is proportional to the stiffness of the material. Its creep resistance is excellent and is substantially superior to most plastics. Its mechanical properties are: Flexural modulus = 650 MPa, Poisson's ratio = 0.4, and Density = 1 150 kg/m^3. A complete family of hemispherical shells and some experiments are shown in Figs. 6 and 7. The manufacturing of these shells was carried out with the aid of machined male and female molds made from cast aluminum alloy. The aluminum alloy molds were machined with considerable precision and then the spherical shells were cast by the "puddling" technique. Each shell was inspected by a polariscope to ensure that no air bubbles were trapped in the shell wall.

Figure 6 Hemispherical shells samples made of polyethylene

Figure 7 Initial buckling of hemispherical shells under uniform suction pressure

RESULTS AND DISCUSSION

In this study, the Southwell predictions are compared with experimentally, and numerically obtained values. The results of the experimental and numerical investigation are summarized in Figs. 8 and 9. Scatter in the experimentally obtained buckling pressures is probably due to variations that existed in the specimens because of the fact that each one was cast separately. The manufacturers did, however, take considerable care during the manufacturing process, and especially with the mix, and for this reason the scatter is very small. Thus, it is likely that the main source of error compared to theory is because of measurement reading errors, and imperfections in material properties. However, the agreement between measured buckling load and Southwell prediction is remarkable. Mostly, the Southwell method tended to yield buckling loads which are slightly higher than those measured and the disparity of buckling load is somewhat difficult to detect. Nevertheless, the predicted loads are reliable to be slightly higher (up to about 17%) than the actual buckling loads encountered. Therefore, a reasonable degree of caution is recommended to be exercised. Scatter in the numerically obtained buckling pressures for axisymmetrical buckling cases are very small and it is most likely due to the assumptions of the finite element solution. For the case of buckling of spherical shell with roller support under uniform pressure, the buckled shape is not axisymmetric anymore (Figure 3). So, once data are collected from the principle axes locations, the answer is acceptable and there is only 13% error otherwise, the deviation from the correct values is considerable.

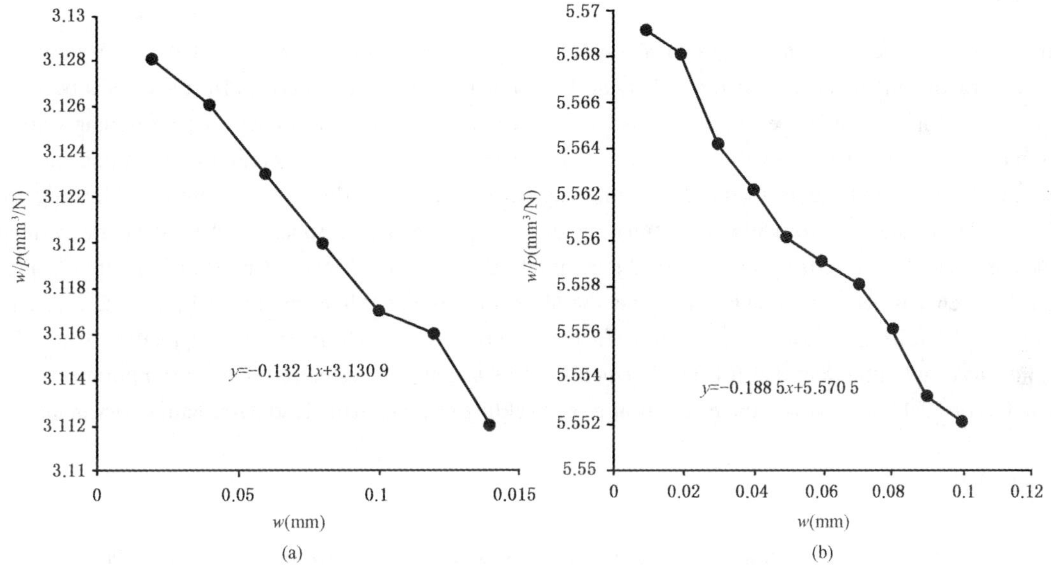

Figure 8 Plot of $\frac{w}{p}$ against w for experimental study (a): (R = 50 mm, t_0 = 0.5 mm) and (b): (R = 75 mm, t_0 = 0.5 mm)

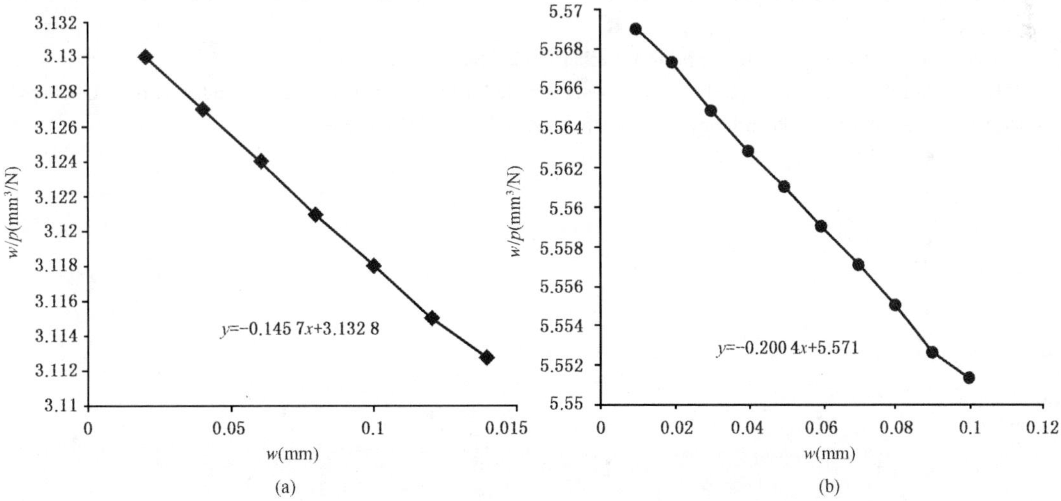

Figure 9 Plot of $\frac{w}{p}$ against w *for numerical study* (*a*): (R=50 mm, t_0=0.5 mm) and (b): (R=75 mm, t_0=0.5 mm) under uniform suction pressure.

For sample (a) in Figure 8, slope of line is equal to $S=-0.1321$ so, $p_{cr}^2 = \dfrac{-4\,Em^3}{3(1-\upsilon^2)S} \rightarrow p_{cr} = 0.088$ MPa and from test buckling pressure is measured $\bar{p} = 0.078$ MPa therefore, error for this case is equal to: $\dfrac{0.088-0.078}{0.078} \times 100 = 12.8\%$ and from a similar calculation, the error for sample (b) is equal to 14.28%. In Figure 9 using the same computations, error is equal to 3.7% and 5.4 % respectively.

CONCLUSION

In this study, the well known graphical method of predicting buckling loads, i. e., the Southwell's nondestructive method is extended to spherical shells and a new formula is derived for the critical buckling load of uniformly compressed spherical shells. The feasibility of this technique in determining critical buckling loads of spherical shells is demonstrated provided that caution is exercised in analyzing the test results. This method may be used in any kind of spherical shells with arbitrary axisymmetrical loading and it provides a valuable procedure for estimating the buckling load of a spherical shell structure without having to conduct a destructive test. In this method, the curve of displacement/load is plotted against displacement, which is a straight line, and the slope of this line when multiplied by a constant value presents the critical buckling load with sufficient accuracy. The expediency of the method lies in its simplicity, and in the fact that it is nondestructive. This technique does not need any assumption as to the number of buckling waves or the exact locality of buckling so long as the loading remains axisymmetric.

REFERENCES

[1] Southwell, R. V., "On the analysis of experimental observations in problems of elastic stability", Proceedings of Royal Society of London (A), 1932, Vol. 135, pp. 601-616.

[2] Timoshenko, S. P., and Woinowsky-Krieger, S. "Theory of Plates and Shells", Mc Graw-Hill Second Edition, 1970, New York, 580p.

[3] Timoshenko, S. P., and Gere, J. M. "Theory of Elastic Stability , Second Edition", McGraw-Hill, 1961, New York, 541p.

[4] Flügge, W., "Stresses in Shells", Springer-Verlag, 1962, Berlin, 499p.

[5] Nayyeri Amiri, S., and Rasheed H. A., "Elastic/Plastic Buckling of Spherical Shells under Various Loadings", LAP Lambert Academic Publishing, 2011, 194p., ISBN-978-3-8454-3985-3.

MOLECULAR DYNAMICS SIMULATION RESULTS FOR BUCKLING OF DOUBLE-WALLED CARBON NANOTUBES WITH SMALL ASPECT RATIOS

* A. N. R. Chowdhury and C. M. Wang

Department of Civil and Environmental Engineering, National University of Singapore, Singapore
* Email: amar@nus. edu. sg

KEYWORDS

Double-walled carbon nanotubes, molecular dynamics, axial buckling, critical load, critical strain, incremental displacement, displacement rate, benchmark results.

ABSTRACT

Researchers using molecular dynamics simulations for buckling analysis of double walled carbon nanotubes (DWCNTs) have furnished somewhat different results, especially for DWCNTs with small aspect ratios ($L/D_{in}<10$ where L is the length and D_{in} the inner tube diameter). In the development of continuum models for buckling of DWCNTs that has van der Waals interaction between walls, it is necessary to have benchmark molecular dynamics (MD) results for calibration. This paper aims to review the existing MD buckling results for DWCNTs with small aspect ratios and to provide an additional comprehensive set of results for benchmarking purpose. For the latter objective, molecular dynamics simulations are conducted on nine sets of DWCNTs of different lengths at 1 K and 300 K. All the DWCNTs chosen for the MD simulation have an inner tube with chiral indices (5, 5) and an outer tube with chiral indices (10, 10). For static buckling analysis using MD simulations, the displacement rate have to be small enough say 0.005 Å/ps respectively, since these parameters influence the critical strain/load significantly as will be shown herein.

INTRODUCTION

First discovered by Iijima[1], carbon nanotubes have triggered an intense research activities because of their superior properties such as high strength, excellent thermal and electrical conductivity. Carbon nanotubes are commonly found nested in one another. The special double-walled carbon nanotubes (DWCNTs) are a synthetic blend of both single walled carbon nanotubes (SWCNTs) and multi walled carbon nanotubes (MWCNTs), and they exhibit the electrical and thermal stability of the latter and the flexibility of the former. Owing to these reasons DWCNTs are very suitable for practical applications. Under axial compressive loading, nanotubes are prone to buckling due to its hollow tubular geometry. Nanotubes with a low aspect ratio (diameter to length ratio <15) buckle in a shell like buckling mode[2-5]. This paper will focus on the buckling of DWCNTs with small aspect ratios (i. e. $L/D_{in}\leqslant 10$) under axial compression.

Researchers have employed molecular dynamics (MD) simulations to investigate the buckling behavior of axially loaded DWCNTs, but the simulation conditions and the type of DWCNT used in reported papers are mostly different. A summary of existing MD simulation results for buckling of DWCNT is given in Table 1.

TABLE 1 EXISTING MD-SIMULATIONS OF DWCNT BY VARIOUS RESEARCHERS

No.	Researchers	Simulation conditions		DWCNT			Critical	
		Potential	T (K)	Length Å	Chiral indices		Load (nN)	Strain
					Outer	Inner		
1	Liew et al. [6]	$REBO^{2nd}$ + L-J	1	60	5, 5	10, 10	172.00	0.060 0
					10, 10	15, 15	202.00	—
					15, 15	20, 20	209.00	—
					20, 20	25, 25	210.00	—
2	Guo et al. [7]	* AFEM with $REBO^{2nd}$ + L-J	0	61.2	5, 5	10, 10	—	0.046
3	Zhang et al. [8]	$REBO^{2nd}$ + L-J	0	75	5, 5	10, 10	94.94^{+}	0.055 5
							83.35*	0.050 2
4	Zhang et al. [9]	$REBO^{2nd}$ + L-J	0	60	5, 5	10, 10	148.20	0.055 3
					15, 15	20, 20	184.00	0.034 1
5	Lu et al. [10]	Tersoff	1	96.03	5, 5	10, 10	—	0.025 0
6	Lu et al. [11]	Tersoff	1	25.00	5, 5	10, 10	—	0.023 7
				37.50	10, 10	15, 15		0.012 1
7	Zhang et al. [12]	$REBO^{2nd}$ + L-J	300	53.2	4, 4	9, 9	—	0.069 3
					9, 9	14, 14		0.049 9
8	Zhang et al. [13]	$REBO^{2nd}$ + L-J	1	65	10, 0	16, 0	—	0.066 1
					10, 0	17, 0		0.063 7
					10, 0	18, 0		0.057 6
					10, 0	19, 0		0.046 1
					11, 0	19, 0		0.055 2
					12, 0	19, 0		0.057 6
					13, 0	19, 0		0.058 2
9	Kulathunga et al. [14]	COMPASS	0	21.97	4, 4	9, 9	—	0.066 1
				43.94				0.063 7
				65.91				0.057 6
				87.88				0.046 1
				109.85				0.055 2
				65.10	3, 3	8, 8		0.057 6
				81.37	5, 5	10, 10		0.047 0
				89.51	6, 6	11, 11		0.066 1
				97.65	7, 7	12, 12		0.063 7

continued

10	Korayem et al.[15]	COMPASS	0	12.03	5, 5	10, 10	—	0.057 7
				14.44				0.055 2
				16.85				0.053 2
				19.28				0.051 7
				21.66				0.050 4
				24.06				0.049 5
				26.47				0.045 3
				28.87				0.045 1
				60.16				0.043 2

Note: + at displacement rate 0.062 25 Å/ps,
* at displacement rate 6.225×10^{-3} Å/ps
* AFEM = Atomistic finite element.
A DWCNT with outer tube indices (10, 10) and inner tube indices (5, 5) will be referred as DWCNT((5, 5),(10, 10))

Table 1 shows that, for a typical DWCNT, the critical strains obtained by different researchers are different. For example, Liew et al.[6] and Zhang et al.[9] used the same potential and temperature in their MD simulations but they obtained different critical strains (0.060 0 and 0.055 3 respectively) for DWCNT((5, 5),(10, 10)) of length 60 Å. An even lower critical strain of 0.043 2 is calculated by Korayem et al.[15] for, DWCNT((5, 5),(10, 10)) of length 60 Å, by using a different potential. For the same DWCNT, Guo et al.[7] obtained a critical strain of 0.046 from atomistic finite element simulation. Therefore, it is evident that different researchers have obtained different critical strains for the same DWCNT. Also we note that Kulathunga et al.[14] reported a critical strain of 0.047 for DWCNT((5, 5), (10, 10)) of length 81 Å but this value is more than the critical strain of a shorter tube [60 Å long DWCNT((5, 5), (10, 10))] as reported by Korayem et al.[15]. From a continuum mechanics perspective, the critical strain ought to decrease with increasing length of the cylindrical tube. Lu et al.[11] reported a critical strain of 0.023 7 for 25 Å long DWCNT ((5, 5),(10, 10)) which is very small when compared to the critical strains calculated by other researchers for longer tubes. All these differences and anomalies need to be examined to filter out the inaccurate or even grossly erroneous results.

Continuum models for analysis of DWCNTs are a suitable alternative to MD simulation as the computational effort is considerably less. In order to develop an effective continuum model of DWCNTs which have van der Waals interaction between walls, one needs accurate benchmark solutions from MD simulation results as there are scarce experimental results. Based on our previous discussions it can be seen that, from Table 1, it is difficult to establish which sets of the different MD results are the ones for use in calibrating the continuum models. Moreover, it can also be seen that most of the simulations in Table 1 are performed near 0 K which is not realistic in practical operating conditions. Because of these reasons, we present a review of MD simulation technique to evaluate the critical strains of DWCNTs and we also present MD simulation results for the buckling of nine sets of axially loaded DWCNTs, with different lengths at 1K and 300 K. The DWCNT consists of an inner tube with chiral indices (5, 5) and outer tube chiral indices (10, 10).

DESCRIPTION OF MD SIMULATION SETUP

MD simulations are conducted at 1K and 300 K by using the molecular dynamics package LAMMPS which was downloaded in May 2011[16]. For the simulations, the AIREBO potential of Stuart et al.[17] is employed. A brief description of the AIREBO potential, as given in the LAMMPS manual, is given by (Plimpton[18])

$$E_{AIREBO} = \sum_{i}^{Natoms}\sum_{j\neq i}^{Natoms}\left[E_{ij}^{REBO} + E_{ij}^{LJ} + \sum_{k\neq i}^{Natoms}\sum_{l\neq i,\, j,\, k}^{Natoms} E_{kijl}^{TORSION}\right] \tag{1}$$

where E^{REBO} is the 2nd generation REBO potential developed by Brenner et al.[19], E^{LJ} is the 12-6 Lennard Jone's potential that takes into account the long range ($\geqslant 2$ Å) interaction and $E^{TORSION}$ is a 4-body potential that allows for the dihedral angle interaction. In the MD simulations, 6 sets of atoms on either ends of a DWCNT are fixed so as to simulate the clamped boundary condition. The length of nanotube is calculated by subtracting the combined length of constrained atoms at both ends, from the total length of the tube. A schematic representation of MD simulation setup is shown in Figure 1.

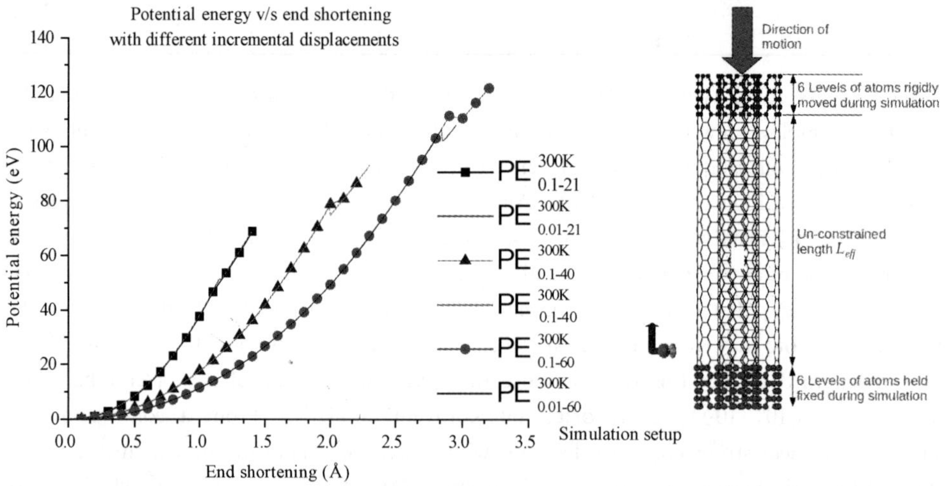

Figure 1 Potential energy versus end shortening plots at 300 K and simulation setup

Different lengths of DWCNT((5, 5),(10, 10)) considered are shown in Table 2. Before initiating the analysis, the total potential energy of the system is minimized using the conjugate gradient method. After that, a displacement controlled buckling analysis is performed by moving one end of the tube incrementally

TABLE 2 DWCNT((5, 5),(10, 10)) LENGTHS AND CRITICAL LOADS AND CRITICAL STRAINS $D_{OUT}/D_{IN}=2$

No.	L (Å)	$\frac{L}{D_{in}}$	Critical load in (nN) and strain							
			At 1 K				At 300 K			
			$P^{1K}_{cr-0.1}$	$\varepsilon^{1K}_{cr-0.1}$	$P^{1K}_{cr-0.01}$	$\varepsilon^{1K}_{cr-0.01}$	$P^{300K}_{cr-0.1}$	$\varepsilon^{300K}_{cr-0.1}$	$P^{300K}_{cr-0.01}$	$\varepsilon^{300K}_{cr-0.01}$
1	20.83	3.07	151.68	0.052 4	157.16	0.0538	149.43	0.052 4	154.42	0.053 8
2	23.28	3.43	147.11	0.052 2	152.17	0.053 5	144.76	0.052 2	151.43	0.053 5
3	28.18	4.16	140.05	0.050 0	150.96	0.052 9	137.04	0.050 0	146.44	0.052 1
4	33.09	4.88	146.93	0.051 5	147.06	0.051 8	133.47	0.048 5	139.37	0.049 7
5	40.44	5.96	145.17	0.052 5	143.60	0.051 0	136.83	0.050 0	136.07	0.049 5
6	47.79	7.05	141.87	0.050 0	139.89	0.049 2	140.86	0.050 0	135.38	0.048 3
7	55.14	8.13	143.57	0.050 9	138.40	0.049 1	130.92	0.047 3	132.81	0.048 0
8	60.05	8.86	141.60	0.050 0	136.64	0.048 5	134.02	0.048 3	131.10	0.047 2
9	67.40	9.94	138.47	0.049 3	135.71	0.046 4	130.51	0.047 7	131.10	0.047 0

Note: minimum simulation time $= N * \Delta t = 2$ pico second, maximum displacement rate $= 0.05$ Å/ps.

by a constant displacement (Δd) of value 0.1 Å or 0.01 Å. These two incremental displacement are used to determine how small value that one has to adopt for accurate results. Equations of motion are integrated for a minimum of 2 000 simulation steps (N) using Verlet algorithm with a time step size of 10^{-15} s (Δt). During the dynamics simulations, a Nose Hoover thermostat (NVT) is used to maintain the simulation temperature at 1 K or 300 K. The effect of Δd and N can be cumulatively ascertained by using a parameter displacement rate which is estimated by dividing Δd by product of N and Δt.

RESULTS AND DISCUSSIONS

Two sets of MD simulations are obtained for each DWCNT. One set is conducted with incremental displacement 0.1Å whereas the other set with 0.01Å. From the simulations, the variations of the total potential energy (PE) with respect to the axial deformation and the variations of the axial load with respect to axial deformation are obtained. In the PE versus deformation plots, a sharp kink is observed which corresponds to the critical point at which the total energy is released from the system and this denotes the occurrence of buckling. Typical PE versus deformation plots for three DWCNTs with different lengths are shown in Figure 1. Note that $PE^{300K}_{0.01-L}$ represents potential energy of DWCNT of length L Å measured at 300 K using an incremental displacement of 0.01 Å.

From the plots, it is observed that the incremental displacement value has no effect on the PE versus end shortening curve before the critical point. The incremental displacement value, however, affects the magnitude of PE at which the critical point occurs, and it also affects the nature of PE versus deformation curve beyond the critical point. The critical load/strain is established as the value at which the axial load experiences a significant drop. For a given DWCNT of length L Å, two critical loads $P^{300K}_{cr-0.1}$ and $P^{300K}_{cr-0.01}$ are obtained with incremental displacements 0.1Å and 0.01Å respectively at 300 K. As an example, typical force deformation plots for three DWCNTs having lengths 21Å, 33Å, and 60Å respectively are shown in Figure 2 where $P^{300K}_{0.1-L}$ corresponds to the force measured on DWCNT of length L Å using an incremental displacement value of 0.1Å at 300 K. From Figure 2, it is evident that the effect of incremental displacement on the force versus deformation plot is similar to its effect on PE versus deformation plot.

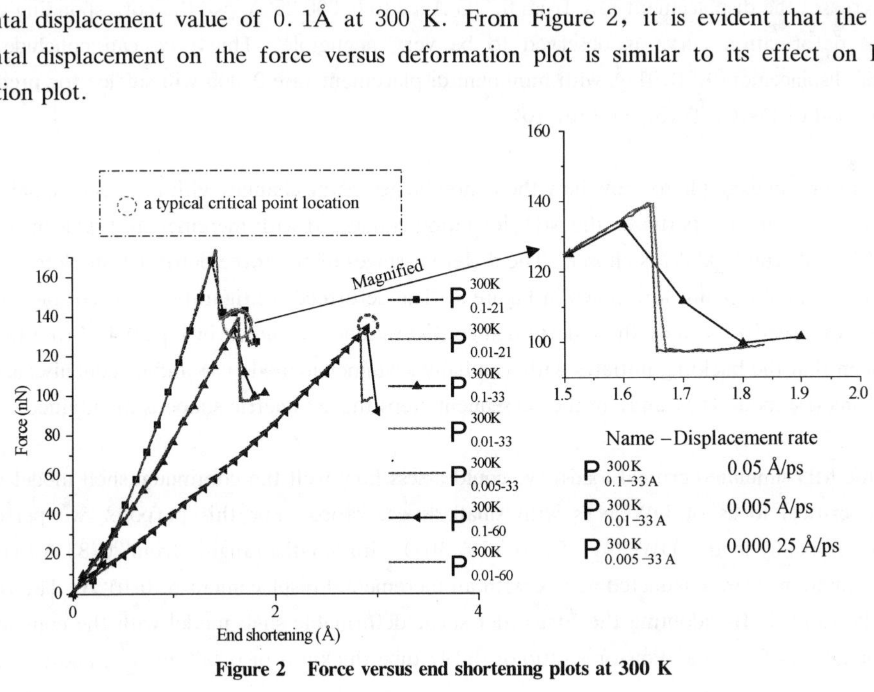

Figure 2　Force versus end shortening plots at 300 K

Critical load values are estimated for DWCNT with different incremental displacements and the values are given in Table 2. From Table 2 it is observed that when an incremental displacement of 0.1Å is used, the critical load obtained may increase or decrease with respect to increasing length. This trend of critical load variation with respect to length contradicts the fact that critical load should decrease with increasing length of the nanotube. If an incremental displacement of 0.01Å is used, it can be seen that the critical load does decrease monotonically with respect to increasing length of the tube. Belytschko et al.[20] observed that if the adopted displacement rate is too high, it will cause considerable overshoot in inter-atomic forces due to stress wave like phenomena. Also, if a larger displacement rate is used, then the force deformation path may skip the critical point during the displacement controlled buckling simulation. Note that in Table 2, the critical strains are obtained by dividing the shortening (which corresponds to the critical load) with the original length.

In our simulations it is observed that the critical loads are significantly affected by the incremental displacement and the number of simulation steps. The combined effect of incremental displacement and the number of simulation steps can be ascertained by using the displacement rate parameter ($\Delta d/N/\Delta t$). To ascertain whether an even smaller displacement rate is necessary, we determine the critical load of DWCNT((5, 5),(10, 10)) at 300 K with a length of 33 Å by using an even lesser incremental displacement of 0.005 Å and displacement rate 0.000 25 Å/ps. The critical load obtained with this displacement rate of 2.5×10^{-4} Å/ps, is 138.952 nN, which is only 0.3% lower than the critical load value 139.365 nN, obtained using an incremental displacement 0.01 Å and displacement rate of 0.005 Å/ps. (Note that MD simulation of 33 Å DWCNT((5, 5),(10, 10)) with a displacement rate of 0.000 25 Å/ps takes almost 7 days to complete whereas it takes only 2 days with a displacement rate of 0.005 Å/ps when the simulations are carried out in a desktop computer). Figure 2 shows a comparison of the force deformation plots for, 33 Å long DWCNT((5, 5),(10, 10)), at different displacement rates. In Figure 2, it can be seen that in the vicinity of the critical point, a significant change in the nature of force versus deformation curve occurs when we lower the displacement rate from 0.05 Å/ps to 0.005 Å/ps. If one further decreases the displacement rate from 0.005 Å/ps to 0.000 25 Å/ps, the corresponding change in load versus deformation curve is observed to be very negligible. Thus, we can conclude that an incremental displacement of 0.01 Å with minimum displacement rate 0.005 will suffice for prediction of the critical load of DWCNT((5, 5),(10, 10)).

Next, we present an example to show how the nanotube geometry changes with increasing axial load near the critical point. For this purpose, the MD simulation conducted with incremental displacement of 0.01 Å on DWCNT of length 33 Å, is chosen. The different stages of the force deformation plots are denoted as a, b, c, d and these points are shown in Figure 3. The deformed configuration at each stage is denoted by a, b, c, d respectively, and, these deformed configurations are shown in Figure 4. From this figure, it can be seen that the buckling initiates with a radially asymmetric mode (a) and it suddenly changes to a radially symmetric mode (b), and, in the subsequent steps the symmetric shape is maintained.

Based on the MD simulated critical loads, we could assess how well the continuum shell model is able to furnish the critical loads of DWCNTs with small aspect ratios. For this purpose, we perform MD simulations on axially loaded DWCNT((5, 5),(10, 10)) with lengths ranging from 20.83 Å to 67.40 Å. The MD simulations were conducted at 1 K with an incremental displacement of 0.01 Å. The results are presented in Table 3. By adopting the first order shear deformable shell model with the commonly used Young's modulus of 5.5 TPa, Poisson's ratio of 0.19, tube thickness of 0.066 nm[21], and the inter-tube

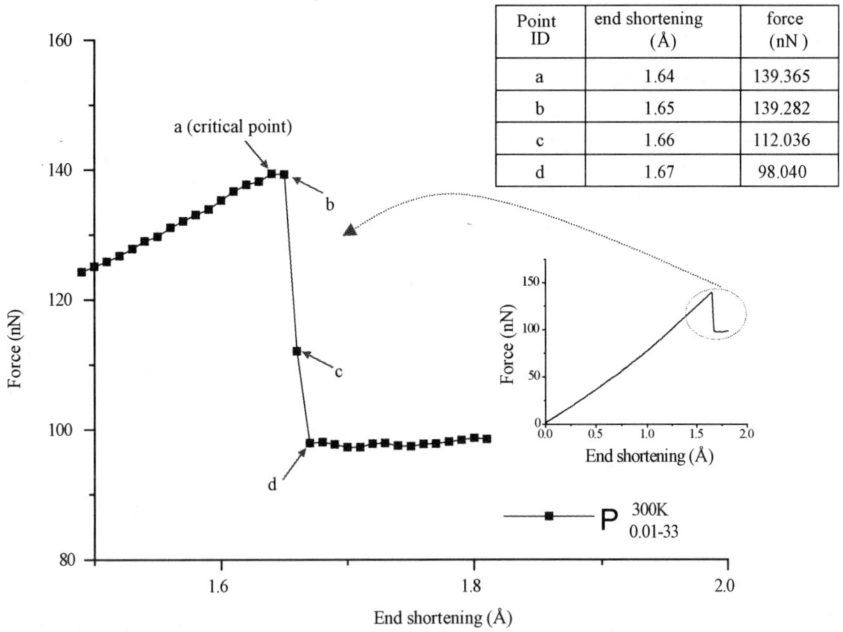

Point ID	end shortening (Å)	force (nN)
a	1.64	139.365
b	1.65	139.282
c	1.66	112.036
d	1.67	98.040

Figure 3 Force versus end shortening plot for DWCNT((5, 5),(10, 10)) of length 33 Å

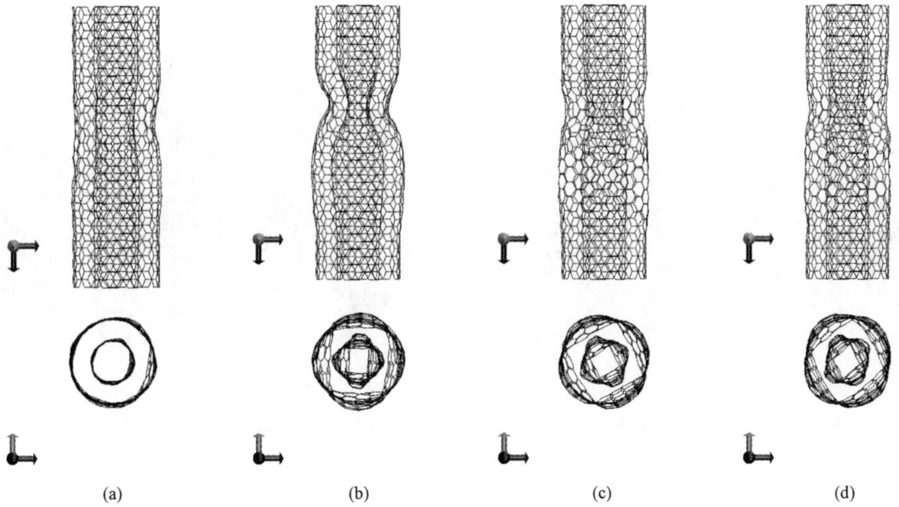

Figure 4 Deformed DWCNT((5, 5),(10, 10)) of length 33 Å at different loading stages

van der Waals interactions being modeled using linear spring elements (assuming van der Waals coefficient $c = 6.199 \times 10^{19}$ N/m^3 [22]), the critical loads obtained using ABAQUS are presented in Table 3. It can be observed that the continuum shell model overestimates the critical load for DWCNT, more so for relatively longer lengths. For the lengths of DWCNTs considered, the maximum difference between the critical loads obtained from MD simulations and the shell model is about 10% when the DWCNTs is relatively long. However, if a Young's modulus of 5 TPa is used instead, the critical load is underestimated (within 8%) for relatively short DWCNTs but the shell model solution comes closer to the

MD results for DWCNTs with longer lengths. So a Young's modulus of 5.25 TPa will furnish results within 5% difference from MD results for the entire range of aspect ratio considered.

TABLE 3 COMPARISON OF CRITICAL LOADS OBTAINED FROM MD SIMULATIONS AND CONTINUUM SHELL MODEL FOR DWCNT((5, 5),(10, 10))

No.	L (Å)	Present MD critical loads (nN)	Other researchers' MD critical loads (nN)	Critical loads obtained from using continuum shell model(nN)		
				E = 5.5 TPa	E = 5.25 TPa	E = 5.0 TPa
1	20.83	157.16	—	155.57	148.07	141.02
2	23.28	152.17	—	155.13	147.62	140.59
3	28.18	150.96	—	154.64	147.29	140.26
4	33.09	147.06	—	154.41	146.73	139.74
5	40.44	143.60	—	152.46	145.19	138.28
6	47.79	139.89	—	151.21	144.03	137.17
7	55.14	138.40	—	150.28	143.00	136.19
8	60.05	136.64	172.00[6] 148.20[9]	149.68	142.44	135.65
9	67.40	135.71	—	149.01	141.86	135.11

A comparison of critical mode-shapes for DWCNT((5, 5),(10, 10)) of length 60 Å, obtained from MD simulation and shell model, is shown in Figure 5. It can be seen that the mode-shapes appear to very similar to each other. Therefore, for buckling analysis of axially loaded DWCNT, one may adopt the first order shear deformable shell theory with the aforementioned elastic parameters.

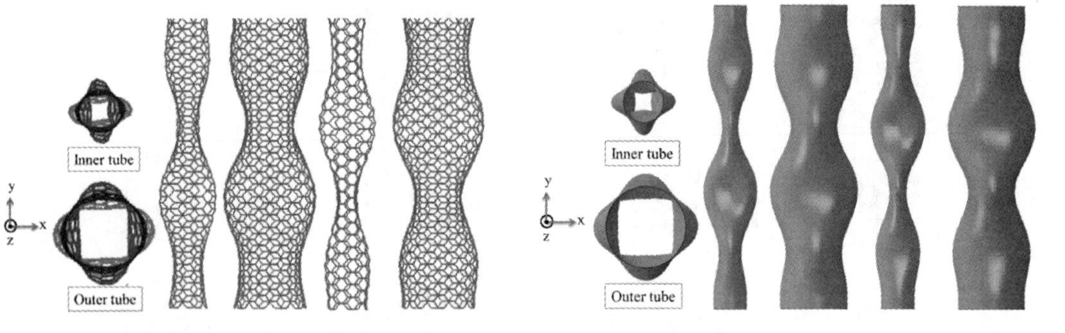

Molecular dynamics (Δd = 0.01 Å, T = 1 K)   Continuum shell theory (E = 5 TPa, μ = 0.19, h = 0.066 nm)

Figure 5 Critical mode-shapes for DWCNT((5, 5),(10, 10) of length 60 Å at 1 K

CONCLUDING REMARKS

In this paper, MD simulations results for the buckling of DWCNTs are reviewed. The existing MD results differ somewhat from each other, especially for DWCNTs with small aspect ratios (L/D_{in} < 10), even though some of them were calculated using the same potential and simulation temperature. The paper also presents additional sets of MD buckling results for DWCNTs with small aspect ratios, simulated at 1K and 300 K and two incremental displacement values. Based on the extensive simulations, we recommend that MD simulations for determining the critical strain of DWCNTs should be performed with a displacement rate of 0.005 Å/ps, along with the simulation parameters mentioned herein. A smaller displacement rate is not necessary. However, one has to adopt a very small time step ($\Delta t \sim 10^{-15}$ s) as we are using an explicit time integration algorithm. From our study, we conclude that the present MD simulation results may be used for calibrating continuum shell model for buckling analysis of DWCNTs. For example, for DWCNTs comprising an inner tube with chiral indices (5, 5) and an outer tube with chiral indices (10, 10), we find that the shear deformable shell model with Young's modulus of about 5.25 TPa, Poisson's ratio of 0.19, tube thickness of 0.066 nm and the van der Waals coefficient $c = 6.199 \times 10^{19}$ N/m^3 can furnish critical loads and mode shapes that are in close agreement (within 5%) with the MD results. Further study is required to ascertain whether the proposed continuum shell model is valid for any size of DWCNTs.

REFERENCES

[1] Iijima, S., "Helical microtubules of graphitic carbon", *Nature*, 1991, 354(7), pp.56-58.

[2] Sears, A. and Batra, R., "Buckling of multiwalled carbon nanotubes under axial compression", *Physical Review B*, 2006, 73(8), pp. 11.

[3] Batra, R. and Sears, A., "Continuum models of multi-walled carbon nanotubes", *International Journal of Solids and Structures*, 2007, 44(22-23), pp. 7577-7596.

[4] Zhang, Y.Y., Wang, C.M., Duan, W.H., Xiang, Y. and Zong, Z., "Assessment of continuum mechanics models in predicting buckling strains of single-walled carbon nanotubes", *Nanotechnology*, 2009, 20(39), pp. 8.

[5] Wang, C.M., Zhang, Y.Y., Xiang, Y. and Reddy, J.N., "Recent studies on buckling of carbon nanotubes", *Applied Mechanics Reviews*, 2010, 63(3), 030804-1 to 030804-18.

[6] Liew, K.M., Wong, C.H., He, X.Q., Tan, M.J. and Meguid, S.A., "Nanomechanics of single and multiwalled carbon nanotubes", *Physical Review B*, 2004, 69(11), pp. 8.

[7] Guo, X., Leung, A.Y.T., Jiang, H., He, X.Q. and Huang, Y., "Critical strain of carbon nanotubes: An atomic-scale finite element study", *Journal of Applied Mechanics*, 2007, 74(2), pp. 347-351.

[8] Zhang, Y.Y., Tan, V.B.C. and Wang, C.M., "Effect of strain rate on the buckling behavior of single-and double-walled carbon nanotubes", *Carbon*, 2007, 45(3), pp. 514-523.

[9] Zhang, Y.Y., Wang, C.M. and Tan, V.B.C., "Examining the effects of wall numbers on buckling behavior and mechanical properties of multiwalled carbon nanotubes via molecular dynamics simulations", *Journal of Applied Physics*, 2008, 103(5), pp. 9.

[10] Lu, J.M., Wang, Y.C., Chang, J.G., Su, M.H. and Hwang, C.C., "Molecular-dynamic investigation of buckling of double-walled carbon nanotubes under uniaxial compression", *Journal of the Physical Society of Japan*, 2008, 77(4), pp. 7.

[11] Lu, J.M., Hwang, C.C., Kuo, Q.Y. and Wang, Y.C., "Mechanical buckling of multi-walled carbon nanotubes: The effects of slenderness ratio", *Physica E: Low-dimensional Systems and Nanostructures*, 2008, 40(5), pp. 1305-1308.

[12] Zhang, C.-L. and Shen, H.-S., "Predicting the elastic properties of double-walled carbon nanotubes by molecular dynamics simulation", *Journal of Physics D: Applied Physics*, 2008, 41(5), pp. 6.

[13] Zhang, Y., Wang, C.M. and Tan, V.B.C., "Mechanical properties and buckling behaviors of condensed double-walled carbon nanotubes", *Journal of Nanoscience and Nanotechnology*, 2009, 9(8), pp. 4870-4879.

[14] Kulathunga, D.D.T.K., Ang, K.K. and Reddy, J.N., "Accurate modeling of buckling of single-and double-walled carbon nanotubes based on shell theories", *Journal of Physics: Condensed Matter*, 2009, 21(43), pp. 8.

[15] Korayem, A.H., Duan, W.H., Zhao, X.L. and Wang, C.M., "Buckling behavior of short multi-walled carbon nanotubes under axial compression load", *International Journal of Structural Stability and Dynamics*, 2012, 12(6).

[16] LAMMPS Molecular Dynamics Simulator, http://lammps.sandia.gov (accessed 7th May, 2010).

[17] Stuart, S.J., Tutein, A.B. and Harrison, J.A., "A reactive potential for hydrocarbons with intermolecular interactions", *The Journal of Chemical Physics*, 2000, 112(14), pp. 6472-6486.

[18] Plimpton, S., "Fast parallel algorithms for short-range molecular dynamics", *Journal of Computational Physics*,

1995, 117(1), pp. 1-19.

[19] Brenner, D. W., Shenderova, O. A., Harrison, J. A., Stuart, S. J., Ni, B. and Sinnott, S. B., "A second-generation reactive empirical bond order (REBO) potential energy expression for hydrocarbons", *Journal of Physics: Condensed Matter*, 2002, 14(4), pp. 783-802.

[20] Belytschko, T., Xiao, S., Schatz, G. and Ruoff, R., "Atomistic simulations of nanotube fracture", *Physical Review B*, 2002, 65(23), pp. 8.

[21] Yakobson, B. I., Brabec, C. J., Bernholc, J., "Nanomechanics of carbon tubes: instabilities beyond linear response", *Physical Review Letters* 76(14), pp. 4.

[22] Ru, C. Q., "Axially compressed buckling of a doublewalled carbon nanotube embedded in an elastic medium", *Journal of the Mechanics and Physics of Solids*, 2001, 49(6), pp. 1265-1279.

ENERGY ABSORPTION OF CARBON NANOTUBE SUBJECTED TO IMPACT LOADS

K.N. Feng, E.J. Hunter, * W.H. Duan and X.L. Zhao

Department of Civil Engineering, Monash University, Clayton, Victoria, Australia, 3800
* Email: wenhui.duan@monash.edu

KEYWORDS

Carbon nanotube, bullet impact, energy absorption, MD simulations.

ABSTRACT

Carbon nanotubes (CNTs) have remarkable resilience to dynamic loads as well as energy absorption capacity due to their excellent mechanical properties. This paper focuses on improving the understanding of energy absorption capacity of CNTs subjected to impact with a diamond projectile referred to as "the bullet". By analyzing the impact-rebound process it is shown that energy absorption in CNTs in local deformation mode is more sensitive to changes in kinetic energy than CNTs in global deformation mode. The bullet is launched at various velocities to create transverse impact at the mid-point of a CNT and the effect of bullet velocities on the energy absorption of CNTs is investigated. The research findings are complementary to the existing research on interfacial strength between CNTs and their surrounding matrix in CNT-nanocomposites, with the aim of optimizing energy absorption capacity of high-performance CNT-nanocomposites.

INTRODUCTION

Since their discovery more than two decades ago[1], carbon nanotubes (CNTs) have maintained the keen attention of the scientific community due to the potential technological advancements they offer in various areas. The Young's modulus for single-walled carbon nanotubes (SWNTs) ranges between 0.4 and 4.15 TPa[2], which makes it conservatively double that of steel, whilst the tensile strength of SWNTs ranges between 10 and 20 GPa making SWNTs between 20 and 40 times more resistant to tensile stresses than structural steel[1, 3]. CNTs can have extremely high aspect ratios of up to 10^4[4] merely by virtue of being able to have nano-sized diameters at the same time as macro-scale lengths in centimeters.

These mechanical and geometric properties mean that CNTs, as nano-reinforcement, can vastly improve the performance of traditional structural materials such as polymers, concretes, metals and ceramics. Li[5] found that when CNTs are admitted at a concentration of 0.5% wt into a concrete paste, the concrete's compressive and tensile strength are enhanced by more than 10%. Lui[6] found that a 1% wt concentration of CNTs would improve the failure strain of an already highly elastic rubbery epoxy by 60%. These results were corroborated by Deng[7] that CNTs of high "persistent length" improve the elastic properties of the composite materials more than shorter ones.

CNT based composites also have excellent energy absorption capacity. Sun[8] compared the performance of epoxy- and ceramic- based nanocomposites whilst varying the amount and type of nanoparticles. Not only was the effect of certain nanoparticles tested, but the shape and aspect ratio, the surface morphology and interfacial slip were all found to be critical factors for producing performent nanocomposites[8]. It was demonstrated that CNTs amongst a myriad of other nanofillerss improve the fracture toughness by up to 79% compared to polymers without nanoparticles[8-9]. Thus far, optimization of CNT based composites has been achieved by reducing interfacial slip between the CNT and the matrix material[4, 7-8], whereas the impact properties of stand-alone CNTs has received limited attention. Guo[10] explored the buckling behavior of CNTs under varying bending loads and demonstrated that energy absorption increases approximately linearly with the angle formed at the central hinge. Sun[11] tested CNTs with varying number of walls with axial impact loads of different magnitudes compare the buckling mode, it was found that the impact force had no clear correlation to buckling mode. Duan[12] showed by testing the collision of a pre-bent CNT with a graphene sheet, that CNTs can withstand impact pressures of up to 10GPa and can store 90% of the kinetic energy associated with impact as elastic strain. Mylavagan[13] has tested the effects of points of impact and end-supports on the energy absorption of CNTs subjected to transverse impacts and found that there is a minimum interval time between consecutive impacts for the CNT to recover[14].

This paper explains in detail how the energy absorption the impact-rebound process of a projectile launched at the center of a SWNT. The local and global deformations associated with the impact-rebound process are to be interpreted in terms of shell and beam modes, respectively. The relationship between the bullet velocity, the deformation mode of the CNT and the energy absorption are explored.

METHODS

A SWNT of radius 7.051 Å, was modeled using Material Studio. The atoms of two ends were fully fixed. A piece of diamond molecule (35.6×35.6×7 Å), which has the highest hardness, was assumed as a bullet striking the central point of SWNT with specific velocities. The interaction between SWNT and diamond bullet were analyzed using Condensed-phase Optimized molecular Potentials for Atomistic Simulation Studies (COMPASS) force field. Two thermodynamic ensembles, which are NVT (the number of particles, the volume and the temperature are constant) and NVE (a fixed number of particles, in a fixed volume and with a fixed total energy) were implemented. NVT ensemble was used to heat up the molecules to the room temperature (298K). Subsequently, NVE was used to achieve the ballistic impact process under the Newton's second law.
The following two equations were adopted to calculate the energy absorption of SWNT:

$$E_{abs} = E_i - E_{tb} = \sum_{bullet} \frac{1}{2} m(v_i^2 - v_{tb}^2) \qquad (1)$$

$$\text{Energy absorption} = \frac{E_{abs}}{E_i} \times 100\% \qquad (2)$$

where E_i is the kinetic energy of the diamond bullet before the impact process and E_{tb} is that after the SWNT-bullet interaction, m is the mass of one bullet atom, v_i and v_{tb} are velocities of one bullet atom before and after the SWNT-bullet interaction.

RESULTS AND DISCUSSION

The Impact-Rebound Process of a Bullet with a (27, 0) SWNT.

Figure 1 shows the kinetic behavior of the bullet and CNT throughout the impact-rebound process, whilst Figure 2 shows the different stages and modes of deformation for the CNT. Figure 1 is divided into four distinct regions corresponding to four different stages of interaction between the bullet and the SWNT. The impact-rebound process of the bullet to the SWNT is essentially symmetrical because the deformation and behavior of the CNT and the bullet in region A are reversed in region D, similarly the processes which occur in region B are undone in region C.

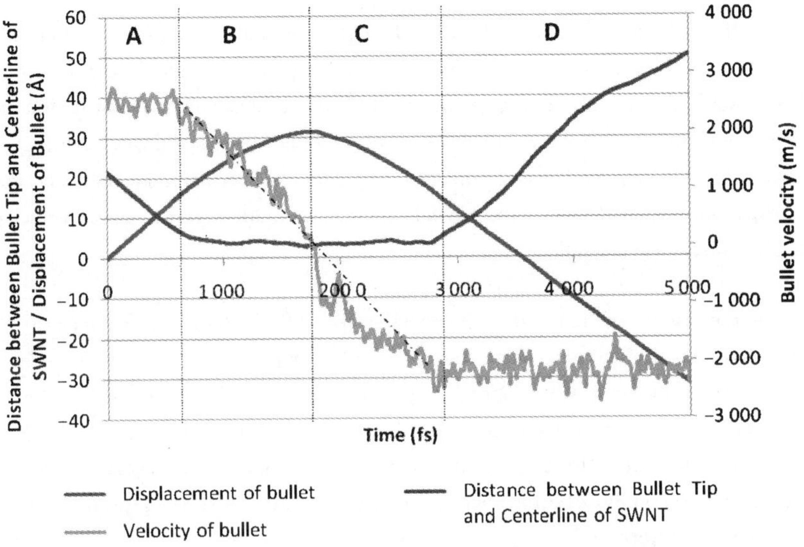

Figure 1 Variation of kinetic properties of bullet during the impact-rebound process when striking the middle of a (27, 0) SWNT

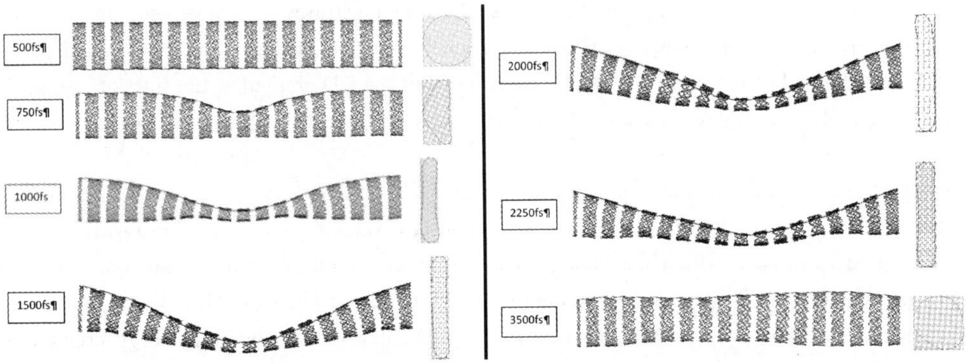

Figure 2 Cross-section and plan view of the CNTs throughout the impact-rebound process

Region A represents the pre-impact velocity and relative distances for the bullet and SWNT. The initial velocity, v_i is constant because no external forces are exerted on the bullet. These changes in region B, where the slowing the bullet indicates that it is interacting with the SWNT in what will be referred to as "impact". Despite the normal use of this term, there is no tactile contact between the bullet and the

SWNT, "impact" is taken to mean that the bullet is so close to the SWNT that very high van der Waals (vdW) repulsive forces are being exerted. Evidence for the strength of these forces is found in Figure 2 at the 750 fs time-point, where the impact-side of the CNT can be seen to be deformed locally.

Figure 2 at the 750 fs time-point, where the impact-side of the CNT can be seen to be deformed locally.

Following this impact at 750 fs, the bullet continues to deform the SWNT; this is shown in Figure 1 as the bullet gets further away from its origin and closer to the centerline of the SWNT. At 1 000 fs the distance between the bullet and the centerline of the SWNT becomes stable at 3.5 Å, which is approximately a third of 10.576 Å, the undeformed radius of the (27, 0) SWNT[14] as shown in Figure 2.

After 1 000 fs and throughout regions B and C, the fact that the bullet remains at a constant 3.5 Å from the centerline of the SWNT shows that no further local deformation occurs. Local deformation (LD) is characterized by shell-like behavior at the point of impact[15]. Subsequent deformation in region B is considered "global deformation" (GD) — these are characterized by beam-like behavior[15], thus the LD-GD sequence predicted by Wegener[16] and confirmed by Jama[15] has been shown to be applicable on a nano-scale. Throughout region B, while the SWNT has deforms globally, the bullet velocity consistently decreases until the bullet stops completely.

Region C — being the reverse of region B — starts when the bullet, albeit at a distance of 3.5 Å from the centerline of the SWNT, begins to rebound and the SWNT begins to return to its undeformed shape. Firstly, after the bullet momentarily stopped there is a negative bullet velocity which increases in magnitude at the same rate as the velocity in region B decreases. This shows that the SWNT continues to exert the same force in the same direction as in region B (see Figure 1 to observe a continuation in the slope of the velocity). Secondly, the distance of the bullet from its starting point begins to decrease, which confirms that it is returning.

Figure 2 at 2 250 fs shows lower magnitude deflections as those at 2000 fs. Thirdly, Figure 1 shows that the distance of about 3.5 Å between the bullet and the center the SWNT remains constant, and that the SWNT is still locally deformed where the bullet sits. This demonstrates the converse of the deformation mode sequence established on a macro-scale by Wegener[15-16] — that during rebound, GD will be undone before LD. At the end of region C the SWNT recovers from the LD created at the start of region B, this can be seen at 3500 fs in Figure 2 because it has completely restored to its pre-impact state.

Region D commences when the bullet moves away from the SWNT and continues to return to and beyond its starting point independently. This is shown by the constant velocity of the bullet in Figure 1 from 3000 fs onwards. The velocity at which the bullet travels from this point onwards reveals one asymmetrical aspect of the impact-rebound process — the kinetic energy balance. The exit velocity, v_{bb} is less than the initial velocity, v_i. The drop in kinetic energy is converted into the thermal energy created by the impact[7, 14].

Effect of Bullet Velocity on Energy Absorption

Figure 3 shows the relationship between different initial velocities of the bullet and the energy absorption of the SWNT. The curve is made up of two main linear sections — one for velocities less than 1 850 m/s and the other for velocities greater than 1 850 m/s. For segment AB, energy absorption increases by 1.9% for an increase of 100 m/s in impact velocity. Whilst for segment BC, energy absorption increases by 0.54% for an increase of 100 m/s of impact velocity. Thus, the energy absorption is four times less sensitive to increases in velocity in the segment BC than in segment AB. The segment AB governs the relationship for velocities that are only able to induce LD on the SWNT. Deformation is considered to be local when the centroid of the most deformed cross-section remains within the bounds of the cross-section of the undeformed SWNT. This mode of deformation can be seen in insert called "segment AB" of Figure 4.

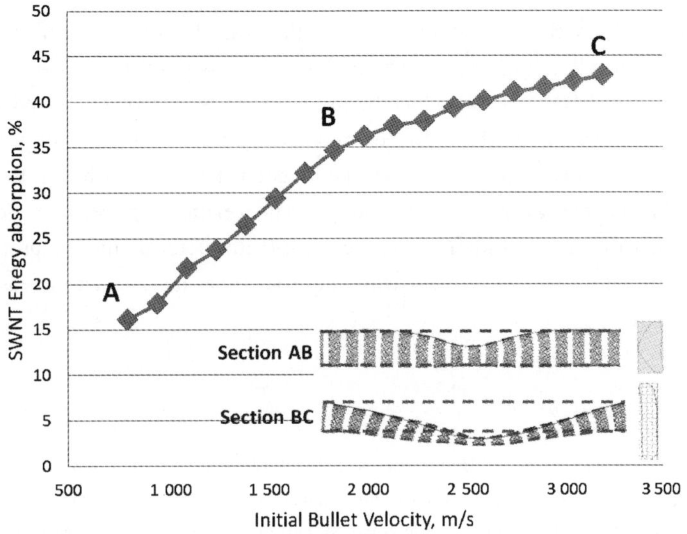

Figure 3 Variation of energy absorption with initial velocities of bullet for a (27, 0) SWNT with L/D ratio of 5 at a temperature of 298K. Segments A to B represent velocities which induce LD only, segments B to C represent velocities which induce the LD-GD sequence.

The culmination of the LD of the first linear segment is a kink at the center of the SWNT. The significance of kinks is that they reduce the bending capacity of the CNT[17-18], and allow for the GD mode to commence. The reduced bending capacity is due to decreased section area that is caused by the small gap between the two walls of the CNT. A central kink acts as a virtual hinge, allowing rotation between both of the sections of the CNT that it links, however the kink itself does not further deform[17].

Subsequent to its formation, only GD takes place in CNT and they occur due to the rotation at the central kink. This rotation can be seen by comparing the time-points of 1 000 fs and 2 000 fs in Figure 2. A visual inspection of the figure shows that in the interval between these two time points, the only deformation to occur is the angle between the top chords left and right of the central kink increases (as predicted by Mylvaganam[17]). Qualitatively, this increase in angle appears be less significant than the early LD which occurred between 750 fs and 1 000 fs of Figure 2. The ease at which the GD takes place as opposed to LD explains the difference by a factor of four between the slopes of segments AB and BC in Figure 4.

The energy absorption ratios found in the simulations for this paper are consistent with those found previously. Mylvaganam[14] found that a (27, 0) SWNT with an length to diameter ratio (L/D) of 7.09 would absorb roughly 10% of the kinetic energy provided by a diamond bullet travelling at 400m/s. This corresponds to the points with the lowest velocities on segment AB of Figure 4. Duan found that the kinetic energy loss for the collision between a pre-bent CNT travelling at velocities of up to 7 000 m/s and a graphene sheet was around 7.5%[12]. This energy absorption for such a velocity appears extremely low when compared to the relationship between velocity and energy absorption presented in Figure 3. However, the discrepancy is due to the fact that pre-bending the CNT locally deforms it and means that impact only triggers the less energy-intensive GD mode.

CONCLUSION

The simulations and analyses carried out have shown that the deformation which takes place in the impact-rebound process is fully elastic. Experiments have demonstrated the converse of the established principle that GD precedes LD — GD is undone before LD for CNTs in rebound. Whilst the undeformed shape of the CNT is recovered and vibrated after impact, the initial kinetic energy in the bullet is not recovered. CNTs when struck with a bullet at their center have been shown to absorb significant proportions of the kinetic energy as thermal energy. The relationship between velocity and energy absorption revealed that the deformation mode is a critical factor determining the proportion of absorbed energy. Energy absorption in the LD segment will react four times more to an increase in kinetic energy than in the GD segment.

REFERENCES

[1] Mylvaganam, K. and Zhang, L.C., "Important issues in a molecular dynamics simulation for characterising the mechanical properties of carbon nanotubes", *Carbon*, 2004, 42(10), pp. 2025-2032.

[2] Guoxin, C. and Xi, C., "Buckling of single-walled carbon nanotubes upon bending: molecular dynamics simulations and finite element method", *Physical Review B (Condensed Matter and Materials Physics)*, 2006, 73 (15), pp. 155435-1.

[3] Xiao, T., et al., "Determination of tensile strength distribution of nanotubes from testing of nanotube bundles", *Composites Science and Technology*, 2008, 68(14), pp. 2937-2942.

[4] Wille, K. and Loh, K. J., "Nanoengineering ultra-high-performance concrete with multiwalled carbon nanotubes", *Transportation Research Record*, 2010(2142), pp. 119-126.

[5] Li, G. Y., Wang, P. M. and Zhao, X., "Mechanical behavior and microstructure of cement composites incorporating surface-treated multi-walled carbon nanotubes", *Carbon*, 2005, 43(6), pp. 1239-1245.

[6] Liu, L. and Wagner, H.D., "Rubbery and glassy epoxy resins reinforced with carbon nanotubes", *Composites Science and Technology*, 2005, 65(11-12), pp. 1861-1868.

[7] Deng, F. , et al., "Elucidation of the reinforcing mechanism in carbon nanotube/rubber nanocomposites", *ACS Nano*, 2011, 5(5), pp. 3858-3866.

[8] Sun, L., et al., "Energy absorption capability of nanocomposites: A review", *Composites Science and Technology*, 2009, 69(14), pp. 2392-2409.

[9] Wei, T. , et al., "A new structure for multi-walled carbon nanotubes reinforced alumina nanocomposite with high strength and toughness", *Materials Letters*, 2008, 62(4-5), pp. 641-644.

[10] Guo, X. , et al., "Bending buckling of single-walled carbon nanotubes by atomic-scale finite element", *Composites Part B*, 2008, 39(1), pp. 202-8.

[11] Sun, C. and Liu, K., "Dynamic column buckling of multi-walled carbon nanotubes under axial impact load", *Solid State Communications*, 2009, 149(11-12), pp. 429-433.

[12] Duan, W.H., Wang, C.M. and Tang, W.X., "Collision of a suddenly released bent carbon nanotube with a circular graphene sheet", *Journal of Applied Physics*, 2010, 107(7), pp. 074303 (7 pp.).

[13] Mylvaganam, K. and Zhang, L.C., "Energy absorption capacity of carbon nanotubes under ballistic impact",

Applied Physics Letters, 2006, 89(12).

[14] Mylvaganam, K. and Zhang, L.C., "Ballistic resistance capacity of carbon nanotubes", *Nanotechnology*, 2007, 18(47).

[15] Jama, H.H., et al., "Numerical modelling of square tubular steel beams subjected to transverse blast loads", *Thin-Walled Structures*, 2009, 47(12), pp. 1523-1534.

[16] Wegener, R. B. and Martin, J. B., "Predictions of permanent deformation of impulsively loaded simply supported square tube steel beams", *International Journal of Mechanical Sciences*, 1985, 27(1-2), pp. 55-69.

[17] Mylvaganam, K., Vodenitcharova, T. and Zhang, L. C., "The bending-kinking analysis of a single-walled carbon nanotube: a combined molecular dynamics and continuum mechanics technique", *Journal of Materials Science*, 2006,41(11), pp. 3341-3347.

[18] Vodenitcharova,T. and Zhang, L.C., "Mechanism of bending with kinking of a single-walled carbon nanotube", *Physical Review B (Condensed Matter)*, 2004, 69(11), pp. 115410-1.

INVESTIGATION ON EFFICIENCY OF WATER TRANSPORT THROUGH SINGLE-WALLED CARBON NANOTUBES

M.Z. Sun, *W.H. Duan and M. Dowman

Department of Civil Engineering, Monash University, Clayton, Victoria 3800, Australia
* Email: wenhui.duan@monash.edu

KEYWORDS

Water transportation, carbon nanotubes, buckling, molecular dynamic simulations.

ABSTRACT

Based on the concept of an energy pump, the transportation of water molecules in a carbon nanotube (CNT) is studied by performing molecular dynamics simulations. The influences of CNT pre-twist angle, temperature, CNT channel length and water mass on energy pump efficiency are investigated. The transportation of one water molecule in a (8, 0) CNT can be described in three stages based on the pre-twist angle: the non-transport stage (between 0 to 65 degrees), the non-monotonous transport stage (between 65.1 to 80 degrees) and the linear-increasing stage (between 80 to 180 degrees). The transport efficiency of water molecules was found to change with temperatures and channel restraint conditions. In a (8, 0) restrained CNT the maximum resultant force and resultant velocity increases with increasing temperature within a range of 1 K to 3 000 K. For a (8, 8) unrestrained CNT, a water molecule was found to be non-transportable beyond 300 K. Three CNT channel lengths are compared with a length of 19.80 nm identified as the fastest and most efficient transporter. Transporting 20 water molecules in a (8, 8) CNT was determined to not be possible due to great fluctuations of the CNT wall and large friction forces between water molecules and the CNT wall.

INTRODUCTION

Carbon nanotubes (CNTs) have generated intense research interest due to their remarkable electrical, mechanical, and thermal properties[1-3]. Research on CNTs indicates that the employment of CNTs for novel microelectromechanical and nanoelectromechanical applications, such as nanorobotic spot-welding, biological cargos delivery and nanopumping devices for atomic transportation[4-6], is a possibility.

The transport of simple and complex fluids in CNTs is important from both a fundamental science and an application point of view[7-9]. Currently water transport through CNTs is being widely investigated to explore the potential of CNTs functioning as microcapillarries in microfluidic systems. Microfluidic systems have been shown to have potential in a diverse array of biological applications, including biomolecular separations, enzymatic assays, the polymerase chain reaction, and immuno-hybridization reactions[10]. The ability of CNTs to transport water molecules can be applied in microfluidic systems such as DNA microarrays, drug screening, optical display technologies, tunable fiber optic waveguides, and environmental monitoring[11].

Research has been undertaken to address issues associated with the nanoscale transportation of water. The transport properties, including the diffusivity, the thermal conductivity, and the shear viscosity, of water confined in single-walled CNTs (SWCNT) have been studied with different diameters[12]. It was found that the diameter of SWCNT makes an important impact on the transport properties and molecular distribution of water. Molecular Dynamics (MD) simulations[13-15] have suggested that CNTs have very low surface friction with respect to fluid flow. These simulations have been confirmed by direct experiments on timescales which show fast flow corresponding to large slip lengths[16-18].

The structure and flow of water inside CNTs have also been examined using MD simulations[19]. Unlike predictions from continuum mechanics, the flow enhancement in subcontinuum systems might not increase monotonically with decreasing flow area. MD simulations were also conducted on a water jet from a (6, 6) CNT that confined water in the form of a single-file molecular chain[20]. The results showed that the water formed nanoscale clusters at the outlet which were released intermittently. The jet breakup was dominated by thermal fluctuations, which led to a strong dependence on the temperature.

Measurements of water flow through membranes in which aligned CNTs served as pores have been reported[21]. It was observed that the water permeability of these nanotube-based membranes were several orders of magnitude higher than those of commercial polycarbonate membranes. The effects of CNT diameter on mass density, molecular distribution, and molecular orientation were identified for both confined and unconfined fluids[22]. A recent study on the single-file water transport through a biomimic water channel consisting of a (6, 6) CNT showed that, with different types of external point charges, water-water interaction influenced the transport rate significantly[23].

The twisting of a small portion of the wall of a CNT to function as an energy pump for possible smooth transportation of water molecules has been considered[24]. The preliminarily research on the influence of energy pump length, restrain condition on CNT channel and number of water molecules on transport efficiency were conducted. However, parameters such as the experimental temperature and the torsional angle of the energy pump were only briefly analysed with more data needed to find optimal levels while important parameters such as pump diameter and CNT length were not considered at all. The focus of this work is to understand in a much greater depth how changes to the torsional angle of the energy pump, the experimental temperature and the CNT length influence the transportation efficiency of one water molecule. The viability of transporting 20 water molecules inside a (8, 8) CNT using the energy pump concept is also investigated.

METHOD

The atomic interaction is modeled using the COMPASS force field (condensed-phased optimized molecular potential for atomistic simulation studies)[25-26]. This force field is the first *ab initio* force field that was parameterized and validated using condensed-phase properties. Research has shown that this force field can be used to describe the mechanical properties of carbon nanotubes[27-28]. In the COMPASS force field, the total potential energy E is expressed as[25]:

$$E = \sum E^{(b)} + \sum E^{(\theta)} + \sum E^{(\phi)} + \sum E^{(\chi)} + \sum E^{(bb')} + \sum E^{(b\theta)} + \sum E^{(b\phi)} + \sum E^{(b'\phi)} + \sum E^{(\theta\theta')} + \sum E^{(\theta\phi)} + \sum E^{(\theta\theta'\phi)} + \sum E^{(vdw)} + \sum E^{(elec)}$$

where b and b' are the lengths of two adjacent bonds, θ and θ' are the adjacent two-bond angles, ϕ is the dihedral torsion angle, and χ is the out of plane angle. The total potential energy may be divided into two categories, namely, (ⅰ) contributions from each of the internal valence coordinates (i. e., $\sum E^{(b)}$, $\sum E^{(\theta)}$, $\sum E^{(\phi)}$), and $\sum E^{(\chi)}$); and (ⅱ) cross-coupling terms between internal coordinates (i. e., $\sum E^{(bb')}$, $\sum E^{(b\theta)}$, $\sum E^{(b\phi)}$, $\sum E^{(b'\phi)}$, $\sum E^{(\theta\theta')}$, $\sum E^{(\theta\phi)}$ and $\sum E^{(\theta\theta'\phi)}$); and (ⅲ) nonbonded interactions (i. e., the van der Waals energy, $\sum E^{(vdw)}$), and the Coulomb electrostatic energy, $\sum E^{(elec)}$). It should be noted that the sum of repulsive and attractive Lennard-Jones terms represents the vdW energy[29].

The transport of a water molecule by CNT is modeled based on the concept of an energy pump[24]. The typical set-up of a CNT in molecular dynamics simulations with an energy pump is shown in Figure 1. The pump is made by restraining the motion of a portion of the CNT. The two ends of this portion, shown as El and E2, have several layers of carbon atoms. The pump is placed in a torsion buckling state by applying a pre-twisted angle to El. The restraint on E2 is removed once pre-buckling of the pump is achieved which allows the propagation of local torsion buckling in the direction from the pump to the channel portion of the CNT. The resultant driving force by the vdW energy interacted between the CNT wall and the molecule through the propagation of the local buckling from the pump is expected to be strong enough to initiate molecular transportation.

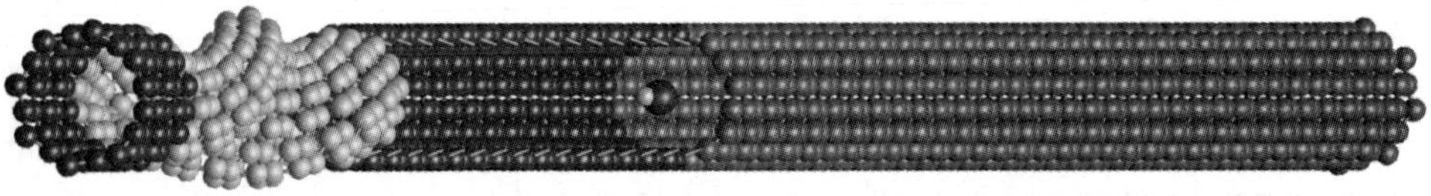

Figure 1 **an energy pump for transportation of one water molecule in a zigzag (8, 0) CNT channel[24]. E1 (blue) and E2 (blue) are fixed, and a pre-twisted angle, 135°, is applied to E1, which results in a torsion buckling of the pump (yellow). Once the restraint on E2 is removed, the potential energy stored in the pump will push the water molecule to travel along the CNT channel (green).**

The initial step of MD simulations involves a geometry optimization process (using the conjugate-gradient method) for the pump portion. The motions of the El and E2 portions are restrained. After this minimization process is complete, the CNT with the twisted pump and water molecule(s) are subjected to an NVT ensemble (i. e., constant volume and constant temperature dynamics) simulation process for 100 ps. This process is undertaken at a prescribed temperature during which the entire system reaches a thermodynamic equilibrium. A NVE (i. e., constant volume/constant energy dynamics) simulation process of another 100 ps will be followed by releasing the E2 to initiate the motion of the water molecule (s). The NVE simulation adopts the Verlet velocity algorithm[30] to integrate the motion of equations for the whole system. This NVE process is the only focus of discussion in the present research. The incremental step in the dynamics simulations is chosen to be 1.0 fs. By calculating the product of the mass and the velocity of the water molecules the amount of momentum that the water molecules possess can be found.. The velocity on the center of mass of the 1/20 water molecules is used as the velocity of the molecules. According to Newton's second law, the rate of change of the momentum of the water molecule is proportional to the resultant force on the water molecule. Therefore, the resultant force applied to water molecules can be obtained through a calculation of the derivative of the momentum with respect to the time.

RESULTS AND DISCUSSION

In the previous paper, only pump length, channel restrain condition and the effect of water mass were investigated[24]. This paper aims to study the effects that some of the other influencing factors, such as pre-twist angle, temperature, and channel length, have on the transport efficiency of a CNT. Also the effect of water mass is further explored. In order to fulfill this objective, 5 scenarios as shown in Table 1 were developed and investigated. Scenario 1 aims at studying how the torsional angle of the energy pump changes the transport efficiency in a restrained CNT channel. Scenario 2 and 3 are used to explore the effect of temperature on the transportation process of a water molecule under different channel restraint conditions. Scenario 4 tries to gain a preliminarily understanding of the effect of channel lengths on the transportation of a water molecule. Scenario 5 is designed to find out the feasibility of transporting 20 water molecules in a (8, 8) CNT. For the study of the water transport process in a restrained channel, such as in scenarios 1 and 2, two parameters are particularly focused on: maximum resultant force and maximum resultant velocity. Maximum force indicates the largest degree of collision between water and the CNT channel and the maximum velocity shows the largest speed achievable at a certain temperature.

Effect of Torsional Angle on Transportation of One Water Molecule

For scenario 1, the temperature is kept constant and the motion of the channel is restrained to remove the influence of environmental temperature and channel fluctuation on the transportation process. Under these conditions, the transportation of a water molecule can be described by four stages[24]. The largest resultant force and velocity experienced by a water molecule takes place in the second stage[24]. Different pre-twist angles from 5 to 180 degrees were examined and the maximum resultant force and velocity that a water molecule experiences after E2 is released were obtained and are presented in Figure 2 and Figure 3, respectively.

TABLE 1 SIMULATION DETAILS IN THIS STUDY

Scenario	(n, m)	Diameter (nm)	Environmental Temperature (K)	Restrain on CNT channel	Number of water molecules	Length of CNT channel (nm)	Pre-twist angle (degrees)
1	(8, 0)	0.626	1	restrained	1	20.26	5-180*
2	(8, 0)	0.626	1-3 000***	restrained	1	17.16	135
3	(8, 8)	1.085	300	unrestrained	1	9.90-26.70**	135
4	(8, 8)	1.085	1	unrestrained	20	17.16	135
5	(8, 8)	1.085	1-3 000***	unrestrained	1	17.16	135

Note:

1. Lengths of E1, E2 for all scenarios are 1.11 and 1.11nm, respectively. For scenario 1, 2, 4 and 5, the CNT length is 25.56 nm. For scenario 1, the length of the pump is 3.1 nm; for other scenarios, pump length is chosen as 6.2 nm.
2. *: Variation of pre-twist angles is 5, 45, 60, 65, 65.1, 65.5, 66, 67, 70, 72, 75, 78, 80, 81, 83, 84, 85, 90, 100, 110, 120, 135, 140, 150, 160, 170 and 180 degrees.
3. **: Variation of CNT channel length is 9.90, 19.80, 26.70 nm.
4. ***: Variation of temperatures is 1, 300, 600, 900, 1 200, 1 500, 2 000, 2 500, and 3 000 K.

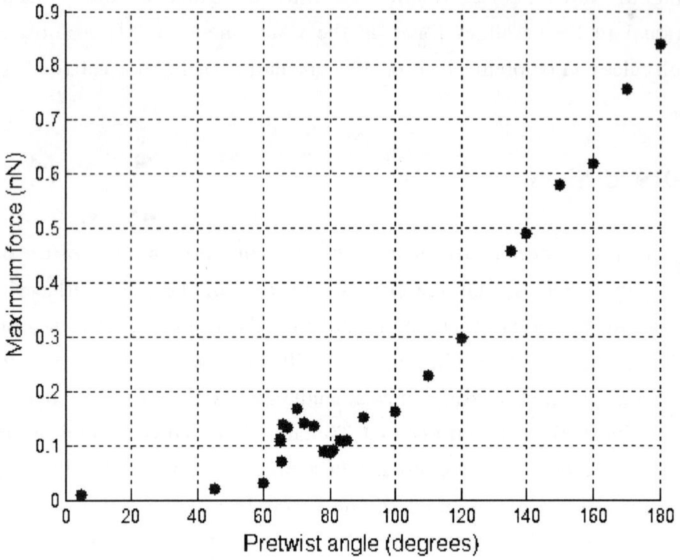

Figure 2 Maximum resultant forces experienced by a water molecule under different pre-twist angles during the transport process.

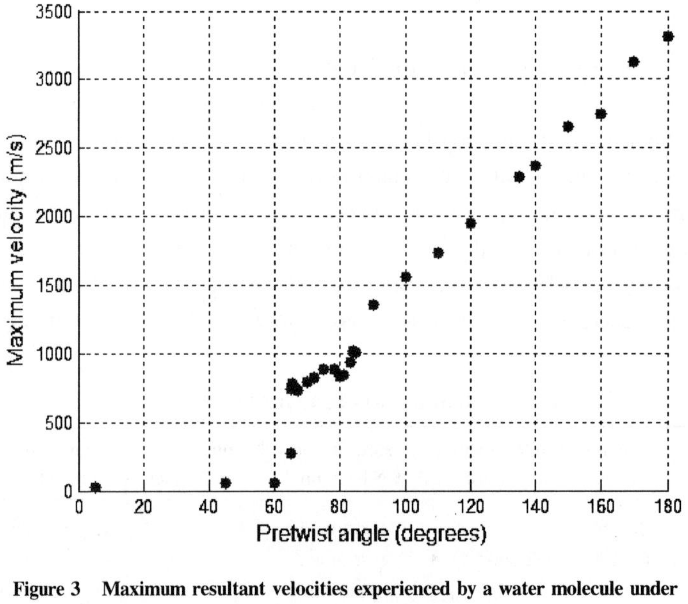

Figure 3 Maximum resultant velocities experienced by a water molecule under different pre-twist angles during transport process.

Based on the data presented in Figure 2 and Figure 3, changes in the maximum resultant force and velocity with increases in the pre-twist angle can be described in three stages. The first stage is between 0 to 65 degrees. At these pre-twist angles the maximum resultant force experienced by the water molecule is nearly zero which results in the water molecule not being adequately accelerated towards the end of the CNT channel after the unlocking of E2. Interestingly, it was found that water transportation can easily be achieved once the pre-twist angle is increased from 65 degrees to 65.1 degrees, as shown in Figure 4. At a pre-twist angle of 65 degrees and after releasing E2, the water molecule can only move back and forth from its original position. However, by increasing the pre-twist angle by just 0.1 degrees, an obvious increase in the resultant velocity and force could be seen from around 11.7 ps after releasing E2. The displacement of water molecule increases steadily and linearly after 11.7 ps as the mean resultant force tends to be zero and velocity tends to be constant.

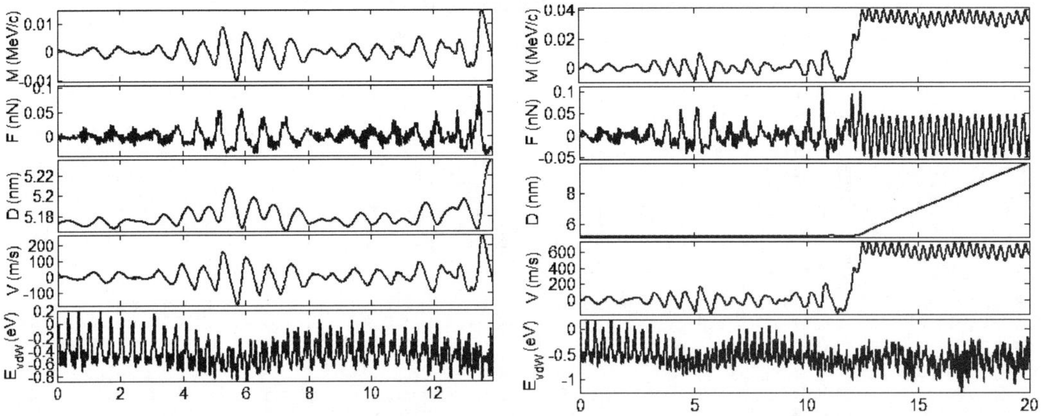

Figure 4 Comparison of the momentum, resultant force, displacement, resultant velocity of the water molecule and vdW energy of the whole system between pre-twist angle of 65 degrees (left) and 65.1 degrees (right).

The second stage occurs from 65.1 to 80 degrees, showing non-monotonously changes in both maximum resultant force and velocity. The maximum force at this stage increases rapidly from nearly 0.105 nN at 65.1 degrees to a peak of 0.168 nN at 70 degrees but drops significantly to 0.088 2 nN at 80 degrees. The maximum velocity at the second stage jumps from 269.48 m/s at 65 degrees to 737.98 m/s at 65.1 degrees and continues to increase but in a more gradual manner to reach a peak of 898.48 m/s at 78 degrees. A slight drop occurs after 78 degrees on Figure 3 with the maximum velocity at 80 degrees decreasing to 804.45 m/s. Although the transport process can be initiated at this stage it was found that the water molecule could not be pumped out of the end of the channel with MD simulations showing the molecule bouncing back when it is neared the end of the channel. Thus the water molecule cannot be successfully transported in this stage.

The third stage is between 80 and 180 degrees and follows a generally linear increasing trend for both the maximum force and the velocity. The maximum force and velocity at 180 degrees was observed to be 0.837 nN and 3 311.16 m/s, respectively. In this stage, a water molecule can be pumped out of the channel without any rebounding when it reaches the end of the CNT.

Based on the above description of Figure 2 and Figure 3, it can be summarized that transportation can only occur within a pre-twist angle range of 80 to 180 degrees. The twisting angle of the pump needs to be kept as large as possible within this allowable range in order to achieve the highest water transport efficiency.

Effect of Temperature on Transportation of One Water Molecule

The thermal effect on the efficiency of water transport is investigated in scenarios 2 and 3 with environmental temperatures investigated between 1K and 3 000 K. This investigation is conducted on both a (8, 0) and a (8, 8) CNT with different restraint conditions on the channel. Higher temperatures can result in stronger thermal motion of the water molecule causing more severe collisions between the water molecule and the CNT channel in both the transverse and the longitudinal directions[24]. As a consequence, the transportation of the water molecule is hindered by larger friction and energy loses as the temperature increases.

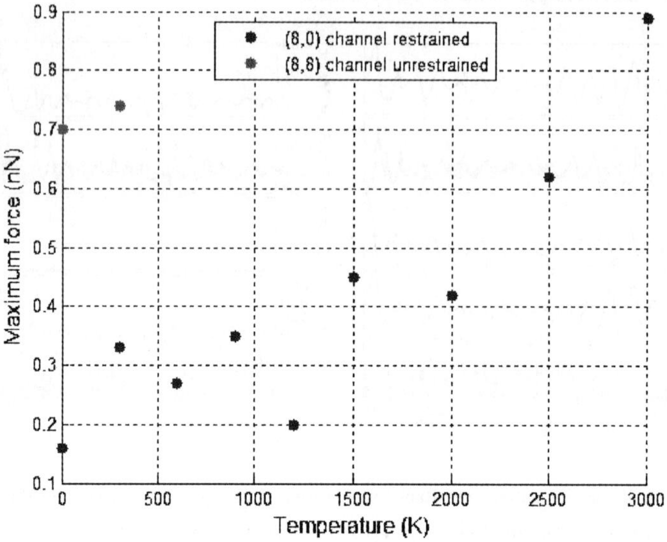

Figure 5 Maximum resultant forces a water molecule experiences travelling through a (8, 0) and a (8, 8) CNT channel under different environmental temperatures

The values of the maximum resultant force and maximum resultant velocity for scenario 2 are presented in Figure 5 and Figure 6, respectively. In Figure 5, it can be observed that when the temperature is increased in increments from 1K to 3 000 K, the maximum force subjected by the water molecule follows a generally increasing trend. This can be explained by the Brownian motion theory which states that the higher the temperature is, the faster the water molecule will move, thereby resulting in a larger impact force when the molecule collides with the CNT wall.

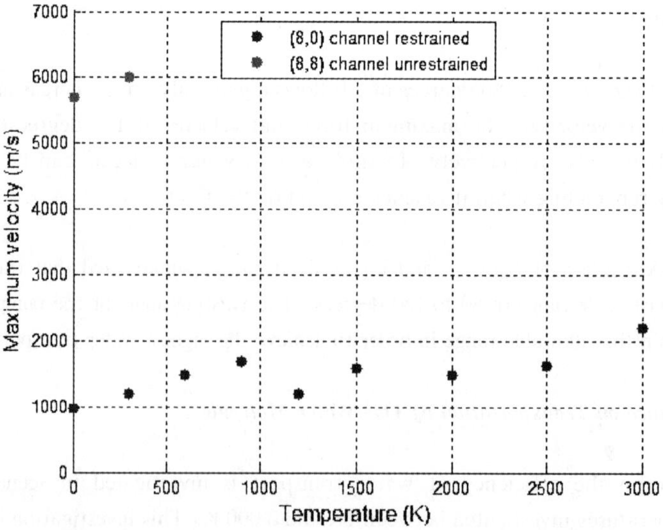

Figure 6 Maximum resultant velocities a water molecule experiences travelling through a (8, 0) and a (8, 8) CNT channel under different environmental temperatures

The maximum velocity achieved by the water molecule also shows an increasing trend as shown in Figure 6. As the channel is restricted from motion environmental temperature is the main driving force for this ascending trend. Owing to the thermal effect, the collisions that water molecule experiences generally become stronger. This leads to the prediction that maximum velocity will increase based on Newton's second law.

For a (8, 8) CNT with channel released, it was observed that the transport of one water molecule was not possible beyond 300K. MD simulations showed that a kink, which is initially confined within the pump, propagates towards the end of the channel soon after E2 is released. The length of this generated kink is almost 4/5 of the length of the CNT channel as shown in Figure 7. It was observed that within the length of the kink the channel becomes so flat that the tube deformed nearly into a sheet shape. This phenomenon greatly increases the friction between water molecule and CNT wall when the water molecule is within the kink. Although the unlocking of E2 initiated the transportation of the water molecule, the motions of water is observed to be greatly retarded by the kink once the kink propagates back from the end of the channel. The water molecule moves much slower and it even bounced back towards the pump under very high temperatures such as 1 500 K and 2 000 K. When the environmental temperature is above 2 500 K the water molecule could even be pumped out of E1. The impact of large friction due to the severely deformed channel and the Brownian motion are therefore seen as the two main causes for the unsuccessful delivery of the water molecule towards the outlet of the CNT.

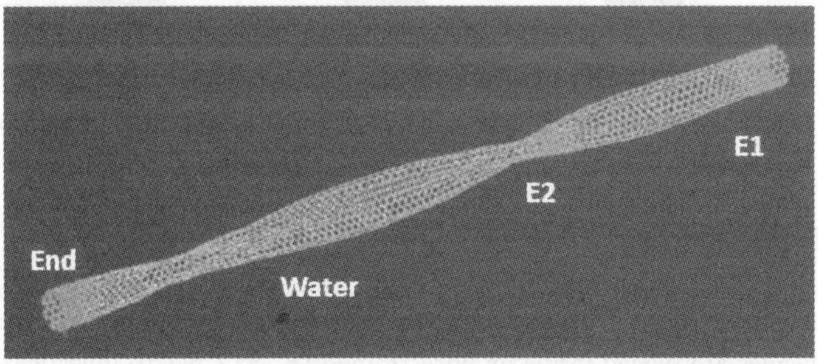

Figure 7 Kink propagation within a (8, 8) CNT in case of an unrestrained channel

Effect of Channel Length on Transportation of One Water Molecule

We now investigate how the length of the CNT channel affects the transport efficiency of one water molecule in a (8, 8) CNT. In scenario 4, the channel is completely unrestrained. Three channel lengths are considered, i.e. 9.90 nm, 19.80 nm and 29.70 nm. Simulations are conducted at 300 K. Variations in the travelling time of water are investigated with Figure 8 showing the values of the time for the different lengths.

From Figure 8 it can be seen that the fastest transporter of the water molecule is the CNT with a channel length of 19.80 nm. In this case, 4.63 ps are needed for the water molecule to be pumped out of the channel. Although the 9.90 nm case has the shortest travelling length, the time it takes for the water to be pumped out of the CNT is 1.3 times that of the fastest case. The slowest transporter is the 29.70 nm CNT, approximately 2 times slower than the fastest case.

This phenomenon means that in an unrestrained channel, length is not the only decisive factor that influences the transport efficiency. Other factors, such as kink propagation should also be taken into consideration when analysing the transportation efficiency. Shorter length means a shorter travel distance. However, the average resultant force may not be adequately large to allow the shortest case to be the fastest transporter. Further efforts are needed to explore the effect of the channel length and are currently underway in our research group.

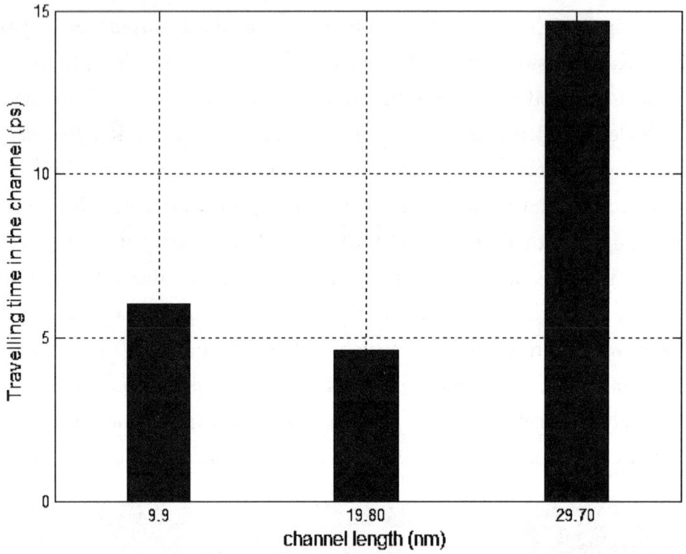

Figure 8 Time water molecule spends travelling in a (8, 8) CNT channel in different channel lengths.

Effect of Water Mass on Transportation Process

In the previous paper, the effect of water mass in a (8, 0) CNT was only briefly explored[24]. In this paper, 20 water molecules are placed inside an unrestrained (8, 8) CNT to investigate the transportation viability. All water molecules are placed in a horizontally straight line with a distance of 0.5 nm between each two water molecules. The torsion angle is set as 135 degrees. When E2 is released, all 20 water molecules can be seen to move gradually towards E1, instead of E2, under the effect of wave propagation. However, it is observed that during the transport process, not all water molecules are in a horizontal line. At some places, two or three water molecules lie in the same vertical plane on their way towards E1. This molecule behaviour is different from the case of using (8, 0) CNT, in which all water molecules are aligned in a straight line during the entire process[24].

This phenomenon means that 20 water molecules cannot be successfully pumped out of the (8, 8) CNT. MD simulation shows that all of them are out of E1 about 9 ps after E2 is unlocked. The mechanism behind this result could be that very large resistance is encountered when 20 water molecules are travelling inside the tube. This resistance may be caused by friction between the water molecules as a whole and the CNT wall. The larger diameter of the (8, 8) CNT when compared to the (8, 0) CNT case may lead to water having less vertical movement restrictions. Therefore, two to three water molecules can lie in the same vertical plane in the CNT, causing a large amount of friction between the water mass and the CNT wall; at the same time, the fluctuation of the kink pumping back from E2 becomes more severe in the (8, 8) CNT, causing a negative influence on the movement of water molecules. The combination of these two facts accumulates in a great resistance between the water molecules and the CNT, resulting in the water mass being pushed out of E1.

CONCLUSION

In view of extending Duan and Wang's research[24], we have performed MD simulations to investigate how CNT pre-twist angles, temperatures, CNT channel length and water mass influence energy pump efficiency. It is found that in order to have successful transportation, the pre-twist angle must be larger than certain threshold, 80 degrees for the case of one water molecule in a (8, 0) CNT. Furthermore, the water molecule is found to have different transport efficiency under different temperatures in CNTs with different channel restrain conditions. In a (8, 0) restrained CNT, general increases in the maximum resultant force and resultant velocity can be observed with incremental increases in temperature from 1K to 3 000 K. However, MD simulation indicates that a water molecule cannot be transported beyond 300 K in a (8, 8) unrestrained CNT. By comparing three CNT channel lengths, the channel length of 19.80 nm is identified as a faster and more efficient transporter. Finally, when transporting 20 water molecules in a (8, 8) CNT, it is observed that the transportation process cannot be successfully completed. Further study is needed using MD simulations to further explore the influence of factors such as channel length and the water mass effect.

REFERENCES

[1] Iijima, S., "Helical microtubules of graphitic carbon", *Nature*, 1991. **354**(6348): pp. 56-58.

[2] Wildoer, J.W.G., et al., "Electronic structure of atomically resolved carbon nanotubes", *Nature*, 1998. **391**(6662): pp. 59-62.

[3] Ball, P., "Roll up for the revolution", *Nature*, 2001. **414**(6860): pp. 142-144.

[4] Dong, L., et al., "Nanorobotic spot welding: Controlled metal deposition with attogram precision from copper-filled carbon nanotubes", *Nano Letters*, 2007. **7**(1): pp. 58-63.

[5] Kam, N.W.S. and Dai, H., "Single walled carbon nanotubes for transport and delivery of biological cargos", *Physica Status Solidi B-Basic Solid State Physics*, 2006. **243**(13): pp. 3561-3566.

[6] Kral, P. and Tomanek, D., "Laser-driven atomic pump", *Physical Review Letters*, 1999. **82**(26): pp. 5373-5376.

[7] Supple, S. and Quirke, N., "Rapid imbibition of fluids in carbon nanotubes", *Physical Review Letters*, 2003. **90**(21).

[8] Ito, T., et al., "A carbon nanotube-based coulter nanoparticle counter (vol 37, pg 938, 2004)", *Accounts of Chemical Research*, 2005. **38**(8): pp. 687-687.

[9] Kalra, A., Hummer, G. and Garde, S., "Methane partitioning and transport in hydrated carbon nanotubes", *Journal of Physical Chemistry B*, 2004. **108**(2): pp. 544-549.

[10] Darhuber, A.A. and Troian, S.M., "Principles of microfluidic actuation by modulation of surface stresses", in *Annual Review of Fluid Mechanics*. 2005. pp. 425-455.

[11] Thorsen, T., Maerkl, S.J. and Quake, S.R., "Microfluidic large-scale integration", *Science*, 2002. **298**(5593): pp. 580-584.

[12] Liu, Y.C. and Wang, Q., "Transport behavior of water confined in carbon nanotubes", *Physical Review B*, 2005. **72**(8).

[13] Longhurst, M.J. and Quirke, N., "The environmental effect on the radial breathing mode of carbon nanotubes in water", *Journal of Chemical Physics*, 2006. **124**(23).

[14] Sokhan, V.P., Nicholson, D. and Quirke, N., "Fluid flow in nanopores: Accurate boundary conditions for carbon nanotubes", *Journal of Chemical Physics*, 2002. **117**(18): pp. 8531-8539.

[15] Skoulidas, A. I., et al., "Rapid transport of gases in carbon nanotubes", *Physical Review Letters*, 2002. **89**(18).

[16] Supple, S. and Quirke, N., "Molecular dynamics of transient oil flows in nanopores I: Imbibition speeds for single wall carbon nanotubes", *Journal of Chemical Physics*, 2004. **121**(17): pp. 8571-8579.

[17] Mattia, D., et al., "Effect of graphitization on the wettability and electrical conductivity of CVD-carbon nanotubes and films", *Journal of Physical Chemistry B*, 2006. **110**(20): pp. 9850-9855.

[18] Miller, S. A., Young, V. Y. and Martin, C. R., "Electroosmotic flow in template-prepared carbon nanotube membranes", *Journal of the American Chemical Society*, 2001. **123**(49): pp. 12335-12342.

[19] Thomas, J. A. and McGaughey, A. J. H., "Water flow in carbon nanotubes: transition to subcontinuum transport", *Physical Review Letters*, 2009. **102**(18): pp. 184502.

[20] Hanasaki, I., Yonebayashi, T. and Kawano, S., "Molecular dynamics of a water jet from a carbon nanotube", *Physical Review E*, 2009. **79**(4).

[21] Holt, J. K., et al., "Fast mass transport through sub-2-nanometer carbon nanotubes", *Science*, 2006. **312** (5776): pp. 1034-1037.

[22] Thomas, J. A. and McGaughey, A. J. H., "Density, distribution, and orientation of water molecules inside and outside carbon nanotubes", *Journal of Chemical Physics*, 2008. **128**(8).

[23] Zuo, G., et al., "Transport properties of single-file water molecules inside a carbon nanotube biomimicking water channel", *Acs Nano*, 2010. **4**(1): pp. 205-210.

[24] Duan, W. H. and Wang, Q., "Water transport with a carbon nanotube pump", *Acs Nano*, 2010. **4**(4): pp. 2338-2344.

[25] Rigby, D., Sun, H. and Eichinger, B. E., "Computer simulations of poly(ethylene oxide): Force field, PVT diagram and cyclization behaviour", *Polymer International*, 1997. **44**(3):pp. 311-330.

[26] Sun, H., "COMPASS: An ab initio force-field optimized for condensed-phase applications-Overview with details on alkane and benzene compounds", *Journal of Physical Chemistry B*, 1998. **102**(38): pp. 7338-7364.

[27] Duan, W. H., et al., "Molecular mechanics modeling of carbon nanotube fracture", *Carbon*, 2007. **45**(9): pp. 1769-1776.

[28] Wang, Q., et al., "Inelastic buckling of carbon nanotubes", *Applied Physics Letters*, 2007. **90**(3).

[29] Jones, J. E., "On the determination of molecular fields-II From the equation of state of a gas", *Proceedings of the Royal Society of London Series A*-Containing Papers of a Mathematical and Physical Character, 1924. **106**(738): pp. 463-477.

[30] Verlet, L., "Computer experiments on classical fluids I. Thermodynamical properties of Lennard-Jones molecules", *Physical Review*, 1967. **159**(1): pp. 98-&.

ON THE APPLICABILITY OF HILBERT-HUANG TRANSFORM FOR ANALYSIS OF A TWO-MEMBER TRUSS IN VIBRATION

* Y.B. Yang, C.T. Chen and K.C. Chang

Department of Civil Engineering, Taiwan University, Taipei, China
President of the Yunlin University of Science and Technology Yunlin, China
* Email: ybyang@ntu.edu.tw

KEYWORDS

Fast Fourier transform, Hilbert-Huang transform, instantaneous frequency, truss, nonlinearity.

ABSTRACT

The Hilbert-Huang transform (HHT) proposed for treating the time-history data of nonlinear dynamic systems comprises two parts: the empirical mode decomposition (EMD) and the Hilbert transform (HT). In this study, focus is placed on the *physical interpretation* of the HHT results for an elastic two-member truss, for which the nonlinear behavior can be *clearly interpreted* at various stages of excitation. The instantaneous frequency of the truss is computed using the HHT or derivative-based HHT. The results obtained by the fast Fourier transform (FFT) are also included for comparison. It is concluded that *the enforcement by the HHT of the data set to be symmetric with respect to the local zero mean in extracting the intrinsic mode functions (IMFs) may result in frequencies that are not physically meaningful*. Care needs to be taken when using HHT for interpreting the frequency contents of a nonlinear structure.

INTRODUCTION

The Hilbert-Huang transform (HHT) proposed for treating the non-stationary data generated by nonlinear systems comprises two major parts[1, 2]: (1) the empirical mode decomposition (EMD): a sifting process by which the data can be decomposed into a collection of intrinsic mode functions (IMF) that admit well-behaved Hilbert transforms; and (2) Hilbert transform (HT): a type of transform by which the instantaneous frequency and amplitude can be calculated for any instant. Since the data set is decomposed according to its characteristic time scales, while the time-varying oscillatory information is revealed through the Hilbert transform, the HHT is considered to be a technique generally suitable for nonlinear data analysis. As useful as it has been proved to be, the HHT still leaves some annoying difficulties unresolved. One drawback is that the instantaneous frequencies calculated from the IMFs, obtained by the EMD with zero average means, are not fully consistent with the physical reality, especially in the analysis of nonlinear structures.

The objective of this paper is to investigate the applicability of the HHT to analyzing the dynamic response of a two-member truss that is sensitive to geometric nonlinearity. This truss is selected because it is so simple that its behavior at each stage can be *physically interpretable*. The dynamic response of this system is computed by the Newmark-β method (with $\beta = 1/4$ and $\gamma = 1/2$). Focus will be placed on interpretation of the instantaneous frequencies obtained by the HHT or derivative-based HHT, along with that from the fast Fourier transform FFT, to see if they are truly reflective of the physical reality of the structure.

HILBERT-HUANG TRANSFORM (HHT)

The HHT comprises two major parts, i. e., the empirical mode decomposition (EMD) and Hilbert transform (HT). For a set of time-history data $X(t)$, the HT can be expressed as

$$\tilde{X}(t) = \frac{1}{\pi} P \int_{-\infty}^{\infty} \frac{X(\tau)}{t-\tau} \mathrm{d}\tau, \tag{1}$$

where P is the Cauchy principal value. With this transform, one can form an analytical set of data as

$$Z(t) = X(t) + \mathrm{i}\tilde{X}(t) = a(t)\mathrm{e}^{\mathrm{i}\theta(t)}, \tag{2}$$

where

$$a(t) = \sqrt{X^2(t) + \tilde{X}^2(t)}, \tag{3}$$

$$\theta(t) = \tan^{-1} \frac{X(t)}{\tilde{X}(t)}. \tag{4}$$

The polar coordinate expression on the right side of Eq. (2) offers a physical interpretation for the data set: $a(t)$ represents the amplitude and $\theta(t)$ the phase of the data set with respect to time. The instantaneous frequency can be defined as

$$\omega(t) = \frac{\mathrm{d}\theta(t)}{\mathrm{d}t}. \tag{5}$$

Although each data set can be processed by the HT to generate the instantaneous frequencies and time-varying amplitudes, there is no guarantee that each of these derived is physically meaningful. According to the definition in Eq. (5), the instantaneous frequency is a single-valued function of time. Only for those data sets containing one single frequency component at each instant can a meaningful instantaneous frequency be generated by the HT.

In generating the instantaneous frequencies, the data set should satisfy the following conditions[1]: (1) It should possess the same or nearly the same numbers of extrema and zero-crossings. (2) It should be symmetric with respect to the *local zero mean*. The functions satisfying these conditions are designated as the *intrinsic mode functions* (IMFs). Since most data sets are not IMFs, they should be processed by the EMD first and decomposed into a limited number of IMFs that admit well-behaved HT. Given a data set $Y(t)$, the algorithm of the EMD can be briefly described as follows:

(1) Identify all the local maxima and minima of the data $Y(t)$, and then form the upper and lower envelopes by interpolating the local maxima and minima, respectively, by cubic spline curves.

(2) Obtain the first component h_1 by subtracting the mean of the upper and lower envelopes from the data $Y(t)$. If h_1 does not satisfy the IMF requirements, then treat it as $Y(t)$ and repeat step (1) until the requirements are satisfied. The first IMF component obtained is designated as c_1.

(3) Subtract c_1 from $Y(t)$ and obtain the residue r_1. If r_1 still contains information of other scale components, then treat it as $Y(t)$ and repeat steps (1)-(2) to obtain the next IMF c_2. Such a process is

repeated until the residue r_n is either too small to be physically meaningful or becomes a monotonic function from which no more IMF can be extracted. Consequently, the data $Y(t)$ is decomposed into n IMFs, c_1 to c_n, plus the final residue, r_n.

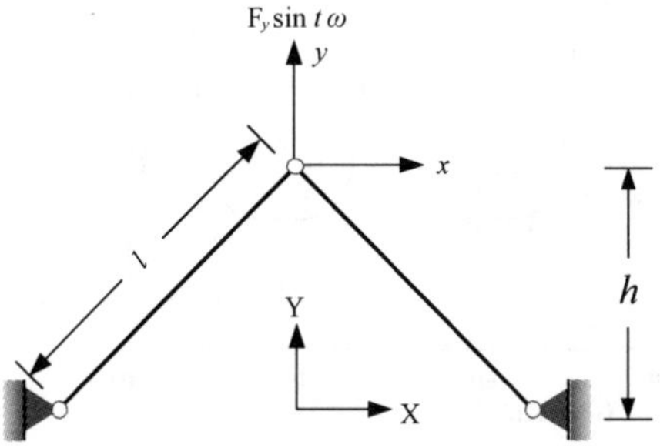

Figure 1 Two-member truss.

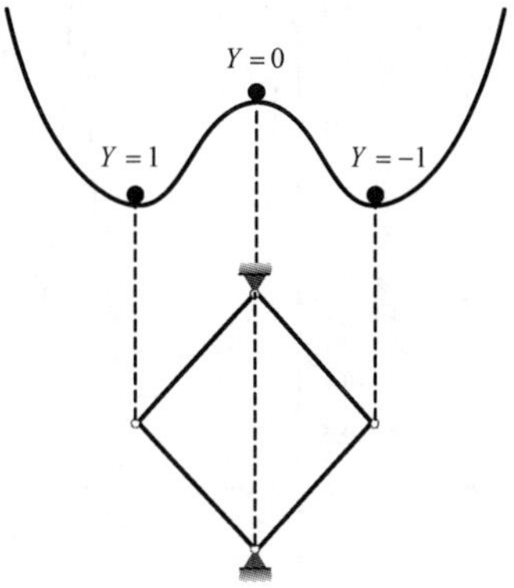

Figure 2 Potential wells.

NONLINEAR DYNAMIC ANALYSIS OF TWO-MEMBER TRUSS

Consider an elastic simple truss with two identical members of length l, height h, cross sectional area A, density ρ, Young's modulus E, and vertical damping ratio C_y (Figure 1). Supposing that the truss is subjected to a harmonic force $F_y \sin\omega t$, where ω is the frequency of excitation, the motion of the truss can be simplified into one vertical degree of freedom (Y) with the governing equation given as[3]

$$\ddot{Y} + \gamma\dot{Y} + kY(Y^2 - 1) = f\sin\omega t\ , \tag{6}$$

where

$$Y = \frac{\mathbf{Y}}{h},\quad \gamma = \frac{C_y}{\rho Al},\quad k = \frac{Eh^2}{\rho l^4},\quad f = \frac{F_y}{\rho Alh}\ . \tag{7a-d}$$

Equation (6) is a Duffing equation. It possesses three states of equilibrium: $Y = \pm 1$ (stationary), and 0 (non-stationary). It is known as a two-well potential problem, with two wells ($Y = 1, -1$) and one hill ($Y = 0$). The former behaves like attractors, and the latter like a repellor (Figure 2). In the following, Eq. (6) is first solved by the Newmark-β method (with $\beta = 1/4$ and $\gamma = 1/2$). The dynamic response will be illustrated via not only the time-history plot, but also the phase plot and Poincare plot (being mapped every forced period $T = 1/f$, where f is the frequency of the external force.) By performing the FFT and HHT to the time-history response solved for the truss, the frequency property will be revealed in the resulting Fourier spectrum and instantaneous frequency plot. In the following, three of the five cases studied for the truss under various excitations will be presented.

Case 1. Undamped Free Vibration ($\gamma = 0$, $k = 0.5$, $Y(0) = 1$, $\dot{Y}(0) = 0.5$, $f = 0$)

The displacement time history and phase plot have been shown in Figures 3 and 4, respectively. From Figure 3, one observes that the response is a periodical one, which is also confirmed by the closed curve in Figure 4. However, the motion is not a regular sine-like wave, due to the involvement of nonlinearity of various levels at different stages of motion for the case. The Fourier spectrum obtained by the FFT in Figure 5 indicates that there are several peaks distributed over the domain. However, none of these peaks represent any frequency of the truss, as will be explained below. This example illustrates that the FFT-based peak-picking method is invalid for picking the frequencies of nonlinear structures for this case.

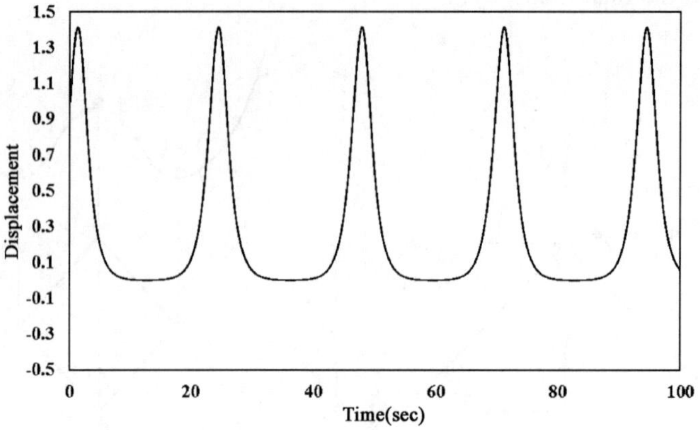

Figure 3 Case 1: Displacement time history.

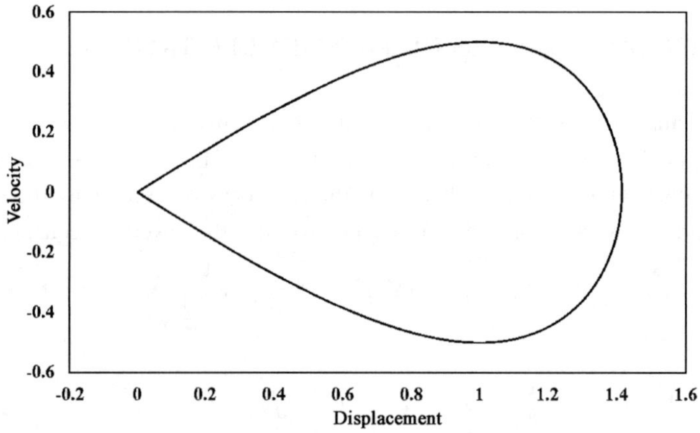

Figure 4 Case 1: Phase plot.

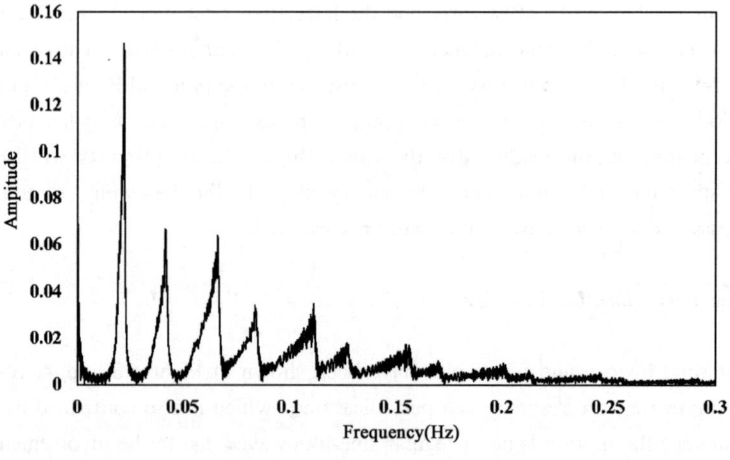

Figure 5 Case 1: Fourier spectrum.

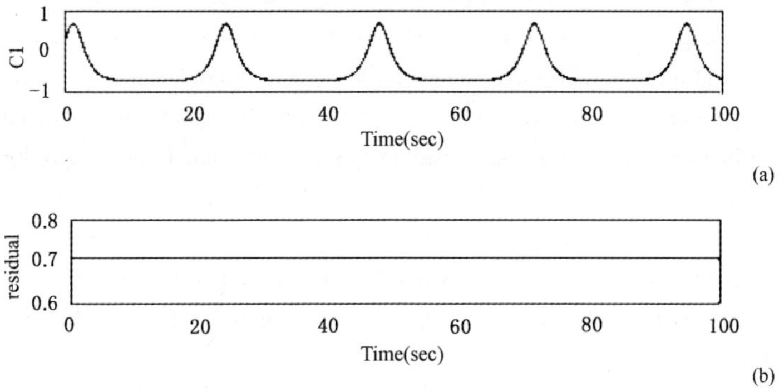

Figure 6. Case 1: IMFs by EMD: (a) c_1, (b) residual.

By processing the time history data with the EMD, only one IMF can be obtained, shown as c_1 in Figure 6. By further performing the HT to c_1, one obtains the instantaneous frequency as the solid line in Figure 7. For this case, the EMD only shifts the time history data $Y(t)$ by a constant of -0.708, as indicated by the residual in Figure 6(b), due to *enforcement* of the data set to be symmetric with respect to the *local zero mean*. Thus, the response for c_1 here is equal to $Y(t)-0.708$.

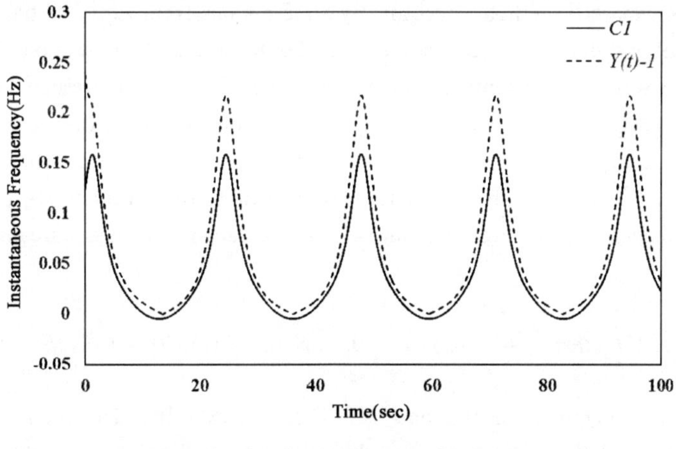

Figure 7 Case 1: Instantaneous frequency.

For this example, the time-history response obtained in Figure 3 represents the free vibration of the single DOF system, which contains only one single frequency, i. e., the *intrinsic frequency*, at each instant. There is no need to perform the EMD to the response. For this reason, the data can be directly processed by the HT once the reference point is selected. Further, for this case the *static equilibrium point*, i. e., $Y=1$, is considered a proper reference point. With this, the original data $Y(t)$ is shifted to $Y(t)-1$ as a new data set. By performing the HT to this new data set, one obtains the instantaneous frequency shown as the dotted line in Figure 7.

Considering only the Duffing Equation (6), one can hardly judge if any instantaneous frequency in Figure 7 is physically meaningful for the system. One way is to look back at the original truss in Figure 1. When in free vibration, the truss oscillates from the static equilibrium point, i. e., $Y=1$. In Figure 3, the displacement time-history shows that the motion of the truss oscillates from $Y=0.0015$ to 1.41. By the fact that the static equilibrium point is $Y=1$, the vibration has an upper amplitude as 0.41, and a lower amplitude as 0.998 5. Clearly, the motion not only changes its *frequency*, but also its *amplitude* in its course. As the EMD has enforced the IMFs to be symmetric with respect to the local *zero mean*, the upper and lower amplitudes of the IMFs are forced to be equal. Evidently, such a restraint has *artificially changed* the physical properties of the system. For this reason, the instantaneous frequency of $Y(t)-1$, i.e., the dotted line in Figure 7, should be considered as the one that is physically more meaningful. It should be noted that that for the two-member truss considered, the equilibrium position can be easily identified, which enables us to make the above considerations. For complicated structures, the equilibrium position will not be readily made available. In that case, it is hard to estimate the level of artificial change brought by the enforcement of zero average means in finding the IMFs.

Considering that the truss tip reaches its highest position with $Y=1.41$, one can express the displacement of the truss tip close to the highest position as

$$Y(t)=0.41\sin\omega t+1\,, \tag{8}$$

and the acceleration of the truss tip as

$$\ddot{Y}(t)=-\omega^2 0.41\sin\omega t\,. \tag{9}$$

The maximum acceleration of the motion occurs at the highest position, i.e., -0.707, as can be calculated by the Newmark-β method. Accordingly, one obtains from Eq. (9) the angular velocity as 1.313 rad/s, and the instantaneous frequency as 0.21 Hz for the highest position of $Y=1.41$. From Figure 7, one observes that the instantaneous frequency of $Y(t)-1$, as represented by the dotted line, oscillates from 0.000 000 05 to 0.21 Hz, while the instantaneous frequency of c_1, as represented by the solid line, oscillates from -0.005 to 0.16 Hz. Evidently, the instantaneous frequency predicted by $Y(t)-1$ is physically more meaningful for this case since it is identical to that calculated by Eq. (9). Moreover, the instantaneous frequency of c_1 even contains a *negative* range, which is unrealistic for engineering structures.

Case 2. Damped Free Vibration ($\gamma=0.084$, $k=0.5$, $Y(0)=1$, $\dot{Y}(0)=0.5$, $f=0$).

The parameters adopted herein are as the same as in Case 1, except that damping is newly included. The displacement response and phase plot have been shown in Figures 8 and 9, respectively. As revealed by Figure 8, the motion decreases with time and finally stops at the static equilibrium point $Y=1$, which is also confirmed by the phase plot of Figure 9. The amplitude of the motion is large enough to induce geometric nonlinear effect at the onset of motion. For this reason, the response does not appear as a regular sine-like motion. However, as the amplitude decreases with time, the geometric nonlinear effect also diminishes, and the response becomes a sine-like motion at the end.

The low-frequency components distributed over the domain of the Fourier spectrum in Figure 10 do *not* represent any physical reality. Using the EMD, two IMFs c_1 and c_2 are constructed in Figure 11. By performing the HT to c_1, the instantaneous frequency is obtained as the solid line in Figure 12. Since the displacement response is of free vibration, it contains only a single frequency component at each instant as discussed in Case 1. Thus, one can shift the data $Y(t)$ to $Y(t)-1$ and perform the HT to the new data set to obtain the instantaneous frequency, which is plotted as the dotted line in Figure 12. As was mentioned in Case 1, the instantaneous frequency of $Y(t)-1$ is *physically more meaningful*, since no enforcement is made of zero average mean. Besides, both the instantaneous frequencies obtained for c_1

and $Y(t)-1$ converge to the frequency of 0.159 Hz, which is considered reasonable since the truss is still in the linear range for small free vibrations.

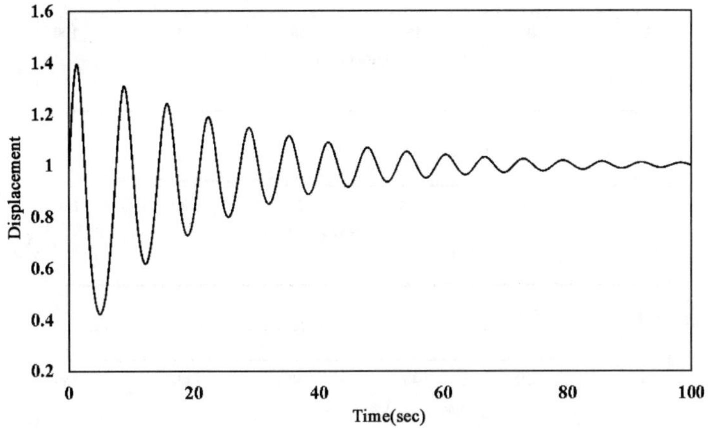

Figure 8 Case 2: Displacement time history.

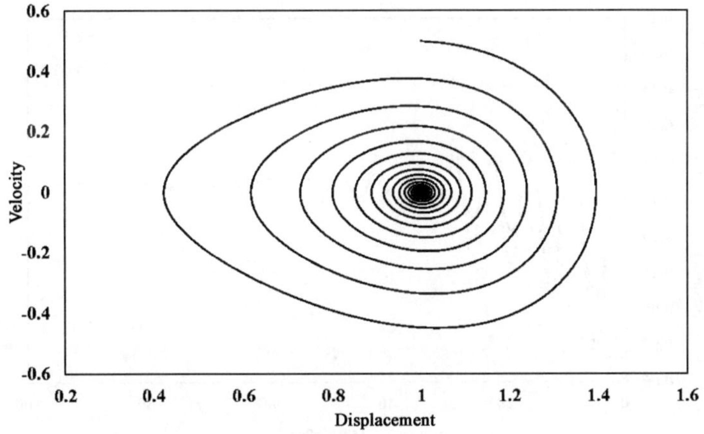

Figure 9 Case 2: Phase plot.

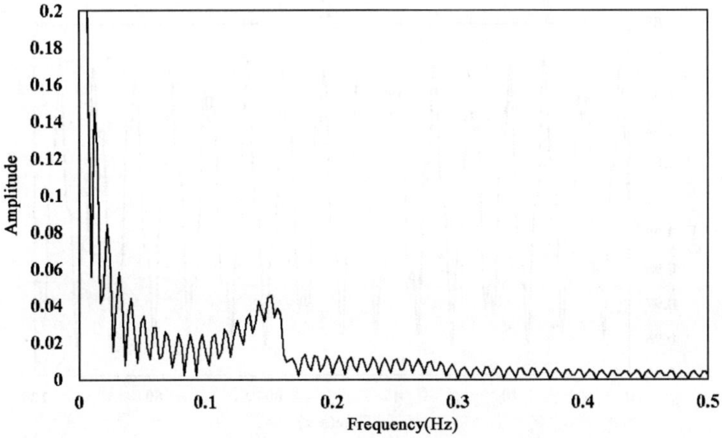

Figure 10 Case 2: Fourier spectrum.

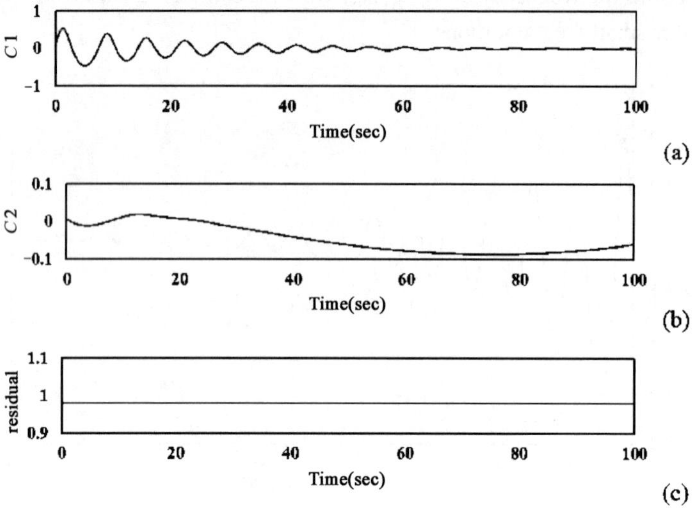

Figure 11 Case 2: IMFs: (a) c_1, (b) c_2, (c) residual.

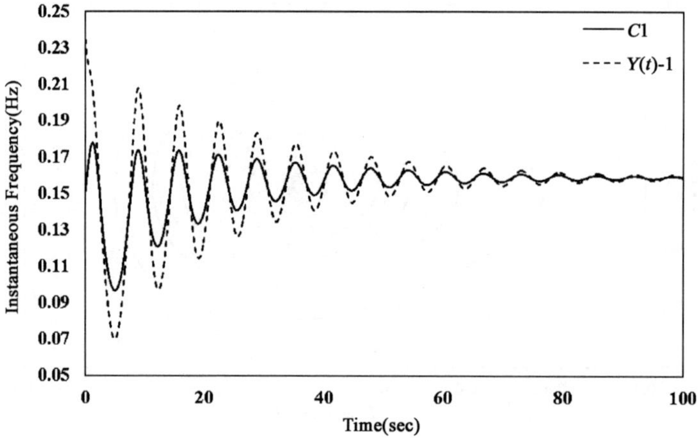

Figure 12 Case 2: Instantaneous frequency.

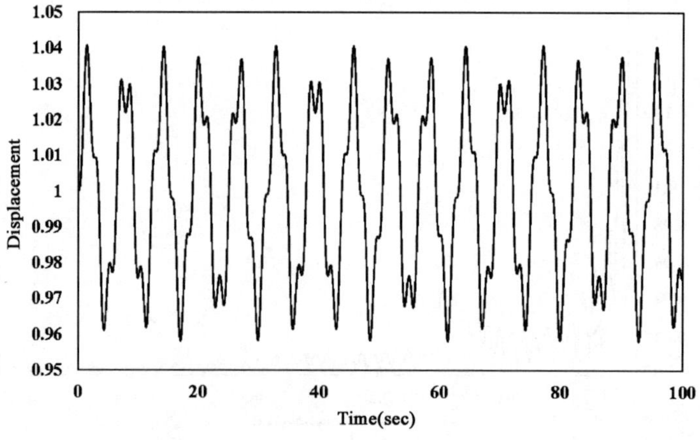

Figure 13 Case 3: Displacement time history.

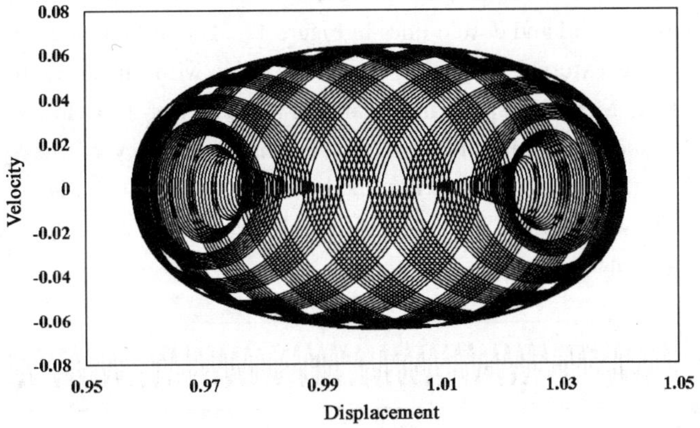

Figure 14 **Case 3: Phase plot.**

Case 3. Undamped Forced Vibration with Small Excitation ($\gamma = 0$, $k = 0.5$, $Y(0) = 1$, $\dot{Y}(0) = 0$, $f = 0.1$, $\omega = 3.4$).

This example is tailored for the forced vibration of the undamped truss with the external force as the vibration source. From the displacement response in Figure 13, one cannot tell whether it is a periodic one or not. However, the regular curves in the phase plot of Figure 14 indicate that the response should be regarded as a periodic motion.

From the Fourier spectrum in Figure 15, one observes that there exist two peaks, 0.159 and 0.541 1 Hz, and that there is a continuous distribution of frequencies below 0.541 1 Hz. As for the truss under a small external force, the response is similar to the one for the truss in Case 2 under damped free vibration. As was mentioned in Case 2, for vibrations of small amplitudes, the natural frequency of the truss is close to 0.159 Hz. Therefore, the peak frequency of 0.159 Hz in Figure 15 represents exactly the intrinsic frequency of the system, and the other peak frequency of 0.541 1 Hz ($= 3.4/2\pi$ Hz) the frequency of the external source. Since only slight nonlinear effect has been induced, the continuous low-frequency distribution over the frequency domain should *not* be interpreted as a physical reality.

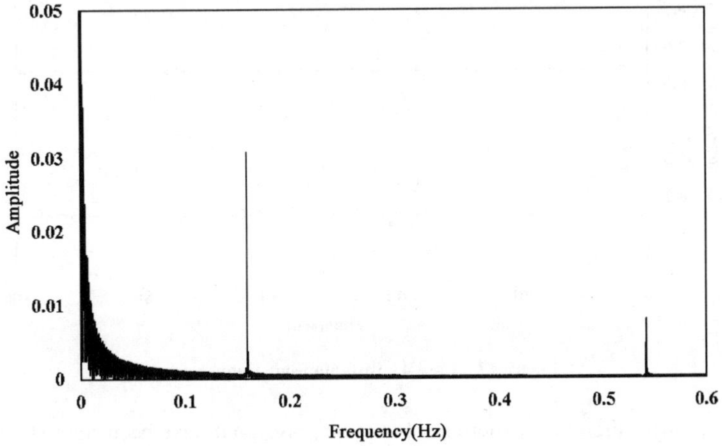

Figure 15 **Case 3: Fourier spectrum.**

Two IMFs c_1 and c_2 obtained by the EMD were plotted in Figure 16. Correspondingly, the instantaneous frequencies are shown as the solid and dotted lines in Figure 17. The instantaneous frequency of c_1 is close to 0.541 1 Hz, i. e., the external excitation frequency, but with slight oscillation, caused by the geometric nonlinear effect. Meanwhile, the instantaneous frequency of c_2 is close to 0.159 Hz, also with slight oscillation, which should be interpreted as the intrinsic frequency of the system. This example shows that the EMD is a nonlinear decomposition method, and the result produced by the HHT is more interpretable than the FFT, since the former generates only two frequency components and both of them possess clear physical meaning.

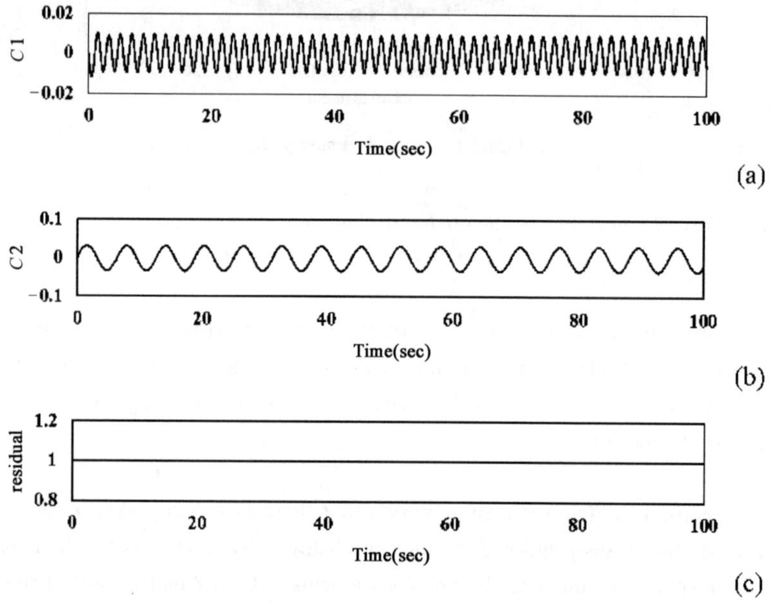

Figure 16 Case 3: IMFs: (a) c_1, (b) c_2, (c) residual.

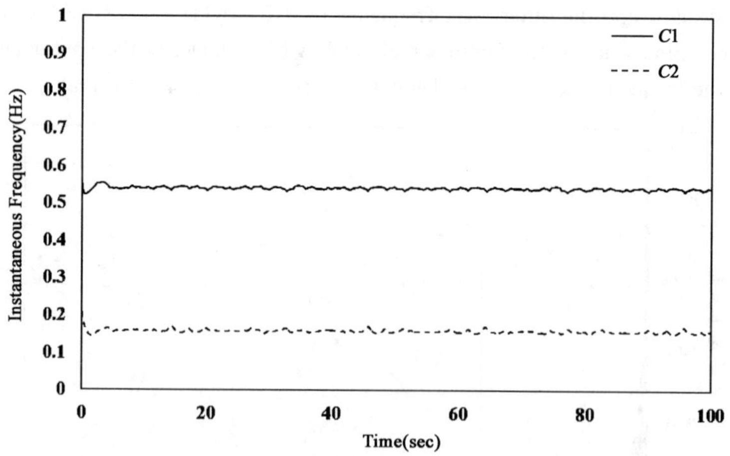

Figure 17 Case 4: Instantaneous frequency.

Owing to the paper length limitation, there are two other cases that have been studied by the authors, but have not been presented herein, i.e., damped forced vibration of the truss with large excitation and with extremely large excitation.

CONCLUSIONS

The following conclusions can be drawn based on the three cases presented herein for the two-member truss: (1) The reference point for EMD is a crucial factor in calculating the instantaneous frequency. The conventional EMD requires the IMFs to be symmetric with respect to the local zero mean. This may be good for fluid or wave problems, but may *artificially change* the physical properties of real structures with *non-zero equilibrium state*, such as the two-member truss presented herein. (2) In general, performing the FFT to the time-history response of a nonlinear structure will yield spectrums that do not truly reflect the physical property of the structure at various stages of loadings. (3) The excitation frequency of a structure plays an important role in the system response, regardless of whether the system is under periodic motion or in chaotic status.

ACKNOWLEDGEMENT

The research reported herein is sponsored in part by the NSC through a research project with Grant No. 98-2221-E-002-106-MY2 to the senior author. Such a financial aid is gratefully acknowledged.

REFERENCES

[1] Huang, N.E., Shen, Z., Long, S.R., Wu, M.C., Shih, H.H., Zheng, Q., Yen, N.-C., Tung, C.C., and Liu, H.H., "The empirical mode decomposition and the Hilbert spectrum for nonlinear and non-stationary time series analysis." *Proc. of the Royal Society of London*, 454, 1998, 903-995.

[2] Huang, N.E., Shen, Z., and Long, S.R., "A new view of nonlinear water waves: the Hilbert spectrum," *Annual Review of Fluid Mechanics*, 31, 1999, 417-457.

[3] Yang, Y.B., and Wu, Y.S., "Chaotic behaviors of a two-member truss," *Journal of Vibration and Control*, 3, 1997, 103-118.

IJSSD Symposium on Progress in Structural Stability and Dynamics
14-16 April 2012, Nanjing, China

DYNAMIC ANALYSIS BY KRIGING-BASED FINITE ELEMENT METHODS

* W. Kanok-Nukulchai and C. Wicaksana

School of Engineering and Technology, Asian Institute of Technology, Thailand
* Email: worsak@gmail.com

KEYWORDS

Dynamic analysis, kriging shape function, Timoshenko beam, mindlin plate.

ABSTRACT

This paper presents dynamic analyses by finite element methods (FEM) with Kriging shape function (Kriging-based finite element methods). A previous study has shown that Kriging-based FEMis able to furnish remarkably accurate solutions for static problems (Plengkhom and Kanok-Nukulchai, 2005). In this study, the application of Kriging-based FEM is enhanced to dynamic problems. One dimensional (Timoshenko beam) and two dimensional (Mindlin plate) dynamic analyses are conducted. Some numerical examples of free and forced vibration are taken to evaluate the accuracy of the method. For each analysis, the accuracy of Kriging-based FEMis compared with the standard FEM. The results show that Kriging-based FEM significantly improve the accuracy when compared to the standard FEM, especially for two dimensional problems.

INTRODUCTION

One drawback of the standard finite element (FEM) is the general limitation of its shape functions due to their construction over the element structure. Meshless method is one of the attempts to overcome this drawback, but it creates another problem. A meshless shape function doesnot have the Kronecker's delta property, and therefore the essential boundary conditions have to be enforced. Based on those two conditions, attempts to investigate a method that combines the advantage of FEM shape function and meshless shape function are conducted. Previous studies[1-2] have found a good method which is called the Kriging-based FEM since the Kriging interpolation is used in the formation of FEM shape functions. These studies have already shown that Kriging-based FEM performs well and it could achieve higher accuracy as compared to the standard finite element method for elasto-static problems. More investigations are, however, needed to evaluate the method for more complex problems. In this study, the use of Kriging-based FEM will be applied for dynamic analysis. This includes free vibration and forced vibration of one-dimensional Timoshenko beam and two-dimensional Mindlin plate problems.

KRIGING INTERPOLATION

In the context of shape function, the Kriging interpolation can be written in the following form:

$$u_i^h(x) = \sum_i^n N^a(x) u_i^a \tag{1}$$

$$N^a(x) = \sum_j^m p_j(x) A_j^a + \sum_k^n r_k(x) B_k^a \tag{2}$$

where p_j is the vector of polynomial basis, and r_k is the vector of correlation function between nodes in the domain of influence. The number of the polynomial basis term is denoted by $m$and the number of nodes in the domain of influence by n. A and B can be written in the following form

$$A = (P^T R^{-1} P)^{-1} P^T R^{-1} \tag{3}$$

$$B = R^{-1}(I - PA) \tag{4}$$

where P is ann × m matrix that collects values of polynomial basis function at given n set of coupling nodes and R is the n × n matrix of correlation functions. Two types of correlation functions are used in this study, i. e. Gauss and quartic spline correlation function. Based on the study, it is found that Gauss correlation function is better than the quartic spline for one-dimensional problems, but for two-dimensional problems, the quartic spline is better than Gauss function with respect to accuracy. The quartic spline correlation function can be written as:

$$R(x_i, x_j) = \begin{cases} 1 - 6d^2 + 8d^3 - 3d^4 & \text{for } d \leqslant 1 \\ 0 & \text{for } d > 1 \end{cases} \tag{5}$$

while the Gauss correlation function can be written as:

$$R(x_i, x_j) = e^{-\left(\theta \frac{r_{ij}}{d_{\max}}\right)^2} \tag{6}$$

where $d = \frac{r_{ij}}{d_{\max}}\theta$, r_{ij} is the distance between two points, θ is correlation parameter and $d_{\max}$ is the maximum distance between two points in the domain of influence. Plengkhom[3] has investigated and proposed that the value of θ has to be set in between two limits to ensure the stability of the method. The bounds are proposed as follows:

Lower bound: $\left| \sum_{i=1}^n N^i - 1 \right| \leqslant 1 \times 10^{-10+m}$, where m is the order of basis function (7)

Upper bound: Det $(R) \leqslant 1 \times 10^{-k}$, where k is the dimension of the problem (8)

The proposed upper and lower bound limits above are the mathematical consideration which guarantees the quality of Kriging interpolation with a rational physical meaning. The lower bound will ensure the consistency property of correlation matrix **R** since too low θ will allow too much influence of all nodes and decreases the non singularity of correlation matrix **R**. The upper bound will ensure the Kriging interpolation's quality. Layered domain of influence concepts will be used herein, and it can be illustrated in Figure 1.

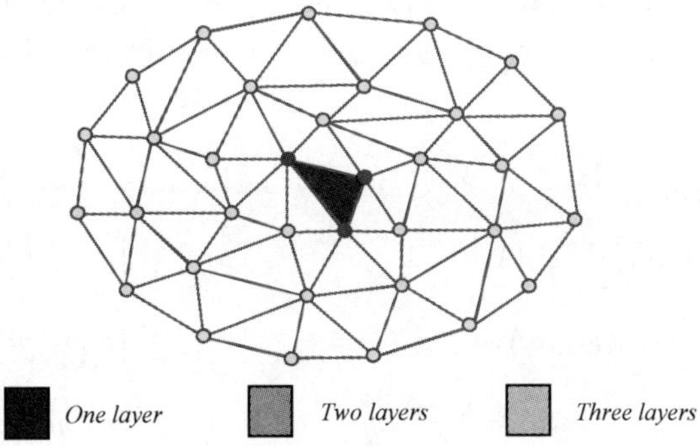

Figure 1 Layered domain of influence.

TIMOSHENKO BEAM

In the Galerkin discretized equation of beams, the unknown variables can be interpolated using the Kriging shape function as

$$u(x) = \sum_n \phi^n u^n = \{\phi\}\{u\} \tag{9a}$$

$$\theta(x) = \sum_n \eta^n \theta^n = \{\eta\}\{\theta\} \tag{9b}$$

The global equations for free vibration of Timoshenko beams can be defined as:

$$\left[\begin{bmatrix} [K_u]_1 & [K_\theta]_1 \\ [K_u]_2 & [K_\theta]_2 \end{bmatrix} - \omega^2 \begin{bmatrix} [M_1] & 0 \\ 0 & [M_2] \end{bmatrix}\right]\begin{bmatrix} u \\ \theta \end{bmatrix} = \begin{bmatrix} 0 \\ 0 \end{bmatrix} \tag{10}$$

where:

$$[K_u]_1 = \int kGA\,\{\phi,_x\}^T\{\phi,_x\}\mathrm{d}x;\quad [K_u]_2 = -\int kGA\,\{\eta\}^T\{\phi,_x\}\mathrm{d}x;$$

$$[K_\theta]_1 = -\int kGA\,\{\phi,_x\}^T\{\eta\}\mathrm{d}x;\quad [K_\theta]_2 = \int kGA\,\{\eta\}^T\{\eta\}\mathrm{d}x + \int EI\,\{\eta,_x\}^T\{\eta,_x\}\mathrm{d}x;$$

$$[M_1] = \int \rho A\,\{\phi\}^T\{\phi\}\mathrm{d}x;\ \text{and}\quad [M_2] = \int \rho I\,\{\eta\}^T\{\eta\}\mathrm{d}x.$$

The bending moment and shear forces can be expressed, respectively as $M = EI\dfrac{\mathrm{d}\theta}{\mathrm{d}x}$ and $Q = \dfrac{\mathrm{d}M}{\mathrm{d}x} - \rho I\dfrac{\mathrm{d}^2\theta}{\mathrm{d}t^2}$.

MINDLIN PLATE

In the Galerkin discretized equation of plates, the unknown variables can be interpolated using the Kriging shape function as

$$u(x, y) = \sum_n \phi^n u^n = \{\phi\}\{u\} \tag{11}$$

$$\theta_x(x, y) = \sum_n \eta^n \theta_x^n = \{\eta\}\{\theta_x\} \tag{12}$$

$$\theta_y(x, y) = \sum_n \xi^n \theta_y^n = \{\xi\}\{\theta_y\} \tag{13}$$

The global equations for free vibration of the Mindlin plate can be defined as:

$$\left[\begin{bmatrix} [K_u]_1 & [K_{\theta_x}]_1 & [K_{\theta_y}]_1 \\ [K_u]_2 & [K_{\theta_x}]_2 & [K_{\theta_y}]_2 \\ [K_u]_3 & [K_{\theta_x}]_3 & [K_{\theta_y}]_3 \end{bmatrix} - \omega^2 \begin{bmatrix} [M_1] & 0 & 0 \\ 0 & [M_2] & 0 \\ 0 & 0 & [M_3] \end{bmatrix}\right]\begin{bmatrix} u \\ \theta_x \\ \theta_y \end{bmatrix} = \begin{bmatrix} 0 \\ 0 \\ 0 \end{bmatrix} \tag{14}$$

where

$$\left.\begin{aligned}
&[K_u]_1 = \int D_s[\{\phi,_x\}^T\{\phi,_x\} + \{\phi,_y\}^T\{\phi,_y\}]\mathrm{d}A;\ [K_{\theta_y}]_1 = -\int D_s\,\{\phi,_y\}^T\{\xi\}\mathrm{d}A;\\
&[K_{\theta_x}]_1 = -\int D_s\,\{\phi,_x\}^T\{\eta\}\mathrm{d}A;\qquad [M_1] = \int \rho h\,\{\phi\}^T\{\phi\}\mathrm{d}A;\\
&[K_u]_2 = -\int D_s\,\{\eta\}^T\{\phi,_x\}\mathrm{d}A;\qquad [M_2] = \int \rho\frac{h^3}{12}\,\{\eta\}^T\{\eta\}\mathrm{d}A;\\
&[K_{\theta_x}]_2 = \int D_b[\{\eta,_x\}^T\{\eta,_x\} + \frac{1-\nu}{2}\,\{\eta,_y\}^T\{\eta,_y\}]\mathrm{d}A + \int D_s\,\{\eta\}^T\{\eta\}\mathrm{d}A;
\end{aligned}\right\}$$

$$\left.\begin{aligned}
&[K_{\theta_y}]_2 = \int D_b[\nu\{\eta,_x\}^T\{\xi,_y\} + \frac{1-\nu}{2}\{\eta,_y\}^T\{\xi,_x\}]dA; \\
&[K_u]_3 = -\int D_s\{\xi\}^T\{\phi,_y\}dA; \qquad [M_3] = \int \rho\frac{h^3}{12}\{\xi\}^T\{\xi\}dA; \\
&[K_{\theta_x}]_3 = \int D_b[\nu\{\xi,_y\}^T\{\eta,_x\} + \frac{1-\nu}{2}\{\xi,_x\}^T\{\eta,_y\}]dA; \\
&[K_{\theta_y}]_3 = \int D_b[\{\xi,_y\}^T\{\xi,_y\} + \frac{1-\nu}{2}\{\xi,_x\}^T\{\xi,_x\}]dA + \int D_s\{\xi\}^T\{\xi\}dA;
\end{aligned}\right\} \tag{15}$$

and bending moments M_x, M_y, twisting moment M_{xy} and shear forces Q_x, Q_y can be expressed as:

$$\left.\begin{aligned}
&M_x = -D_b\left(\frac{d\theta_x}{dx} + \nu\frac{d\theta_y}{dy}\right); \qquad Q_x = \frac{dM_x}{dx} + \frac{dM_{xy}}{dy} + \frac{1}{12}\rho h^3\frac{d^2\theta_x}{dt^2}; \\
&M_y = -D_b\left(\frac{d\theta_y}{dy} + \nu\frac{d\theta_x}{dx}\right); \qquad Q_y = \frac{dM_y}{dy} + \frac{dM_{xy}}{dx} + \frac{1}{12}\rho h^3\frac{d^2\theta_y}{dt^2}; \\
&\text{and } M_{xy} = -D_b\frac{1-\nu}{2}\left(\frac{d\theta_x}{dy} + \frac{d\theta_y}{dx}\right)
\end{aligned}\right\} \tag{16}$$

FREE VIBRATION RESULTS

In this section, free vibration analysis of Timoshenko beam and Mindlin plate are conducted. The results are compared among various layered domain of influence and basis function. The comparison of the numerical results with analytical solutions[3] and the percentage errors can be seen in Tables 1 and 2.

TABLE 1 NON-DIMENSIONALIZED FREQUENCIES OF SIMPLY SUPPORTED TIMOSHENKO BEAM

Mode	Analytical solution	Finite Element	Kriging		
		Quadratic	Quadratic		Cubic
			2 layers	3 layers	3 layers
1	3.141 59	3.141 60 (0.000 35)	3.141 55 (0.001 27)	3.141 62 (0.000 72)	3.141 61 (0.000 56)
2	6.283 19	6.283 19 (0.000 14)	6.283 01 (0.002 79)	6.283 14 (0.000 68)	6.283 16 (0.000 48)
3	9.424 78	9.424 90 (0.001 33)	9.424 26 (0.005 48)	9.424 61 (0.001 82)	9.424 63 (0.001 60)
4	12.566 37	12.567 03 (0.005 21)	12.565 35 (0.008 11)	12.565 89 (0.003 83)	12.565 94 (0.003 39)
5	15.707 96	15.710 13 (0.013 80)	15.706 50 (0.009 32)	15.706 87 (0.006 97)	15.706 98 (0.006 25)
6	18.849 56	18.855 15 (0.029 67)	18.848 16 (0.007 43)	18.847 34 (0.011 77)	18.847 57 (0.010 53)
7	21.991 15	22.003 44 (0.055 89)	21.990 99 (0.007 00)	21.987 07 (0.018 56)	21.987 48 (0.016 68)
8	25.132 74	25.156 84 (0.095 90)	25.136 15 (0.013 56)	25.125 92 (0.027 16)	25.126 53 (0.024 70)
9	28.274 33	28.317 74 (0.153 51)	28.285 03 (0.037 84)	28.263 50 (0.038 32)	28.264 54 (0.034 63)
10	31.415 93	31.489 07 (0.232 83)	31.439 50 (0.075 04)	31.399 71 (0.051 63)	31.401 36 (0.046 38)

Note: Numbers in parenthesis shown percentage error compared with analytical solution

TABLE 2 NON-DIMENSIONALIZED FREQUENCIES OF SIMPLY SUPPORTED MINDLIN PLATE

Mode	Analytical solution	Finite Element	Kriging		
		Quadratic	Quadratic		Cubic
			2 layers	3 layers	3 layers
1	19.739 21	20.047 58 (1.562 2)	19.732 99 (0.031 5)	19.694 06 (0.228 7)	19.696 32 (0.217 3)
2	49.348 02	50.131 42 (1.588 5)	49.402 24 (0.109 9)	49.276 56 (0.144 8)	49.255 66 (0.187 2)
3	49.348 02	51.498 32 (4.357 4)	49.478 67 (0.246 7)	49.312 02 (0.073 0)	49.256 27 (0.185 9)
4	78.956 84	83.685 65 (5.989 1)	79.169 84 (0.269 8)	78.677 57 (0.353 7)	78.739 53 (0.275 2)
5	98.696 04	102.847 13 (4.205 9)	99.146 65 (0.456 6)	98.757 84 (0.062 6)	98.580 97 (0.116 6)
6	98.696 04	103.357 73 (4.723 3)	99.304 85 (0.616 8)	98.910 75 (0.217 5)	98.604 12 (0.093 1)
7	128.304 86	134.683 74 (4.971 7)	128.741 22 (0.340 1)	127.789 95 (0.401 3)	127.940 91 (0.283 7)
8	128.304 86	134.854 98 (5.105 1)	129.210 11 (0.705 5)	128.040 41 (0.206 1)	127.969 72 (0.261 2)
9	167.783 27	177.398 98 (5.731 0)	169.378 83 (0.951 0)	168.328 92 (0.325 3)	167.954 88 (0.102 3)
10	167.783 27	178.639 02 (6.470 1)	169.814 81 (1.210 8)	168.924 96 (0.680 5)	168.086 70 (0.180 8)

Note: Numbers in parenthesis shown percentage error compared with analytical solution

Table 1 shows that the Kriging-based FEM with 2 and 3 layers perform better than the standard FEM for one dimensional problem. For two-dimensional problems, the Kriging-based FEM performs much better than the standard FEM.

FORCED VIBRATION RESULTS

Timoshenko Beam

Forced vibration of a simply supported Timoshenko beams subjected to moving load is tested here. The material property, dimensions and loadings are as follows: $E = 2\times 10^9\,\mathrm{kg/m^2}$; $h/L = 1/100$; $L = 10$ m; $b = 1$ m; $M = 240$ kg/m; no damping; load $P = 100$ N moving loads with $v = 10$ m/s.

The results of the numerical analysis are compared with the analytical solution. The analytical solution to this problem is taken from Ref. [4]. The numerical results are in very good agreement with the analytical solutions as shown in Figure 3. In order to determine the accuracy, the error norm of displacement, moment and shear are calculated at the peak for comparison. Error norm can be defined as the integration of error over the whole domain, so that it represents the average error over the whole domain. For thin beams, in order to eliminate the effect of shear locking, a reduced integration is implemented in the analysis. The analysis for linear finite element and linear Kriging-based FEM with one layer shows exactly the same result, and it is shown in the comparison. Table 3 compares the results obtained from quadratic FEM and Kriging-based FEM with quadratic and cubic basis function.

TABLE 3 PERCENTAGE ERROR NORM (%) AT T=1.105 SEC

	FEM	Quadratic	Cubic
	Quadratic	2 layers	3 layers
Displacement	0.086 00	0.079 50	0.086 30
Moment	0.809 80	0.719 30	0.641 40
Shear	5.330 40	5.109 30	4.178 30

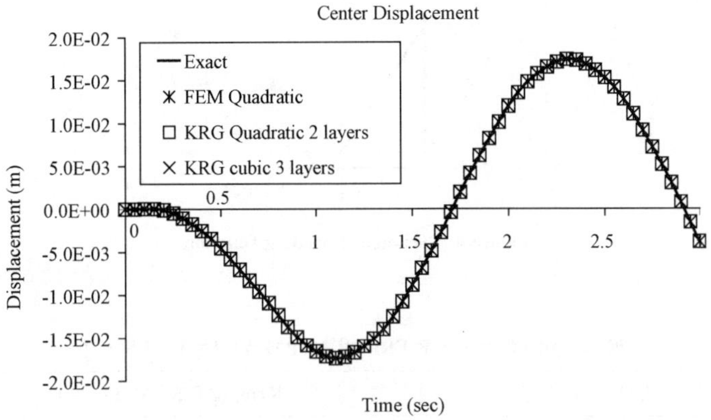

Figure 2 Center displacement time history.

Figure 3 and Table 3 show that the Kriging-based FEM is better than standard FEM, especially with regard to moment and shear distributions.

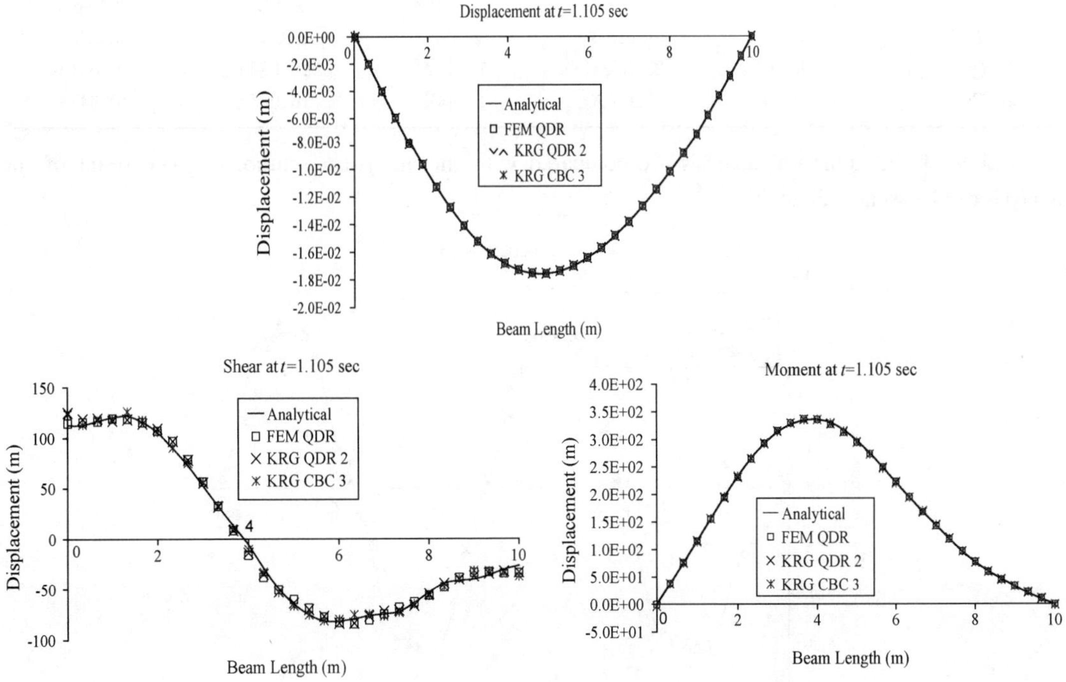

Figure 3 Displacement, moment and shear at t=1.105 second.

Mindlin Plate

Consider next the forced vibration analysis of simply supported Mindin plate subjected to a point load at the center of the plate. The plate was modeled by rectangular elements, with the following parameters: $E = 2\times 10^9$ kg/m^2; $h = 0.1$ m; $Lx = Ly = 10$ m; $h/L = 1/100$ (thin plate); and $M = 240$ kg/m^2. No damping is considered here. The loading functions a triangular pulse force as shown in Figure 4.

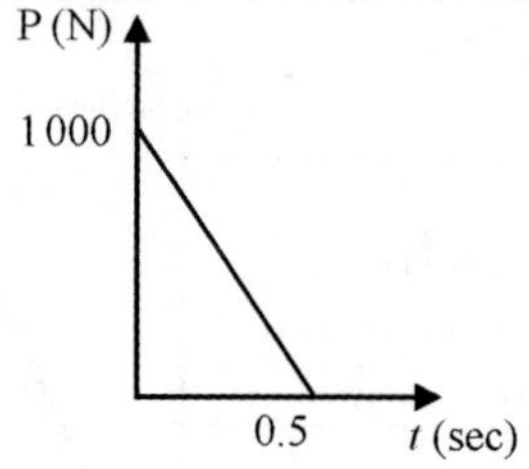

Figure 4 Triangular loading function

TABLE 4 PERCENTAGE ERROR NORM AT T=1.55 SEC (%)

	FEM	Kriging (25×25 nodes)			
	Quadratic	Quadratic		Cubic	
		2 layers	3 layers	3 layers	4 layers
U	3.012 5	1.527 1	0.538 6	0.464 6	0.511 2
Mx	17.205 0	9.062 7	3.409 0	2.457 3	2.789 3
My	17.205 0	9.062 7	3.409 0	2.457 3	2.789 3
Mxy	8.823 6	6.247 8	2.332 3	3.053 8	3.073 9
Qx	45.488 0	23.062 0	17.447 0	10.083 0	10.641 0
Qy	45.488 0	23.062 0	17.447 0	10.083 0	10.641 0

The results of the numerical analysis are compared with the analytical solution[5]. The result of the analysis can be seen in Figure 5.

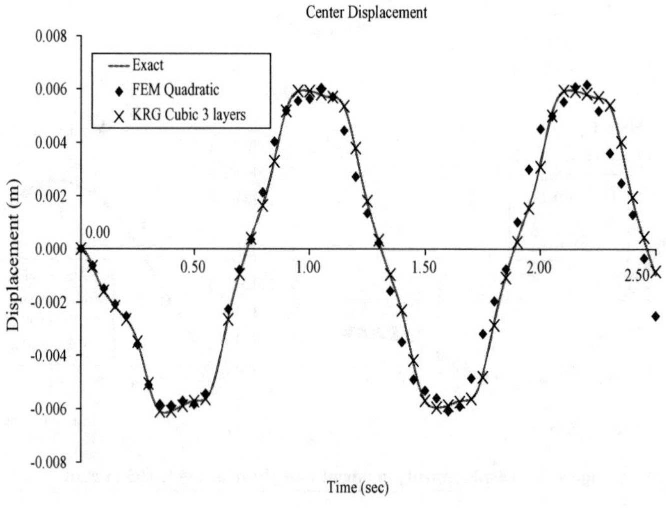

Figure 5 Center displacement time history

Table 4 and Figures 5-7 confirm that the Kriging-based FEM is more accurate than the standard FEM. Analysis with cubic basis function with three layers of domain of influence is the most accurate one.

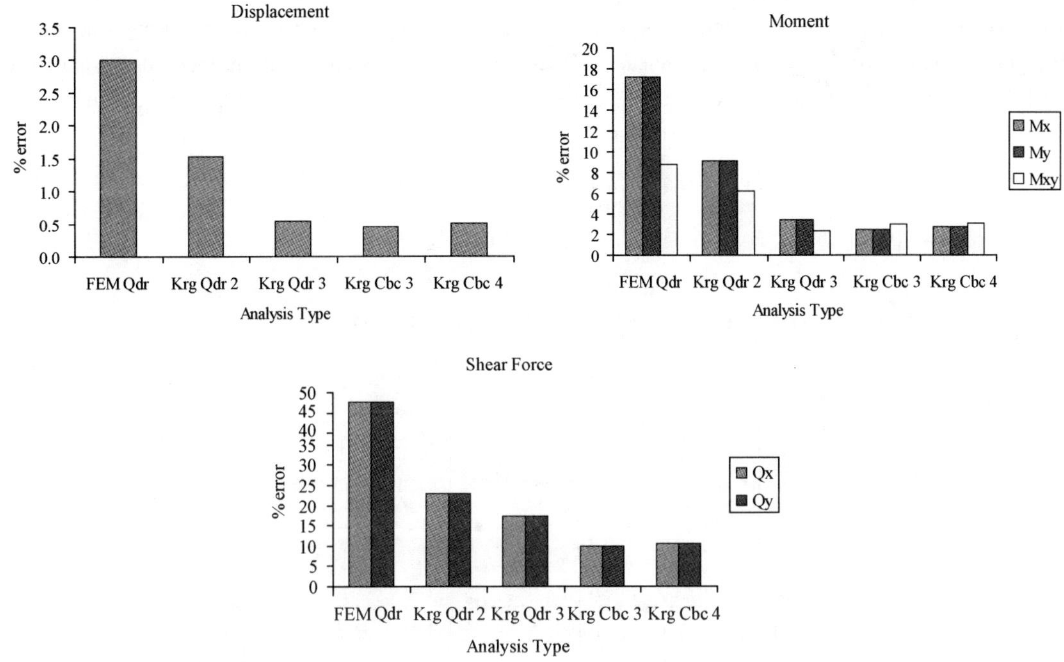

Figure 6 Percentage error norm of displacement moment and shear at $t=1.55$ sec

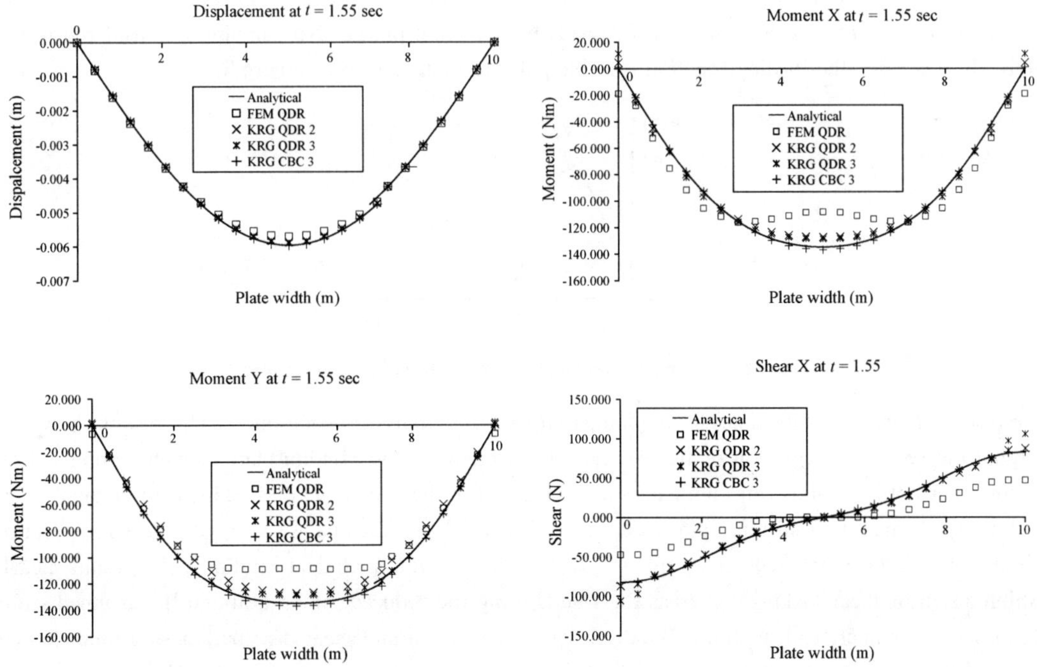

Figure 7 Displacement, moment and shear at middle section at $t=1.55$ sec

The next example problem will discuss about attempts to eliminate the shear locking phenomenon[6] in the Kriging-based FEM for Mindlin plate problems. The two methods that will be implemented are the reduced integration method and hybrid plate formulations based on the assumed strain method[7]. Reduced integration is a well-known method that can be simply implemented by using one Gauss integration point for integrating the shear stiffness term. The assumed strain methods assume the variation of the bending and shears strain (usually using a polynomial term) within the element. The stiffness matrix for the hybrid plate formulation can be expressed as:

$$[K] = [H_b]^T [G_b]^{-T}[H_b] + [H_s]^T [G_s]^{-T}[H_s] \tag{15}$$

where: $[G_b] = \int_\Omega [B_b]^T[D_b][B_b]\mathrm{d}\Omega$; $[G_s] = \int_\Omega [B_s]^T[D_s][B_s]\mathrm{d}\Omega$;

$[H_b] = \int_\Omega [B_b]^T[D_b]L_b[\phi]\mathrm{d}\Omega$; $[H_s] = \int_\Omega [B_s]^T[D_s]L_s[\phi]\mathrm{d}\Omega$;

$$L_b[\phi] = \begin{bmatrix} 0 & \dfrac{\partial\phi}{\partial x} & 0 \\ 0 & 0 & \dfrac{\partial\phi}{\partial y} \\ 0 & \dfrac{\partial\phi}{\partial y} & \dfrac{\partial\phi}{\partial x} \end{bmatrix}; \qquad L_s[\phi] = \begin{bmatrix} \dfrac{\partial\phi}{\partial x} & -\phi \\ \dfrac{\partial\phi}{\partial y} & -\phi \end{bmatrix}; \text{ and}$$

$$[B_b] = \begin{bmatrix} 1 & 0 & 0 & x & 0 & 0 & y & 0 & 0 \\ 0 & 1 & 0 & 0 & x & 0 & 0 & y & 0 \\ 0 & 0 & 1 & 0 & 0 & x & 0 & 0 & y \end{bmatrix} \qquad [B_s] = \begin{bmatrix} 1 & 0 \\ 0 & 1 \end{bmatrix}.$$

The problem considered is a simply supported rectangular Mindlin plate with a point load at the center of the plate. Rectangular elements are used. The problem parameters are defined as follows: $E = 2\times10^9$ kg/m^2; $h = 0.001$ m; $Lx = Ly = 10$ m; and $h/L = 1/10\,000$ (thin plate). No damping is considered in this study. In this case, the loading function is a sine pulse force as shown in Figure 8.

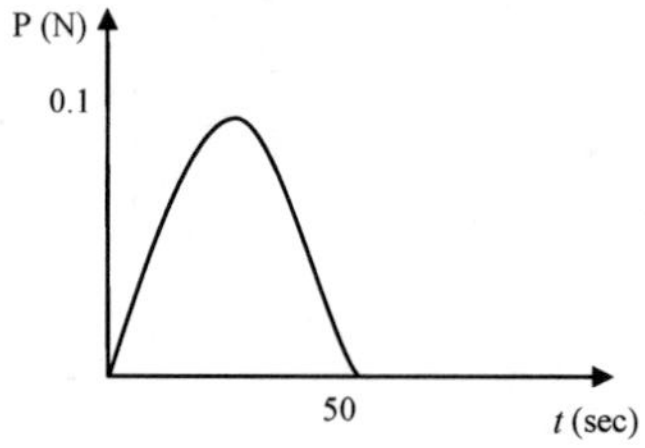

Figure 8 A sine pulse loading function applied at the plate center.

The results of the numerical analysis are compared with the analytical solution[6]. The results show that without the reduced integration or the assumed strain method, the Mindlin plate element exhibits shear locking, but both methods can eliminate shear locking for the rectangular Mindlin plate element. The center displacement time history (plotted in Figure 9) reveals that the analysis without reduced integration vibrates with a different frequency (stiffer) as compared to the analytical solution. The results clearly exhibit a sign of shear locking, whereas the results using the reduced integration and the assumed strain are closer to the analytical solution. The displacement, moment and shear distributions at a time step (in this case it is chosen to be at peak center displacement at $t = 54$ sec) are presented in Figure 10.

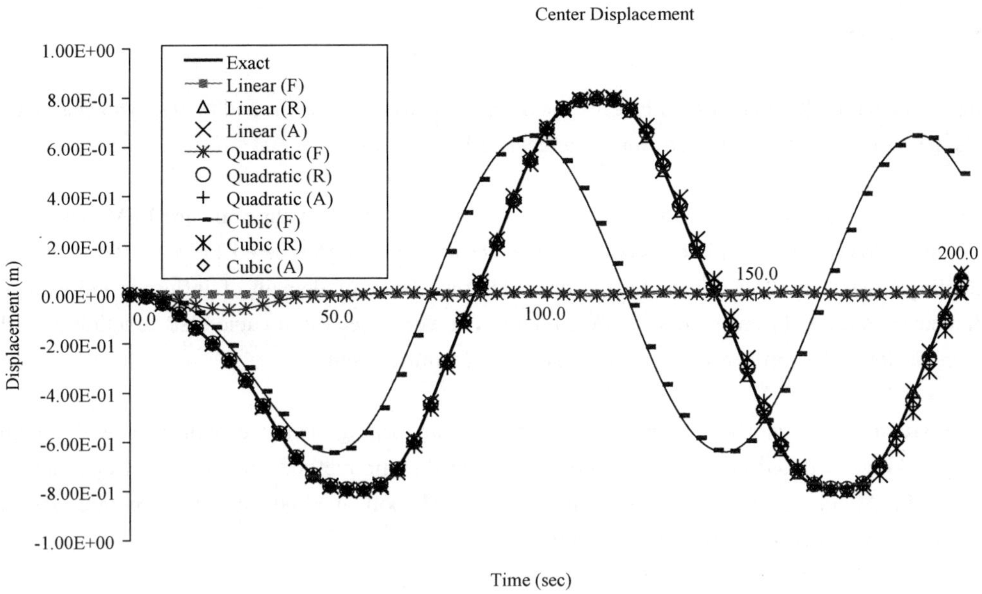

Figure 9 Displacement time histories, in which F=Full integration; R=Reduced integration; A=Assumed strain.

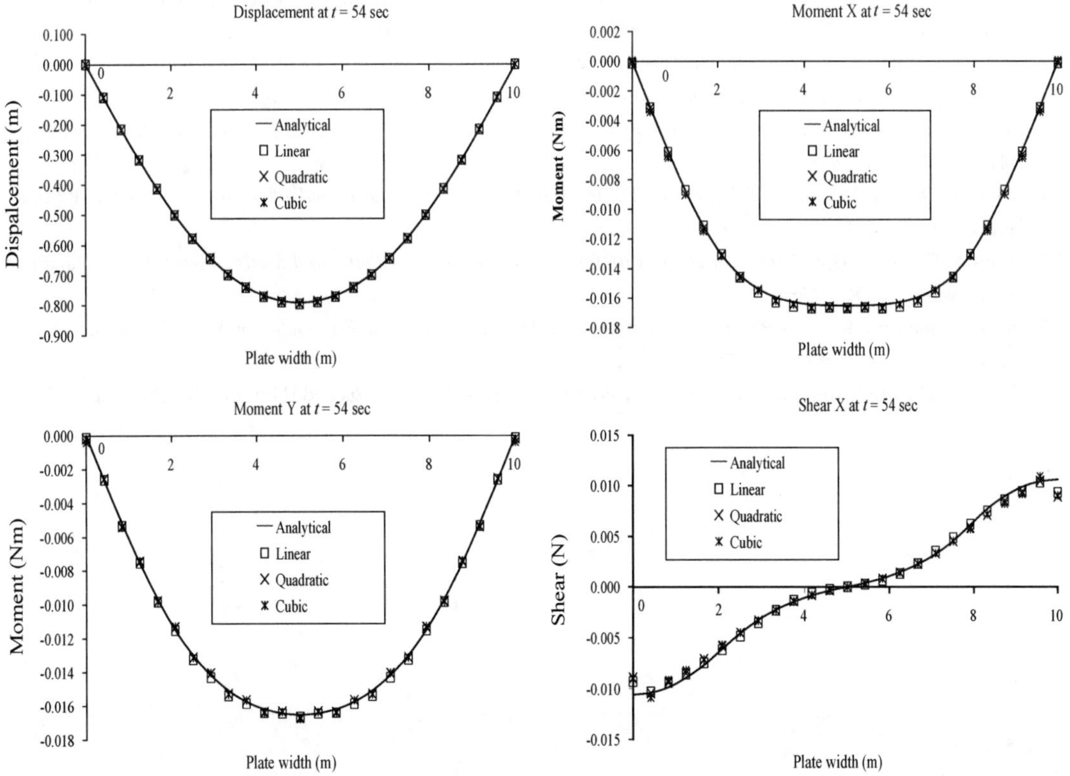

Figure 10 Moment and shear at middle section at t=54 sec (assumed strain method).

CONCLUSIONS

We have applied the Kriging-based FEM for the dynamic analyses of beams and plates. The characteristic and the performance of the method can be summarized as follows:

1. For dynamic analysis, the Kriging-based FEM is more accurate than the standard FEM especially for two-dimensional (Mindlin plate) problems. The accuracy of Kriging-based FEM can be improved by adopting a high basis function and using one additional layer of domain of influence more than the minimum required. From the case analyzed herein, it can be seen that cubic basis function with three layers of domain of influence can achieve the most accurate result.

2. High basis function can help reduce the effect of shear locking, but it cannot completely eliminate shear locking. Reduced integration method and assumed strain method can completely eliminate shear locking for Mindlin plate with rectangular elements, but both methods are only accurate when used with low basis functions (linear and quadratic).

REFERENCES

[1] Tongsuk, P., and Kanok-Nukulchai, W., "Further Investigation of Element Free Galerkin Method Using Moving Kriging Interpolation", *International Journal of Computational Methods*, 2004, 1(2), pp. 345-366.

[2] Plengkhom, K. and Kanok-Nukulchai, W., "An Enhancement of Finite Element Method with Moving Kriging Shape Functions", *International Journal of Computational Methods*, 2005, 2(4), pp. 451-475.

[3] Blevins, R. D., *Formulas for Natural Frequency and Mode Shape*, Robert E Krieger Publishing Company, Malabar, Florida, 1986.

[4] Chopra, A K., *Dynamics of Structures Theory and Application to Earthquake Engineering*, Second Edition, Prentice-Hall, 2001.

[5] Hinton, E., *Numerical Methods and Software for Dynamic Analysis of Plates and Shells*, First Edition, Pineridge Press Swansea U.K., 1988.

[6] Saranyasoontorn, K., *On Elimination of Shear Locking In the Element Free Galerkin Method*, Master Thesis, Asian Institute of Technology, Thailand, 1999.

[7] Kwon, Y.W, and Bang, H., *The Finite Element Method Using Matlab*, Second Edition, CRC Press LLC, 2000.

ON MODE ORTHOGONALITY OF COMPLEX STRUCTURES

* W.Q. Chen[1], Y.Q. Guo[2], Y.H. Pao[3]

[1]Department of Engineering Mechanics, Zhejiang University,
Yuquan Campus, Hangzhou 310027, China
[2]Key Laboratory of Mechanics on Disaster and Environment in Western China,
Ministry of Education, and School of Civil Engineering and Mechanics Lanzhou
University, Lanzhou 730000, China
[3]College of Civil Engineering and Architecture, Zhejiang University
Zijingang Campus, Hangzhou 310058, China
* Email: chenwq@zju.edu.cn

KEYWORDS

Complex structures, free vibration, mode orthogonality, symplectic inner product.

ABSTRACT

The orthogonality of vibration modes plays a central role in determining analytically the transient response of an elastic system subject to general time-varying excitations. A novel method, based on the system of first-order differential equations governing one-dimensional continuum models, is presented for deriving the orthogonality relations of vibration modes of a complex three-dimensional elastic frame. The members in the frame can have non-uniform cross-sections, but the centre lines must be straight, and the joints (that connect the members together to form the network of structure) can be either perfectly rigid or pinned; and further, they may be elastically supported and have concentrated masses. The derived relations can be used to investigate the transient responses of framed structures subjected to distributed/concentrated loads applied on members or concentrated loads applied at joints.

INTRODUCTION

The importance of orthogonality of vibration modes in structural dynamic analysis is much highlighted by its ability of obtaining transient responses of structures efficiently and accurately for arbitrarily dynamic stimuli. In fact, with the help of normal mode expansion, the governing equations for the unknown time-varying functions are so simple that they allow closed-form solutions to be obtainable. This is very attractive from the theoretical viewpoint, in addition to the numerical efficiency associated with analytical expressions. The orthogonality of vibration modes for discrete multi-degree-of-freedom systems, such as the multiple-mass-spring systems and the ones used to approximately model the practical elastic structures by the finite element method (FEM), has been well recognized and established[1-3]. The orthogonality relations in the vibration theory of continuum models have also been presented for simple structures such as rods, beams, strings, plates and shells, and even three-dimensional elastic bodies[4, 5]. Generally, the orthogonality of vibration modes is closely related to the self-adjoint property of the differential operator of the governing differential equations of the one-, two- or three-dimensional elastic continua[6]. These equations are usually expressed in terms of displacements with high-order differential operators, which are at least of order 2, such as the one for the longitudinal vibration of a rod or for the transverse vibration of a stretched string.

However, little attention has been paid to the orthogonality relations for free vibration of complicated planar and space frames based on one-dimensional continuum models for the constituent members. This is because the FEM-based discrete models have been widely employed for approximate dynamic analyses of structures. On the other hand, the derivation of such relations is not routine, and great care must be taken even with an appropriate way of presentation. We have noted that Chan and Williams[7] presented the orthogonality of vibration modes of planar frames with continuum members, based however on a discrete model by a limit analysis, i. e. letting the degree-of-freedom of the discrete system tend to infinite such that a continuum model is recovered. Guo et al.[8] derived the orthogonality relations directly based on the continuum model using the Betti's reciprocity theorem. However, the derivation is lengthy and very cumbersome.

The objective of this paper is to present a novel method to deduce the orthogonality relations of vibration modes of complicated framed structures. The departure point is the system of first-order differential equations governing the dynamic state of each member in the frame. This obviously differs from all the previous analyses[1-8]. With the concept of symplectic inner product, two important relations are then obtained for each member, which are essential to the orthogonality of vibration modes of the whole frame. By taking into account the joint coupling equations, which are the equilibrium conditions of forces and compatibility conditions of displacements at joints, we present two orthogonality relations of vibration modes of a complicated frame. One involves the mass properties of the frame whereas the other is associated with its stiffness properties.

CONTINUUM MODELS FOR STRUCTURAL MEMBERS

A three-dimensional framed structure in a space can be seen as a structural network, comprising a certain number of slender members, which are connected together at their ends by joints of various types. All slender members in the frame are assumed to be perfect carriers of four types of one-dimensional waves, including the longitudinal, torsional and two flexural waves in the two principal planes, when the frame is subjected to dynamic excitations. These waves propagate independently along the member, but mode switching takes place at the joints. Thus, a longitudinal wave in a member may become longitudinal, flexural and/or torsional in other members when transmitting through the joints as well as in the same member when it is reflected by the end joints, and *vise versa*. The dynamic response of the frame is therefore determined by a very complicated procedure of propagation of four different types of waves in members and scattering of these waves at joints, accompanied with certain times of mode switching. Nevertheless, the basic theories of one-dimensional wave propagation in a space member are rather simple, as listed in Ref. [8]. These equations can be arranged in the form of a set of first-order differential equations and may be written in matrix form (in absence of external loads) as

$$\frac{\partial}{\partial x}\boldsymbol{v} = \boldsymbol{A}\boldsymbol{v} \tag{1}$$

where $\boldsymbol{v} = [u_x, u_y, u_z, \phi_x, \phi_y, \phi_z, F_x, F_y, F_z, M_x, M_y, M_z]^T = [\tilde{\boldsymbol{u}}^T, \tilde{\boldsymbol{F}}^T]^T$ is the state vector, with $\boldsymbol{u}$ and $\boldsymbol{F}$ being the generalized displacement (translational and rotational displacements) and generalized force (forces and moments) vectors, respectively, and

$$\boldsymbol{A}=\begin{bmatrix}\boldsymbol{A}_{11} & \boldsymbol{A}_{12}\\ \boldsymbol{A}_{21} & \boldsymbol{A}_{22}\end{bmatrix},\boldsymbol{A}_{11}=\begin{bmatrix}0&0&0&0&0&0\\0&0&0&0&0&1\\0&0&0&0&1&0\\0&0&0&0&0&0\\0&0&0&0&0&0\\0&0&0&0&0&0\end{bmatrix},\boldsymbol{A}_{12}=\left\langle\frac{1}{EA},\frac{1}{\kappa GA},\frac{1}{\kappa GA},\frac{1}{GI_x},\frac{1}{EI_y},\frac{1}{EI_z}\right\rangle$$

$$\boldsymbol{A}_{21}=\left\langle\rho A\frac{\partial^2}{\partial t^2},\rho A\frac{\partial^2}{\partial t^2},\rho A\frac{\partial^2}{\partial t^2},\rho I_x\frac{\partial^2}{\partial t^2},\rho I_y\frac{\partial^2}{\partial t^2},\rho I_z\frac{\partial^2}{\partial t^2}\right\rangle=\boldsymbol{m}\frac{\partial^2}{\partial t^2},\ \boldsymbol{A}_{22}=-\boldsymbol{A}_{11}^T \tag{2}$$

where $\boldsymbol{m}=\langle\rho A,\rho A,\rho A,\rho I_x,\rho I_y,\rho I_z\rangle$. Here $\langle\cdot\rangle$ is the diagonal matrix, $\kappa=\pi^2/12$ is the shear correction factor in the Timoshenko beam theory, and all other quantities bear the conventional meanings. The 12×12 matrix $\boldsymbol{A}$ is the system matrix of a member in a space. Obviously, we have

$$\frac{\partial\tilde{\boldsymbol{u}}}{\partial x}=\boldsymbol{A}_{11}\tilde{\boldsymbol{u}}+\boldsymbol{A}_{12}\tilde{\boldsymbol{F}} \tag{3}$$

TWO ESSENTIAL RELATIONS FOR FREE VIBRATION

Assuming that the length of a structural member is l, we introduce the following symplectic inner product

$$[\boldsymbol{v}_1,\boldsymbol{v}_2]=\int_0^l\boldsymbol{v}_1^T\boldsymbol{J}\boldsymbol{v}_2\,\mathrm{d}x=\int_0^l(\tilde{\boldsymbol{u}}_1^T\tilde{\boldsymbol{F}}_2-\tilde{\boldsymbol{F}}_1^T\tilde{\boldsymbol{u}}_2)\mathrm{d}x=\int_0^l(\tilde{\boldsymbol{u}}_1^T\tilde{\boldsymbol{F}}_2-\tilde{\boldsymbol{u}}_2^T\tilde{\boldsymbol{F}}_1)\mathrm{d}x \tag{4}$$

where $\boldsymbol{J}$ is the unit symplectic matrix with $\boldsymbol{J}^T=-\boldsymbol{J}$ and $\boldsymbol{J}^2=-\boldsymbol{1}_{12\times12}$, and $\boldsymbol{1}$ the identity matrix. For free vibration problem, the system matrix $\boldsymbol{A}$ keeps unaltered, except for $\boldsymbol{A}_{21}$, which now reads

$$\boldsymbol{A}_{21}=-\omega^2\boldsymbol{m} \tag{5}$$

It can be shown that

$$[\boldsymbol{v}_1,\boldsymbol{A}\boldsymbol{v}_2]=\int_0^l\boldsymbol{v}_1^T\boldsymbol{J}\boldsymbol{A}\boldsymbol{v}_2\,\mathrm{d}x=\int_0^l(-\omega^2\tilde{\boldsymbol{u}}_1^T\boldsymbol{m}\tilde{\boldsymbol{u}}_2-\tilde{\boldsymbol{F}}_1^T\boldsymbol{A}_{12}\tilde{\boldsymbol{F}}_2+\tilde{\boldsymbol{u}}_1^T\boldsymbol{A}_{22}\tilde{\boldsymbol{F}}_2-\tilde{\boldsymbol{F}}_1^T\boldsymbol{A}_{11}\tilde{\boldsymbol{u}}_2)\mathrm{d}x \tag{6}$$

Since $\boldsymbol{A}_{12}$ and $\boldsymbol{A}_{21}$ are symmetric matrices and $\boldsymbol{A}_{11}^T=-\boldsymbol{A}_{22}$, i.e.

$$\tilde{\boldsymbol{u}}_1^T\boldsymbol{A}_{22}\tilde{\boldsymbol{F}}_2=-\tilde{\boldsymbol{F}}_2^T\boldsymbol{A}_{11}\tilde{\boldsymbol{u}}_1=-\phi_{z1}F_{y2}-\phi_{y1}F_{z2},\tilde{\boldsymbol{u}}_2^T\boldsymbol{A}_{22}\tilde{\boldsymbol{F}}_1=-\tilde{\boldsymbol{F}}_1^T\boldsymbol{A}_{11}\tilde{\boldsymbol{u}}_2=-\phi_{z2}F_{y1}-\phi_{y2}F_{z1} \tag{7}$$

we have $[\boldsymbol{v}_1,\boldsymbol{A}\boldsymbol{v}_2]=[\boldsymbol{v}_2,\boldsymbol{A}\boldsymbol{v}_1]$, which means that $\boldsymbol{A}$ is a Hamiltonian matrix.

We now consider two vibration modes with frequencies ω_i and ω_j. Thus

$$\frac{\mathrm{d}}{\mathrm{d}x}\boldsymbol{v}_k=\boldsymbol{A}_k\boldsymbol{v}_k\qquad(k=i,\ j) \tag{8}$$

Here the system matrix $\boldsymbol{A}_k$ is the same as before except that $\boldsymbol{A}_{21}k=-\omega_k^2\boldsymbol{m}$.

By multiplying Eq. (8) for $k=i$ with $\boldsymbol{v}_j^T\boldsymbol{J}$ and integrating over the length of the member, we get

$$\int_0^l\boldsymbol{v}_j^T\boldsymbol{J}\frac{\mathrm{d}\boldsymbol{v}_i}{\mathrm{d}x}\mathrm{d}x=\int_0^l\boldsymbol{v}_j^T\boldsymbol{J}\boldsymbol{A}_i\boldsymbol{v}_i\,\mathrm{d}x\quad\text{or}\quad\left[\boldsymbol{v}_j,\frac{\mathrm{d}\boldsymbol{v}_i}{\mathrm{d}x}\right]=[\boldsymbol{v}_j,\boldsymbol{A}_i\boldsymbol{v}_i] \tag{9}$$

Similarly, we have

$$\int_0^l\boldsymbol{v}_i^T\boldsymbol{J}\frac{\mathrm{d}\boldsymbol{v}_j}{\mathrm{d}x}\mathrm{d}x=\int_0^l\boldsymbol{v}_i^T\boldsymbol{J}\boldsymbol{A}_j\boldsymbol{v}_j\,\mathrm{d}x\quad\text{or}\quad\left[\boldsymbol{v}_i,\frac{\mathrm{d}v_j}{\mathrm{d}x}\right]=[\boldsymbol{v}_i,\boldsymbol{A}_j\boldsymbol{v}_j] \tag{10}$$

The subtraction of the above two equations leads to

$$\int_0^l \left(\boldsymbol{v}_j^T \boldsymbol{J} \frac{\mathrm{d}\boldsymbol{v}_i}{\mathrm{d}x} - \boldsymbol{v}_i^T \boldsymbol{J} \frac{\mathrm{d}\boldsymbol{v}_j}{\mathrm{d}x} \right) \mathrm{d}x = -(\omega_i^2 - \omega_j^2) \int_0^l \left(\sum_{k=x,y,z} \rho A u_{ki} u_{kj} + \sum_{k=x,y,z} \rho I_k \phi_{ki} \phi_{kj} \right) \mathrm{d}x \tag{11}$$

By integration by parts, we obtain

$$(\boldsymbol{v}_j^T \boldsymbol{J}\, \boldsymbol{v}_i) \Big|_0^l - \int_0^l \left(\frac{\mathrm{d}\boldsymbol{v}_j^T \boldsymbol{J}}{\mathrm{d}x} \boldsymbol{v}_i + \boldsymbol{v}_i^T \boldsymbol{J} \frac{\mathrm{d}\boldsymbol{v}_j}{\mathrm{d}x} \right) \mathrm{d}x = -(\omega_i^2 - \omega_j^2) \int_0^l \tilde{\boldsymbol{u}}_j^T \boldsymbol{m}\, \tilde{\boldsymbol{u}}_i \mathrm{d}x \tag{12}$$

The second integral term in the left-hand side vanishes because $\boldsymbol{J}^T = -\boldsymbol{J}$, and hence we have

$$(\omega_i^2 - \omega_j^2) \int_0^l \tilde{\boldsymbol{u}}_j^T \boldsymbol{m}\, \tilde{\boldsymbol{u}}_i \mathrm{d}x = (\tilde{\boldsymbol{F}}_j^T \tilde{\boldsymbol{u}}_i - \tilde{\boldsymbol{u}}_j^T \tilde{\boldsymbol{F}}_i) \Big|_0^l = (\tilde{\boldsymbol{u}}_i^T \tilde{\boldsymbol{F}}_j - \tilde{\boldsymbol{u}}_j^T \tilde{\boldsymbol{F}}_i) \Big|_0^l \tag{13}$$

Equation (9) can be rewritten as

$$\int_0^l \left(\tilde{\boldsymbol{u}}_j^T \frac{\mathrm{d}\tilde{\boldsymbol{F}}_i}{\mathrm{d}x} - \tilde{\boldsymbol{F}}_j^T \frac{\mathrm{d}\tilde{\boldsymbol{u}}_i}{\mathrm{d}x} \right) \mathrm{d}x = \int_0^l (-\omega_i^2 \tilde{\boldsymbol{u}}_j^T \boldsymbol{m}\, \tilde{\boldsymbol{u}}_i - \tilde{\boldsymbol{F}}_j^T \boldsymbol{A}_{12} \tilde{\boldsymbol{F}}_i + \tilde{\boldsymbol{u}}_j^T \boldsymbol{A}_{22} \tilde{\boldsymbol{F}}_i - \tilde{\boldsymbol{F}}_j^T \boldsymbol{A}_{11} \tilde{\boldsymbol{u}}_i) \mathrm{d}x \tag{14}$$

By integration by parts, we arrive at

$$(\tilde{\boldsymbol{u}}_j^T \tilde{\boldsymbol{F}}_i) \Big|_0^l - \int_0^l \left(\frac{\mathrm{d}\tilde{\boldsymbol{u}}_j^T}{\mathrm{d}x} \tilde{\boldsymbol{F}}_i + \tilde{\boldsymbol{F}}_j^T \frac{\mathrm{d}\tilde{\boldsymbol{u}}_i}{\mathrm{d}x} \right) \mathrm{d}x = \int_0^l (-\omega_i^2 \tilde{\boldsymbol{u}}_j^T \boldsymbol{m}\, \tilde{\boldsymbol{u}}_i - \tilde{\boldsymbol{F}}_j^T \boldsymbol{A}_{12} \tilde{\boldsymbol{F}}_i + \tilde{\boldsymbol{u}}_j^T \boldsymbol{A}_{22} \tilde{\boldsymbol{F}}_i - \tilde{\boldsymbol{F}}_j^T \boldsymbol{A}_{11} \tilde{\boldsymbol{u}}_i) \mathrm{d}x \tag{15}$$

In terms of Eq. (3), the integral on the left-hand side can be written as

$$\int_0^l \left(\frac{\mathrm{d}\tilde{\boldsymbol{u}}_j^T}{\mathrm{d}x} \tilde{\boldsymbol{F}}_i + \tilde{\boldsymbol{F}}_j^T \frac{\mathrm{d}\tilde{\boldsymbol{u}}_i}{\mathrm{d}x} \right) \mathrm{d}x = \int_0^l (2 \tilde{\boldsymbol{F}}_j^T \boldsymbol{A}_{12} \tilde{\boldsymbol{F}}_i + \tilde{\boldsymbol{F}}_j^T \boldsymbol{A}_{11} \tilde{\boldsymbol{u}}_i + \tilde{\boldsymbol{u}}_j^T \boldsymbol{A}_{11}^T \tilde{\boldsymbol{F}}_i) \mathrm{d}x \tag{16}$$

Thus, on noticing $\boldsymbol{A}_{11}^T = -\boldsymbol{A}_{22}$, Eq. (14) becomes

$$\int_0^l (-\omega_i^2 \tilde{\boldsymbol{u}}_j^T \boldsymbol{m}\, \tilde{\boldsymbol{u}}_i + \tilde{\boldsymbol{F}}_j^T \boldsymbol{A}_{12} \tilde{\boldsymbol{F}}_i) \mathrm{d}x = (\tilde{\boldsymbol{u}}_j^T \tilde{\boldsymbol{F}}_i) \Big|_0^l \tag{17}$$

The two relations given in Eqns. (13) and (17) play a central role in the demonstration of orthogonality of vibration modes of a farmed structure, which will be shown in the following section.

ORTHOGONALITY OF VIBRATION MODES OF 3D FRAMED STRUCTURES

Now we turn to establish the orthogonality relations of vibration modes of a 3D framed structure. It is assumed that the frame has a total of m members, which are connected at their ends through a total of n joints. Each joint is labeled by either a numeral or a capital letter, and each member is labeled by the two numerals or capital letters of the joints at two ends. By denoting the number of members connected to joint J as m^J, we have $\sum_{J=1}^{n} m^J = 2m$ due to the fact that every member has two ends.

For a three-dimensional framed structure, in addition to the governing equations given in Eq. (8) which should be satisfied by each member, we further need to consider the joint coupling equations at each joint, which include the equilibrium conditions of forces and the compatibility considerations of displacements[8, 9]. For the sake of a global analysis, a global coordinate system $\boldsymbol{X} = (X, Y, Z)$ is used, along with a local coordinate system $\boldsymbol{x} = (x, y, z)$ for each member. Denote the transformation matrix between the local coordinate system $\boldsymbol{x}$ and the global one $\boldsymbol{X}$ as $\boldsymbol{T}_0(\boldsymbol{x}, \boldsymbol{X})$, with the elements being directional cosines between respective axes, i. e. $\boldsymbol{T}_{0\,ij} = \cos(x_i, X_j)$. Then we have $\boldsymbol{t}(\boldsymbol{X}) = \boldsymbol{T}_0(\boldsymbol{x}, \boldsymbol{X})\boldsymbol{t}(\boldsymbol{x})$, where $\boldsymbol{t}$ is any physical vector with three components. Thus the transformation matrix for the generalized displacement vector is the same as that for the generalized force vector, i.e.

$$T_u(x, X) = T_F(x, X) = \begin{bmatrix} T_0(x, X) & 0 \\ 0 & T_0(x, X) \end{bmatrix} \equiv T(x, X) \tag{18}$$

Since T_0 is orthogonal, i. e. $T_0 T_0^T = 1$, T is also orthogonal, and we have $TT^T = 1$.

For simplicity, we assume that any joint in the frame is rigid, and its connection to members is perfect. The joint however, may be supported by linear and/or rotational elastic springs and may be with a concentrated mass. Since a local coordinate system is used for each member, some members may connect to a joint at the 0-end and some at the l-end. We thus assume that to the joint J, there are m^{J_1} members connected at the l-end and m^{J_2} members at the 0-end, and hence $m^{J_1} + m^{J_2} = m^J$. Suppose that all positive directions of the generalized displacements and forces of the joint are coincident with the positive directions of axes of the global coordinates (X, Y, Z). Then, we can see that the forces exerted on joint J by member IJ, which is connected to joint J at its l-end, are, if expressed in the global coordinates

$$\widetilde{F}_J^{IJ} = -T(x^{IJ}, X)\, \widetilde{F}^{IJ}(l) \tag{19}$$

On the other hand, the forces exerted on joint J by member JK are

$$\widetilde{F}_J^{JK} = T(x^{JK}, X)\, \widetilde{F}^{JK}(0) \tag{20}$$

In the above, the subscript J indicates the joint under consideration. When it is identical to the first of the double superscripts, it implies that the member is connected to the joint at the 0-end; otherwise the member is connected to the joint at the l-end. The equations of equilibrium of forces of the joint in the state of free vibration can thus be written as

$$\sum_{\eta_1=1}^{m^J 1} \widetilde{F}_J^{\eta_1 J} + \sum_{\eta_2=1}^{m^J 2} \widetilde{F}_J^{J\eta_2} + \widetilde{f}^J = (K^J - M^J \omega^2)\, \widetilde{U}^J \tag{21}$$

where $\widetilde{U} = [U_x, U_y, U_z, \Phi_x, \Phi_y, \Phi_z]^T$ is the generalized displacement vector of the joint, whose components are evaluated in the global coordinate system, and K and M are the stiffness and inertial mass matrices of the joint, both being symmetric and semi-positive definite. $\widetilde{f}^J$ is the force vector of the joint — it is zero when no constraint is imposed on the joint. Generally, when one component of $\widetilde{U}^J$ is fixed, then the corresponding component of $\widetilde{f}^J$ is unknown. Anyway, we have always the relation $(\widetilde{U}^J)^T \widetilde{f}^J = 0$, which is to be employed in the following derivation.

The substitution of Eqns. (13) and (20) into Eqn. (21) gives

$$-\sum_{\eta_1=1}^{m_1^J} T(x\eta_1^J, X)\, \widetilde{F}\eta^J(l) + \sum_{\eta_2=1}^{m_2^J} T(x^{J\eta_2}, X)\, \widetilde{F}^{J\eta_2}(0) + \widetilde{f}^J = (K^J - M^J\omega^2)\, \widetilde{U}^J \tag{22}$$

The compatibility conditions of displacements at the joint are

$$\widetilde{u}_J^{\eta_1 J} = \widetilde{u}_J^{J\eta_2} = \widetilde{U}^J \quad (\eta_1 = 1, 2, \cdots, m^{J_1};\ \eta_2 = 1, 2, \cdots, m^{J_2}) \tag{23}$$

Since the transformations of displacements between local coordinates and global coordinates are

$$\widetilde{u}^J_{IJ} = T(x^{IJ}, X)\, \widetilde{u}^{IJ}(l),\ \widetilde{u}^{JK}{}_J = T(x^{JK}, X)\, \widetilde{u}^{JK}(0) \tag{24}$$

we can rewrite Eqn. (23) in the form

$$T(x^{\eta_1 J}, X)\, \widetilde{u}^{\eta_1 J}(l) = T(x^{J\eta_2}, X)\, \widetilde{u}^{J\eta_2}(0) = \widetilde{U}^J \quad (\eta_1 = 1, 2, \cdots, m^{J_1};\ \eta_2 = 1, 2, \cdots, m^{J_2}) \tag{25}$$

The equilibrium equations of forces, Eqn. (21), and the compatibility conditions, Eqn. (25), are referred as the joint coupling equations. The joint coupling equations are valid for any frequency of free vibration, i.e.

$$-\sum_{\eta_1=1}^{m^J 1} T(x^{\eta_1 J}, X)\widetilde{F}_k^{\eta J}(l)+\sum_{\eta_2=1}^{m^J 2} T(x^{J\eta_2}, X)\widetilde{F}_k^{J\eta_2}(0)+\widetilde{f}_k^J=(K^J-M^J\omega_k^2)\widetilde{U}_k^J \tag{26}$$

$$T(x^{\eta_1 J}, X)\widetilde{u}_k^{\eta_1 J}(l)=T(x^{J\eta_2}, X)\widetilde{u}_k^{J\eta_2}(0)=\widetilde{U}_k^J \quad (\eta_1=1, 2, \cdots, m^{J_1};\ \eta_2=1, 2, \cdots, m^{J_2}) \tag{27}$$

for $k=1, 2, 3, \cdots$. Since T is orthogonal, we have from Eq. (27)

$$\widetilde{u}_k^{\eta_1 J}(l)=T^T(x^{\eta_1 J}, X)\widetilde{U}_k^J,\quad \widetilde{u}_k^{J\eta_2}(0)=T^T(x^{J\eta_2}, X)\widetilde{U}_k^J \tag{28}$$

Equation (19) is valid for each member in the frame. Summation over all m members gives rise to

$$(\omega_i^2-\omega_j^2)\sum_{p=1}^{m}\int_0^{l_p}\widetilde{u}_{pj}^T m_p \widetilde{u}_{pi}\,dx=\sum_{p=1}^{m}(\widetilde{u}_{pi}^T\widetilde{F}_{pj}-\widetilde{u}_{pj}^T\widetilde{F}_{pi})\Big|_0^{lp} \tag{29}$$

where a single subscript p is alternatively used to denote the sequential number of the member in the total m members. Thus, the pth member may have two ends, say I and J. In this case, we have $l_p=l^{IJ}$, $\widetilde{u}_{pi}=\widetilde{u}_i^{IJ}$, etc. The right hand side of Eq. (29) can be written as

$$\begin{aligned}&\sum_{p=1}^{m}(\widetilde{u}_{pi}^T\widetilde{F}_{pj}-\widetilde{u}_{pj}^T\widetilde{F}_{pi})\Big|_0^{l_p}\\&=\sum_{J=1}^{n}\Big(\sum_{\eta_1}^{m^J 1}\{[\widetilde{u}_i^{\eta_1 J}(l)]^T\widetilde{F}_j^{\eta_1 J}(l)-[\widetilde{u}_j^{\eta_1 J}(l)]^T\widetilde{F}_i^{\eta_1 J}(l)\}-\\&\sum_{\eta_2}^{m^J 2}\{[\widetilde{u}_i^{J\eta_2}(0)]^T\widetilde{F}_j^{J\eta_2}(0)-[\widetilde{u}_j^{J\eta_2}(0)]^T\widetilde{F}_i^{J\eta_2}(0)\}\Big)\end{aligned} \tag{30}$$

By multiplying both sides of Eq. (26) for $k=j$ with $(\widetilde{U}_i^J)^T$, we obtain

$$-\sum_{\eta_1=1}^{m^J 1}[T^T(x^{\eta_1 J}, X)\widetilde{U}_i^J]^T\widetilde{F}_j^{\eta J}(l)+\sum_{\eta_2=1}^{m^J 2}[T^T(x^{J\eta_2}, X)\widetilde{U}_i^J]^T\widetilde{F}_j^{J\eta_2}(0)=(\widetilde{U}_i^J)^T(K^J-M^J\omega_j^2)\widetilde{U}_j^J \tag{31}$$

where we have made use of the relation $(\widetilde{U}_i^J)^T\widetilde{f}_j^J=0$. In view of Eq. (28), we get

$$-\sum_{\eta_1=1}^{m^J 1}[\widetilde{u}_i^{\eta_1 J}(l)]^T\widetilde{F}_j^{\eta J}(l)+\sum_{\eta_2=1}^{m^J 2}[\widetilde{u}_i^{J\eta_2}(0)]^T\widetilde{F}_j^{J\eta_2}(0)=(\widetilde{U}_i^J)^T(K^J-M^J\omega_j^2)\widetilde{U}_j^J \tag{32}$$

Similarly, we can derive

$$-\sum_{\eta_1=1}^{m^J 1}[\widetilde{u}_j^{\eta_1 J}(l)]^T\widetilde{F}_i^{\eta J}(l)+\sum_{\eta_2=1}^{m^J 2}[\widetilde{u}_j^{J\eta_2}(0)]^T\widetilde{F}_i^{J\eta_2}(0)=(\widetilde{U}_j^J)^T(K^J-M^J\omega_i^2)\widetilde{U}_i^J \tag{33}$$

Subtracting Eq. (32) with Eq. (33), and noticing the symmetry of stiffness and mass matrices, we get

$$\begin{aligned}&-\sum_{\eta_1=1}^{m^J 1}\{[\widetilde{u}_i^{\eta_1 J}(l)]^T\widetilde{F}_j^{\eta J}(l)-[\widetilde{u}_j^{\eta_1 J}(l)]^T\widetilde{F}_i^{\eta J}(l)\}\\&+\sum_{\eta_2=1}^{m^J 2}\{[\widetilde{u}_i^{J\eta_2}(0)]^T\widetilde{F}_j^{J\eta_2}(0)-[\widetilde{u}_j^{J\eta_2}(0)]^T\widetilde{F}_i^{J\eta_2}(0)\}=(\omega_i^2-\omega_j^2)^T M^J\widetilde{U}_i^J\end{aligned} \tag{34}$$

This can be used in Eq. (30), resulting in

$$\sum_{p=1}^{m}(\widetilde{u}_{pi}^T\widetilde{F}_{pj}-\widetilde{u}_{pj}^T\widetilde{F}_{pi})\Big|_0^{lp}=-(\omega_i^2-\omega_j^2)\sum_{J=1}^{n}(\widetilde{U}_j^J)^T M^J\widetilde{U}_i^J \tag{35}$$

Hence, Eq. (29) becomes

$$(\omega_i^2-\omega_j^2)\left[\sum_{p=1}^{m}\int_0^{l_p}\tilde{\boldsymbol{u}}_{pj}^{\mathrm{T}}\, m_p\, \tilde{\boldsymbol{u}}_{pi}\,\mathrm{d}x+\sum_{J=1}^{n}(\tilde{\boldsymbol{U}}_j^J)^{\mathrm{T}}\boldsymbol{M}^J\,\tilde{\boldsymbol{U}}_i^J\right]=0 \tag{36}$$

For $\omega_i\neq\omega_j$, we obtain

$$\sum_{p=1}^{m}\int_0^{l_p}\tilde{\boldsymbol{u}}_{pj}^{\mathrm{T}}\, m_p\, \tilde{\boldsymbol{u}}_{pi}\,\mathrm{d}x+\sum_{J=1}^{n}(\tilde{\boldsymbol{U}}_j^J)^{\mathrm{T}}\boldsymbol{M}^J\,\tilde{\boldsymbol{U}}_i^J=0\quad(\omega_i\neq\omega_j) \tag{37}$$

This is the orthogonality relation with respect to the mass property for the three-dimensional frame with perfectly rigid joints, which can be elastically supported and with concentrated masses.
The summation of Eq. (17) over all m members gives

$$\sum_{p=1}^{m}\int_0^{l_p}(-\omega_i^2\,\tilde{\boldsymbol{u}}_{pj}^{\mathrm{T}}\,\boldsymbol{m}_p\,\tilde{\boldsymbol{u}}_{pi}+\tilde{\boldsymbol{F}}_{pj}^{\mathrm{T}}\,\boldsymbol{A}_{p12}\,\tilde{\boldsymbol{F}}_{pi})\mathrm{d}x=\sum_{p=1}^{m}(\tilde{\boldsymbol{u}}_j^{T}\,\tilde{\boldsymbol{F}}_i)\Big|_0^{l}=-\sum_{J=1}^{n}(\tilde{\boldsymbol{U}}_j^J)^{\mathrm{T}}(\boldsymbol{K}^J-\boldsymbol{M}^J\omega_i^2)\,\tilde{\boldsymbol{U}}_i^J \tag{38}$$

where Eq. (33) has been used. It is immediately seen that

$$\sum_{p=1}^{m}\int_0^{l_p}\tilde{\boldsymbol{F}}_{pj}^{\mathrm{T}}\,\boldsymbol{A}_{p12}\,\tilde{\boldsymbol{F}}_{pi}\,\mathrm{d}x+\sum_{J=1}^{n}(\tilde{\boldsymbol{U}}_j^J)^{\mathrm{T}}\boldsymbol{K}^J\,\tilde{\boldsymbol{U}}_i^J=\omega_i^2\sum_{p=1}^{m}\int_0^{l_p}\tilde{\boldsymbol{u}}_{pj}^{\mathrm{T}}\,\boldsymbol{m}_p\,\tilde{\boldsymbol{u}}_{pi}\,\mathrm{d}x+\omega_i^2\sum_{J=1}^{n}(\tilde{\boldsymbol{U}}_j^J)^{\mathrm{T}}\boldsymbol{M}^J\,\tilde{\boldsymbol{U}}_i^J \tag{39}$$

In view of Eq. (37), we have

$$\sum_{p=1}^{m}\int_0^{l_p}\tilde{\boldsymbol{F}}_{pj}^{\mathrm{T}}\,\boldsymbol{A}_{p12}\,\tilde{\boldsymbol{F}}_{pi}\,\mathrm{d}x+\sum_{J=1}^{n}(\tilde{\boldsymbol{U}}_j^J)^{\mathrm{T}}\boldsymbol{K}^J\,\tilde{\boldsymbol{U}}_i^J=0\quad(\omega_i\neq\omega_j) \tag{40}$$

which is the orthogonality relation with respect to the stiffness property.
The method outlined above applies equally to three-dimensional frames with pinned-joints or even semi-rigidly connected joints[10]. Owing to the limitation of space for the paper, these are omitted and will be presented in another work.

CONCLUSIONS

We have obtained the orthogonality relations for free vibration of complicated framed structures with different types of joints, which can either be elastically supported or have concentrated masses. A novel derivation based on the system of first-order differential equations is presented. It is found that the symplectic structure of the system matrix plays a similar role to the self-adjoint property of the differential operator in the conventional analysis based on high-order differential equations. This novel method allows us to investigate the orthogonality of eigenfunctions of more complicated problems in a unified way. Notice that, the frame under consideration is complicated, not only because there are many members, either uniform or non-uniform, which are connected by joints, but also because of the diversity of joint types. The two orthogonality relations of vibration modes can be employed along with the mode expansion to study the transient responses of three-dimensional frames subjected to general external time-varying stimuli, including distributed/concentrated loads applied on members, concentrated loads applied at joints and prescribed motions of joints.

ACKNOWLEDGEMENTS

The work was supported by the National Natural Science Foundation of China (Nos. 11090333, 10725210), the National Basic Research Program of China (No. 2009CB623200), and the Fundamental Research Funds for Central Universities (No. 2011XZZX002).

REFERENCES

[1] Rayleigh, L., *The Theory of Sound*. New York: Dover, 1945.
[2] Clough, R. W., and Penzien, J., *Dynamics of Structures*, 3rd ed. Berkeley: Computers & Structures, 2003.
[3] Timoshenko, S., Young, D. H., and Weaver Jr, W., *Vibration Problems in Engineering*, 4th ed. New York: Wiley, 1974.
[4] Meirovitch, L., *Analytical Methods in Vibrations*. London: Macmillan, 1967.
[5] Love, A. E. H., *A Treatise on the Mathematical Theory of Elasticity*, 4th ed. London: Cambridge University Press, 1927.
[6] Courant, R., and Hilbert, D., *Methods of Mathematical Physics*, Vol. 1, London: Interscience Publishers, 1953.
[7] Chan, K. L., and Williams, F. W., "Orthogonality of modes of structures when using the exact transcendental stiffness matrix method", *Shock and Vibration*, 7, 2000, 23-28.
[8] Guo, Y. Q., Chen, W. Q., and Pao, Y. H., "Dynamic analysis of space frames: The method of reverberation-ray matrix and the orthogonality of normal modes", *Journal of Sound and Vibration*, 317, 2008, 716-738.
[9] Pao, Y. H., and Chen, W. Q., "Elastodynamic theory of framed structures and reverberation-ray matrix analysis", *Acta Mechanica*, 204, 2009, 61-79.
[10] Chen, W. F., Goto, Y., and Liew, J. Y. R., *Stability Design of Semi-Rigid Frames*. New York: Wiley, 1996.

A BROAD FREQUENCY VIBRATION ANALYSIS OF BUILT-UP STRUCTURES WITH MODAL UNCERTAINTIES

* H.A. Xu[1], W.L. Li[2]

[1] Department of Virtual Product Development, Volvo Construction Equipment, Shippensburg, PA 17257, USA
[2] Department of Mechanical Engineering, Wayne State University, Detroit, MI 48202, USA
* Email: dw0588@wayne.edu

KEYWORDS

Fourier Spectrum Element Method, built-up structures, vibration, uncertain modal parameters, Monte Carlo Simulation.

ABSTRACT

A Fourier Spectrum Element Method (FSEM) is proposed for the vibration analysis of built-up structures. The basic idea of FSEM is to treat a complex structure as an assembly of a number of fundamental structural components such as beams, plates, and shells. The primary variables, usually the displacements, over each component are sought as a modified Fourier series expansion which is guaranteed to be uniformly convergent at any desired rate for any boundary and coupling conditions. The Fourier coefficients are considered as the generalized coordinates and determined using the Rayleigh-Ritz method. Mathematically, this Fourier series method does not involve any assumption or an introduction of any artificial model parameters, and it is broadly applicable to the whole frequency range which is usually divided into low, mid, and high frequency regions. The mesh-less and grid-free representation of the subsystems makes FSEM particularly attractive and useful for statistical analyses and parametric studies. As an example, this method is used to study the vibration characteristics of a coupled beam-plate structure with uncertain modal parameters. It is shown that the spatial-and frequency-averaging processes may not be desired for the mid-frequency analysis because the important dynamic characteristic of a system tends to be completely wiped out by them.

INTRODUCTION

A dynamic system typically exhibits distinctively different response characteristics in different frequency ranges. At low frequencies, finite element analysis (FEA) is the most commonly used tool for the dynamic analysis of complex structures. It is widely believed that the low frequency limit of FEA methods is only limited by computing resources. However, there are actually some other intrinsic reasons that make FEA methods ineffective, even useless, at high frequencies, regardless of computer resource considerations[1]. While the idea of dividing a structure into a large number of elements is theoretically necessary to represent its geometry, it is clearly not suitable for high frequency applications because the presence of inevitable manufacturing errors or variations tends to make deterministic FEA results erratic and unreliable.

At high frequencies, a response spectrum tends to become smooth without strong modal showings; thus, deterministic methods are usually no longer useful. Instead, the Statistical Energy Analysis (SEA) method has emerged as a major technique for the analyses of complex dynamic systems at high frequencies[2]. In an SEA model, the original dynamic system is divided into a number of subsystems (or mode groups) on which the solution variables are usually the frequency-and space-averaged energy levels. The coupling between any two subsystems is described by the so-called Coupling Loss Factor (CLF) that basically regulates the energy flows through the junction. Theoretically, the CLFs are typically calculated based on wave formulations for the interaction of two semi-infinite subsystems. Consequently, the modal details of the individual sub-systems are simply ignored, which often results in a smoothened estimate of the model variables such as the wave transmission coefficients or coupling loss factors. The flaw of this process may be seen by the fact that the CLFs are usually assumed to be independent of the modal overlap, which is shown not to be the case for small modal overlap factors[3, 4]. The issues and concerns related to the irregularities and varieties of the SEA solutions have also been investigated by other researchers[5-7].

Between the low-frequency and the high-frequency ranges, there is a wide mid-frequency range that is critical to the engineering design of dynamic systems because the dominant excitation bands often fall in this region in real-world dynamic problems. However, the mid-frequency range is not clearly defined in practice. In fact, the medium frequency region for a "uniform" structure implies where the structure shows a highly volatile and uncertain behaviour in response to an excitation or minor changes. Medium-frequency problems are also referred to as a dynamic system that exhibits mixed coherent global and incoherent local motions[8, 9] or consists of subsystems with distinctively different modal densities[10]. In some sense, it is accepted that the mid-frequency range is where the conventional deterministic methods such as FEA are not appropriate, but the SEA assumptions are not yet fulfilled.

In an endeavor to address the critical mid-frequency problems, a number of analysis methods have been developed for predicting the vibrations and energy flows between various structural components such as beam and plates. Among them are the dynamic stiffness methods[11-13], spectral element method[14], and receptance methods[15-18] to name a few. However, there are some issues and concerns associated with each of them, which, for instance, include: a) the standard form of mode shape equations tend to break down after the first dozen modes or so due to the numerical instability of the hyperbolic functions for large wavenumbers; and b) the slow convergence of the Green's functions expressed in the form of modal expansion. In addition, most existing methods are based on the modal data for each structural component. An immediate implication is that a modification or adaptation of the formulations or solution procedures is required to account for the variety of the boundary conditions. However, a more serious concern is that the modal properties for each component have to be calculated from the presumed boundary conditions which can be significantly different from the actual coupling conditions in a system environment and hence make the modal data useless.

Because of the aforementioned problems and concerns, it is of great interest to both researchers and application engineers to develop a sophisticated and robust method which is capable of reducing modelling efforts, simplifying model input data, and improving the accuracy and reliability of the prediction. A Fourier series method was previously proposed for the dynamic analysis of built-up structures consisting of beams[19, 20], plates[21, 22], and beam and plates[23, 24]. In this paper, this method is applied to a coupled beam-plate structure with consideration of the uncertainties with model parameters in the mid-frequency range.

MATHEMATICAL FORMULATION

To illustrate the current modelling strategy, Figure 1 shows how a coupled beam-plate structure is naturally decomposed into a number of basic structural components in the forms of beams and plates. In the current method, the primary solution variables (or displacements) over each beam are invariably expressed as an improved Fourier series expansion in the form of

$$f(x) = \sum_{m}^{\infty} a_m \cos\frac{m\pi}{L}x + \sum_{m=1}^{M_0} b_m \sin\frac{n\pi}{L}x, \qquad 0 \leqslant x \leqslant L \tag{1}$$

where M_0 is an even integer. The sine terms in the Eq. (1) are used to ensure the conditioned solution function (represented by the cosine series), when it is periodically extended, to have $C^{M0/2+1}$ continuity. The expansion coefficients b_m are only dependent on the values of the odd derivatives of $f(x)$ at the ends of the interval, and calculated from

$$b_m = \sum_{k=1}^{Q}\left[f^{(2k-1)}(\pi) + (-1)^m f^{(2k-1)}(0)\right] \frac{\sum\limits_{\substack{1\leqslant j_1<\cdots<j_{Q-k}\leqslant Q \\ j_1,\cdots,j_{Q-k}\neq i}} (-1)^{k+1} x_{j_1}^2 \cdots x_{j_{Q-k}}^2}{2x_i^{\alpha_m} \prod\limits_{\substack{j=1 \\ j\neq 1}}^{Q} x_j^2 - x_i^2} \tag{2}$$

where $\alpha_m = 1$ and $x_i = 2i-1$ for $m = -M_0+1, -M_0+3, \cdots, -1$; $\alpha_m = 2$ and $x_i = 2i$ for $m = -M_0, -M_0+2, \cdots, -2$. Mathematically, it has been proved that the Fourier series given in Eq. (1) has a polynomial convergence rate

$$a_m = 0(M^{-(M_0+1)}) \qquad \text{for} \quad M_0 \leqslant N \tag{3}$$

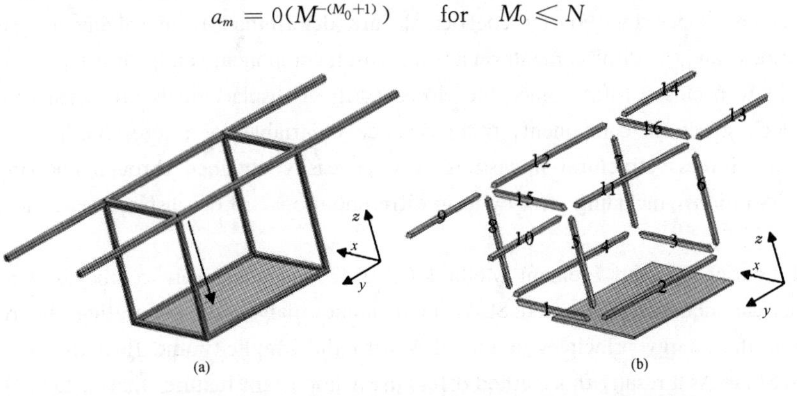

Figure 1 A built-up structure is naturally divided into a number of structural components

The beam governing equation demands that the fourth derivative exists over the entire length of the beam. Accordingly, we need to set $M_0 = 4$ in seeking for a strong X^3 solution, or $M_0 = 2$ for X^1 solution in a weak formulation. If M_0 is set equal to 4, then any secondary variables (e.g., bending moments and shear forces) of interest can be obtained analytically from the term-by-term differentiation of the series solution.

The new Fourier series given in Eq. (1) has laid a solid foundation for developing a spectral method to obtain a highly accurate and uniformly convergent solution throughout the entire domain including boundaries. For the two-dimensional version of Eq. (1), by considering a plate of length a and width b as an example, the displacement function in the x and y-directions can be expressed as

$$w(x, y) = \bar{w}(x, y) - \sum_{i=1}^{4}[\xi_b^i(y)\alpha_i(x) + \xi_a^i(x)\beta_i(y)] \tag{4}$$

where $\bar{w}(x, y) = \sum_{m}^{\infty}\sum_{n=0}^{\infty} A_{mn}\cos\lambda_m x\cos\lambda_n y$ with $\lambda_m = m\pi/a$, $\lambda_n = n\pi/b$,
$\alpha_0(x) = w'(x, y)|_{y=0}$, $\alpha_1(x) = w'(x, y)|_{y=b}$, $\alpha_3(x) = w'''(x, y)|_{y=0}$, $\alpha_4(x) = w'''(x, y)|_{y=b}$,
$\beta_0(y) = w'(x, y)|_{x=0}$, $\beta_1(y) = w'(x, y)|_{x=a}$, $\beta_3(y) = w'''(x, y)|_{x=0}$, $\beta_4(y) = w'''(x, y)|_{x=a}$ and $\xi_a^i(x)$ (or $\xi_b^i(y)$) represent a set of closed-form sufficiently smooth functions defined over $[0, a]$ (or $[0, b]$). The term "sufficiently smooth" implies that third-order derivatives of these functions exist and are continuous at any point on the plate.

Once the displacements are constructed over each structural component, the final system equations can be derived from the Rayleigh-Ritz procedure[23]. As a result, only the neighbouring components are directly coupled together in the final system, as manifested in the sparseness of the stiffness and mass matrices. It should be noted that since the displacement functions are constructed sufficiently smooth in the current model, the weak and strong solutions are essentially the same mathematically. In seeking weak solutions, the Lagrangian L can be expressed as

$$L = V - T = \sum_{i=1}^{N}V_i - \sum_{i=1}^{N}T_i \tag{5}$$

where V_i and T_i are the potential and kinetic energy for the ith structural component, respectively. Readers are referred to Refs. [23] and [24] for details about them.

The final system of equations can be derived by minimizing the Lagrangian against all the unknown Fourier coefficients. Once the Fourier coefficients are determined from solving a system of linear algebraic equations, and the displacements on any structural component can be directly obtained from Eq. (1) or Eq. (4) in a closed form. Since the closed form of displacements are constructed adequately smooth over each structural component, other secondary variables of interest (e. g., stresses, strain energies, internal forces, structural intensities) can be easily obtained through applying appropriate mathematical operations, including term-by-term differentiations, to the displacement functions.

The proposed Fourier Spectral Element Method (FSEM) has effectively combined two of the most powerful numerical methods, FEA and SEA, in dynamic analysis by integrating the robustness and sophistication of the energy principles in the FEA with the simplicity and flexibility of the modeling strategy in the SEA. As a result, this method offers many important features that include: 1) the number of DOFs is substantially smaller than that in an FEA model for the same spatial resolution; 2) the model parameters are explicitly included in the final system which facilitates optimization or sensitivity study against any model variable of interest; 3) the FSEA is better suited for statistical and uncertainty analysis; d) there interactions between any two subsystems are faithfully accounted for in the forms of the displacement compatibility or force equilibrium conditions, rather than some artificial model parameters such as the coupling loss factors in a SEA model. Some of these claims will become evident from the numerical examples given below.

RESULTS AND DISCUSSION

To demonstrate some of the unique advantages of the proposed method, we consider a built-up structure as shown in Figure 1(b). The structure can be divided to 8 beams and 1 plate which are rigidly connected at each joint. The material properties for the plate and beams are given as: $E = 200$ GPa, $\rho = 7\,872$ kg/m^3, and $\mu = 0.29$. Each of the beam has a rectangular cross section $S = 0.02 \times 0.02$ m^2 and the lengths of these beams are specified as: beams 1, 3, 15 & 16 = 0.4 m; beams 2, 4, 11, & 12 = 0.6 m; and beams 5, 6, 7, 8, 9, 10, 13, & 14 = 0.5 m. The plate has a thickness $h = 0.002$ m, length $a = 0.6$m, and width $b = 0.4$ m.

For any given set of model parameters, the dynamic characteristic of this structure can be determined theoretically using any well-established analytical or numerical methods such as FEA. An FEA model typically works well in a low frequency, but it typically becomes unreliable as frequency increases. To better understand this issue, suppose that we are interested in the responses of the structure in a broad frequency range up to 1 500 Hz. The above specified model parameters should be considered as the nominal (perfect) values. Strictly speaking, however, some degree of uncertainty is unavoidably associated with any model parameter due to engineering and manufacturing errors and other factors. It is generally known that such uncertainties tend to have more remarked effects on the responses at higher frequencies. Thus, a critical question is if the results obtained based on the normal values are of practical significance, especially at higher frequencies. To account for the model uncertainties, we assume that the properties for each component vary randomly or in any other statistical manner specified. It is further assumed that the modal variables be mutually independent and uncorrelated. For instance, a model variable χ can be specified as

$$\chi = \bar{\chi}(1 + \varepsilon_\chi) \tag{6}$$

where $\bar{\chi}$ is the nominal value, and ε_χ represents a random variable following the Gaussian distribution with a zero mean and a standard deviation of ε_χ.

In this study, the Monte Carlo Simulation (MCS) is employed to obtain the ensemble statistics of the system responses for a collection of samples created according to the probability density function specified for each uncertain variable. For example, if a population of N responses, $\widetilde{\Psi} = \{\Psi_1, \Psi_2, \cdots \Psi_i, \cdots \Psi_N\}$, the mean value and the variance of the responses can be respectively estimated from

$$E(\boldsymbol{\Psi}_i) = \frac{1}{N}\sum_{i=1}^{N}\boldsymbol{\Psi}_i \tag{7}$$

$$\sigma^2(\Psi_i) = \frac{1}{N-1}\left[\sum_{i=1}^{N}\Psi_i^2 - N \times E^2(\Psi_i)\right] \tag{8}$$

The response Ψ_i may represent any quantity of interest such as displacements, shear forces, strain energy, etc.

As a numerical example, suppose that the structure has a completely free boundary condition and is excited by a unit point force applied at point (0.4a, 0.6b) in z direction, as shown in Figure 1(b). A response point is selected on the plate at (0.5a, 0.3b). Plotted in Figure 2 are the calculated velocity responses of 100 samples generated by assuming that the plate thickness is Gaussian with a standard deviation equal to 5% of the normal value. It is seen that in the low frequency range, approximately up to 200 Hz, all the response curves tend to show a similar and coherent behavior, indicating the predicted behavior is not sensitive to the small perturbations to the model variables and the results obtained based on the nominal values are considered representative. However, as frequency increases, the responses become increasingly uncorrelated and unpredictable. As a consequence, any effort in trying to select or specify a better (deterministic) set of model variables is deemed to be practically meaningless. Instead, one should pay attention to the statistical significance of the results such as the ensemble-average of the responses represented by the red line in Figure 2. While the mean response appears rather similar to each individual response in the low frequency range, such correlations are clearly absent at higher frequencies. It is seen that the systems collectively exhibit a resonance-like behavior in a fairly wide frequency range, and the ensemble mean response can be better used to describe the statistical behaviors of the structure even though their design may be specified in a deterministic manner. Like resonant responses to individual modes, the peaks on the mean-response curve potentially identify the troubling frequencies for the given design with taking into consideration of design and manufacturing uncertainties. On the other hand, those peaks may indicate a more serious design problem because they have much wider bandwidths than their modal counterparts. In the statistical energy analysis (SEA) method, the ensemble mean is obtained simply through frequency averaging, instead of processing the response samples in a statistical way.

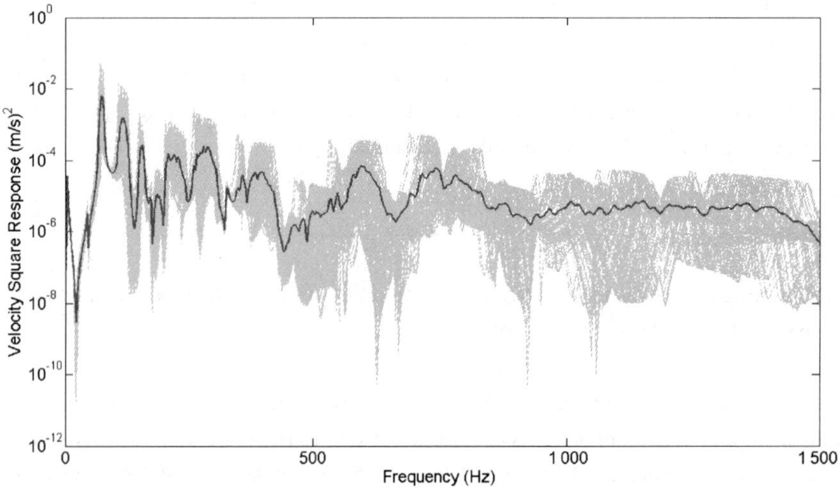

Figure 2 Velocity responses of a built-up structure to a concentrated load, red line: mean; grey line: 100 individual responses

The statistical energy analysis (SEA) also involves another averaging process based on the assumption that the response fields are essentially diffuse over each subsystem. This presumption does not only cause a loss of spatial resolution, but also is generally not true in many cases, especially in the mid-frequency range. To illustrate this point, the ensemble averaged kinetic energy distributions for the structure are plotted in Figure 3 for a two selected frequencies. As anticipated, while the mean kinetic energy (density) tends to become uniformly distributed at high frequencies, it clearly exhibits some strong spatial characteristics over certain structural members in the mid frequency range. Because the primary variables, displacements, are obtained in the closed form and sufficiently smooth over each structural component, other variables of interest such as the structural intensity fields and the power flows through junctions can be readily calculated through appropriate mathematical operations[23-25]. In the foregoing discussions, we loosely divide a broad frequency range into three zones: low, mid, and high. In reality, however, the boundaries between these different frequency regions are usually case-dependent and cannot be clearly defined *a priori*.

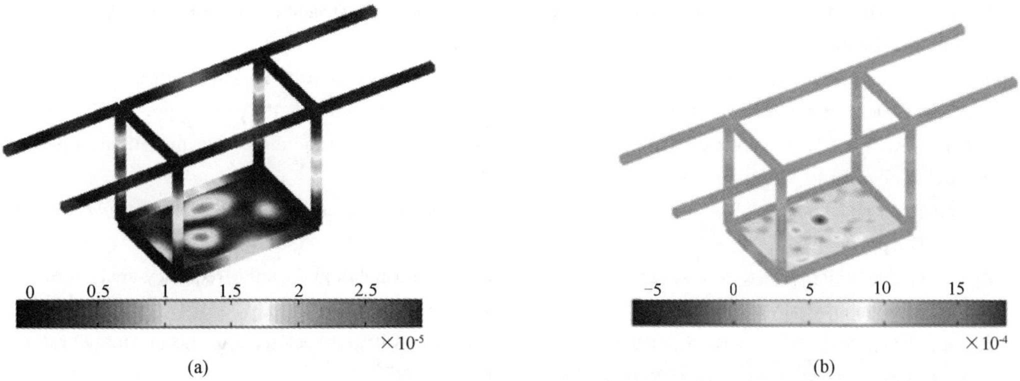

Figure 3 Mean kinetic energy distribution of the structure excited by a unit force applied on the plate in *z* direction at point (0.4a, 0.6b): (a) 328Hz; (b) 868 Hz.

CONCLUSIONS

The Fourier Spectral Element Method (FSEM) is presented for the broad frequency vibration analysis of built-up structures with modal uncertainties. In this method, a built-up structure is divided into a number of basic structural components. Over each structural component, the primary solution variables (typically, the displacements) are sought as a continuous function in the form of an improved Fourier series expansion. Since this method does not involve any assumptions, simplifications or artificial model parameters (such as the coupling loss factors in the SEA method), it is generally applicable to a broad frequency range including the so-called mid frequency region. Further, because the current model is mesh-less, it is particularly suited for statistical analyses and parametric studies. This method is demonstratively applied to study the dynamic characteristic of a coupled beam-plate structure. It is shown that the ensemble-average is an interesting and meaningful quantity to describe the behavior for a group of "identical" dynamic systems. The Monte-Carlo simulations provide a powerful means to extract the statistical characteristics of complex dynamic systems in a broad frequency range. In comparison, the spatial-and frequency averaging procedures can easily deteriorate the resolution of the prediction and cause a loss of information such as the important resonance-like characteristic exhibited by the systems in the mid-frequency range.

REFERENCES

[1] Rabbiolo, G., Bernhard, R. J. and Milner, F. A., "Definition of a high-frequency threshold for plates and acoustical spaces", *Journal of Sound and Vibration*, 2004, 277, pp. 647-667.

[2] Lyon, R. H. and DeJong, R. G., *Theory and Application of Statistical Energy Analysis* (2nd Edition), Butterworth-Heinemann, Boston, 1995.

[3] Mace, B. R. and Rosenberg, J., "The SEA of two coupled plates: an investigation into the effects of system irregularity", *Journal of Sound and Vibration*, 1999, 212, pp. 395-415.

[4] Wester, E. C. N. and Mace, B. R., "Statistical energy analysis of two edge-coupled rectangular plates: ensemble averages", *Journal of Sound and Vibration*, 1996, 193, pp. 793-822.

[5] Craik, R. J. M., Steel, J. A. and Evans, D. I., "Statistical energy analysis of structure-borne sound transmission at low frequencies". *Journal of Sound and Vibration*, 1991, 144, pp. 95-107.

[6] Fahy, F. J., Statistical Energy Analysis: an overview, in: Keane, A. J., Price, W. G. (Eds.), *Statistical Energy Analysis: A Critical Overview, with Applications in Structural Dynamics*, Cambridge University Press, Cambridge, 1997, pp. 1-18.

[7] Fahy, F. J. and Mohamed, A. D., "A study of uncertainty in application of SEA to coupled beam and plate systems, Part I: computational experiments". *Journal of Sound and Vibration*, 1992, 158, pp. 45-67.

[8] Langley, R. S., "Analysis of power flow in beams and frameworks using the direct-dynamic stiffness method", *Journal of Sound and Vibration*, 1990, 136, pp. 439-452.

[9] Shorter P. J. and Langley, R., Vibro-acoustic analysis of complex systems, *Journal of Sound and Vibration*, 2005, 117, pp. 85-95.

[10] Zhao, X. and Vlahopoulos, N., A basic hybrid finite element formulation for mid-frequency analysis of beams connected at an arbitrary angle, *Journal of Sound and Vibration*, 2004, 269, pp. 135-164.

[11] Bercin, A. N. and Langley, R. S., Application of the dynamic stiffness technique to the inplane vibrations of plate structures, *Computers and Structures*, 1996, 59, pp. 869-875.

[12] Langley, R. S., "Analysis of power flow in beams and frameworks using the direct-dynamic stiffness method", *Journal of Sound and Vibration*, 1990, 136, pp. 439-452.

[13] Doyle, J. F., *Wave Propagation in Structures*, Springer, Berlin, 1989.

[14] Ahmida, K. M. and Arruda, J. R. F., "Spectral element based prediction of active power flow in Timoshenko beams", *International Journal of Solids and Structures*, 2001, 38, pp. 1669-1679.

[15] Keane, A. J. and Price, W. G., "A note on the power flowing between two conservatively coupled multi-modal sub-system", *Journal of Sound and Vibration*, 1991, 144, pp. 185-196.

[16] Keane, J., "Energy flows between arbitrary configurations of conservatively coupled multi-modal elastic subsystems", *Proceedings of the Royal Society of London A*, 1992, 436, pp. 537-568.

[17] Beshara, M. and Keane, A. J., "Vibrational power flows in beam networks with compliant and dissipative joints", *Journal of Sound and Vibration*, 1997, 203, pp. 321-339.

[18] Farag, N. H. and Pan, J., "Dynamic response and power flow in three-dimensional coupled beam structures. I. *Analytical modeling*". *Journal of the Acoustical Society of America*, 1997, 102, pp. 315-325.

[19] Li, W. L. and Xu, H. A., "An exact Fourier series method for the vibration analysis of multi-span beam systems", *Journal of Computational and Nonlinear Dynamics*, 2009, 4(2), pp.1-9.

[20] Xu, H. A. and Li, W. L., "Dynamic behavior of multi-span bridges under moving loads with focusing on the effect of the coupling conditions between spans", *Journal of Sound and Vibration*, 2008, 312(4-5), pp. 736-753.

[21] Du, J. T., Li, W. L., Jin, G. Y., Yang, T. J. and Liu, Z. G., "An analytical method for the in-plane vibration analysis of rectangular plates with elastically restrained edges", *Journal of Sound and Vibration*, 2007, 306, pp. 908-927.

[22] Li, W. L., Zhang, X. F., Du, J. T. and Liu, Z. G. "An exact series solution for the transverse vibration of rectangular plates with general elastic boundary supports", *Journal of Sound and Vibration*, 2009, 321, pp. 254-269.

[23] Xu, H. A., Du, J. T. and Li, W. L. "Vibrations of rectangular plates reinforced by any number of beams of arbitrary lengths and placement angles", *Journal of Sound and Vibration*, 2010, 329, pp. 3759-3779.

[24] Xu, H. A. and Li, W. L. "Vibration and power flow analysis of periodically reinforced plates", *Science China*, 2011, 54, pp. 1141-1153.

[25] Li, W. L. and Xu, H. A., "Vibration and power flow analyses of built-up structures in a broad frequency range", *IMAC XXVII*, Orlando, Florida, 2009.

ESTIMATION OF DYNAMIC RESPONSE OF STRUCTURAL ELEMENTS SUBJECT TO BLAST AND IMPACT ACTIONS USING A SIMPLE UNIFIED APPRAOCH

* Y. Yang, R. Lumantarna, N. Lam, L.H. Zhang and P. Mendis

Department of Infrastructure Engineering, Melbourne School of Engineering, University of Melbourne

* Email: y.yang3@student.unimelb.edu.au

KEYWORDS

Blast, impact, modeling, protective structures, performance indicator, damage assessment.

ABSTRACT

Glazing facades, concrete panels and other structural, and non-structural, components are used to protect occupants and contents in buildings in the extreme events of blast and impact. The dynamic response behaviour of structural components to blast and impact actions is often analysed by specialised finite element packages, such as program *LS-DYNA*. This type of simulation is often time-consuming and labour intensive to operate and yet the generated results involving inelastic material behaviour do not always match with observations from physical experimentations (which can also be very expensive to conduct). This paper presents a unified methodology for providing rapid assessment of the response behaviour of structural components subject to both blast and impact actions assuming linear elastic behaviour. The proposed methodology circumvents the need for a full blown finite element analysis. It could be expanded further to model post-damage behaviour thus making significant contributions towards disaster mitigation. Description of calculation methods ranging from simple hand calculation, two-degree-of-freedom spring-mass modelling and finite element modelling techniques will be covered herein.

INTRODUCTION

Extreme loads on structures, such as accidental explosion, intentional explosion attack or projectile impact with vandalism intent, could cause severe structural damage. Recently, there is an increased focus on dynamic behavior assessment of structures and their elements when subject to blast and impact. Topics warranting research include global dynamic behavior of target structure and local indentation around the area of contact (induced by impact action)[1].

An iso-damage curve (also known as pressure-impulse curve) based on deflection criterion, as shown in Figure 1, has been used extensively in damage assessment when structures and their elements are subject to blast loads. In general, when subject to a varying pressure and impulse combination, the response of a structure is governed by the natural period T_n of the structure and load duration T_d. In a blast event, when T_n is much smaller than T_d, the blast load can be reasonably idealized as quasi-static loading which dissipates over a relatively long duration. On the other hand, when T_n is much greater than T_d, the blast load can be treated as impulsive loading which dissipates rapidly over time. The quasi-static and impulsive asymptotes in Figure 1 can be established by using the following energy conservation equations for quasi-static loading and impulsive loading respectively.

$$Fx_{\max} = \frac{1}{2}K_{eff}x_{\max}^2 \tag{1}$$

$$\frac{I^2}{2M_{ef}} = \frac{1}{2}K_{ef}x_{\max}^2 \tag{2}$$

where F is the peak blast load, $x_{\max}$ is the maximum allowable deflection of the structure, K_{eff} is the effective stiffness of the structure, I is the peak reflected impulse and M_{eff} is the effective mass of the structure.

The calculation procedure used to establish pressure-impulse curve is illustrated by Figure 2.

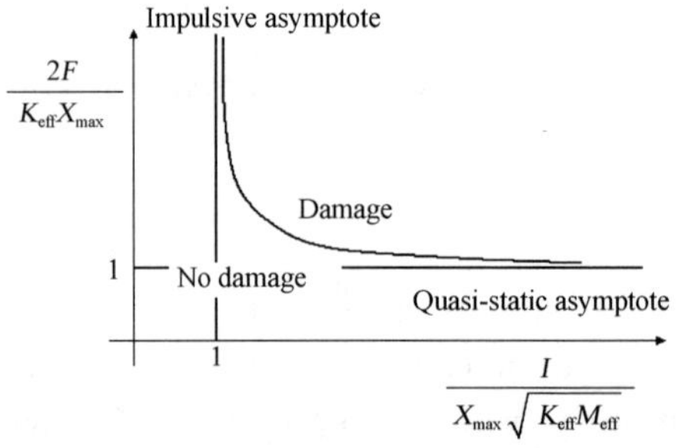

Figure 1 Generic non-dimensional pressure-impulse curve

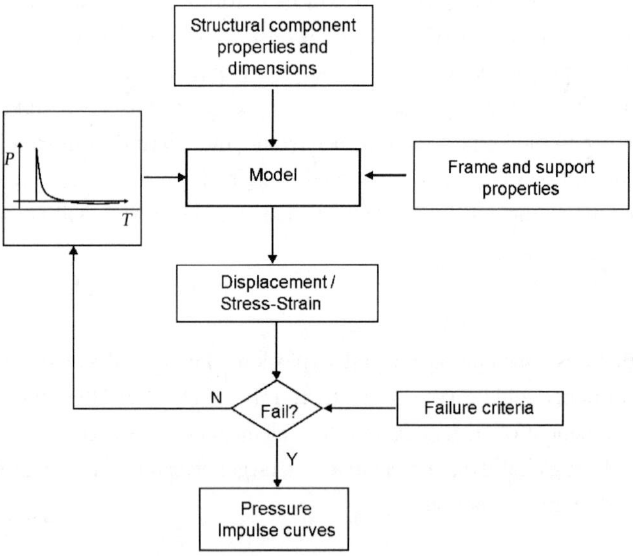

Figure 2 Calculation procedure used in establishing pressure-impulse curve

Several physical experiments have been conducted to study the blast dynamic response of structure[2-4]. However, these experimental approaches are too costly to establish a pressure-impulse curve as a large number of pressure-impulse combination tests is required. Finite element modeling package, such as *LS-*

DYNA is a potentially available alternative to simulate experimental studies and to extrapolate existing experimental data. Numerous finite element models have been developed to establish structure performance indicators[5,6]. One should note that finite element modeling exercise is very time-consuming and labour intensive to undertake. Furthermore, in order to accurately simulate the material behavior, non-linear finite element modeling often requires significant calibration to experimental data. Thus, there is a clear need of developing simple, inexpensive and effective calculation procedures which are easy for end-users to comprehend. Simplified model based on an elastic single-degree-of-freedom (SDOF) system was introduced to estimate the dynamic response of structures subject to blast loadings[7,8]. Although the SDOF model is incapable of capturing the higher mode phenomenon caused by excitation, this approach has been widely used as a first order structure design method for blast action. The aforementioned techniques in blast analysis have also been used by many researchers to study flying object impact induced dynamic behavior of structures and their components[9-11]. However, no parallel body of well-shared knowledge is available amongst engineers for understanding how structure performance is controlled by impact parameters, such as impactor mass and impacting velocity.

This paper presents a unified modelling methodology using SODF idealisation which could be used for estimating impact and blast induced response of elastic material behaviour, and so allows quick assessment of structural components response without having to undertake rigorous finite element analyses. Application of the unified approach for structure damage assessment will be demonstrated by a case study based on a simply supported beam. Other means with varying degree of rigour and complexity will also be introduced.

UNIFIED APPROACH BASED ON SDOF

The proposed unified approach in modeling both blast and impact actions is to idealize a target structure as a SDOF system. It is well established that a SDOF model, which is characterized by the generalized mass (m^*) and generalized stiffness (k^*), can be used to simulate the dynamic behavior of a distributed mass system such as a beam and a plate. The generalized properties m^* and k^* are essentially dependent on the deflected shape caused by external loads. For example, the parameter values of m^* and k^* for beam structures can be obtained from Eqns. (3) and (4), respectively. Table 1 shows the m^* and k^* values with varying boundary conditions and types of load.

$$m^* = \int_0^l m(x)\psi(x)^2 \mathrm{d}x \tag{3}$$

$$k^* = \int_0^l EI\psi''(x)^2 \mathrm{d}x \tag{4}$$

where $m(x)$ is the mass per unit length, l is the length, E is the Young's Modulus, I is the second moment of area and $\Psi(x)$ is the deflection shape function.

TABLE 1 VALUES OF m^* AND k^* WITH VARYING BOUNDARY CONDITIONS AND LOAD TYPES

Boundary Condition	Load type	Deflection shape function	m^*	k^*
Fixed-Fixed	Blast	$\Psi(x)=16[(x/l)^4-2(x/l)^3+(x/l)^2]$	0.4 ml	$205EI/l^3$
	Impact	$\Psi(x)=4[3(x/l)^2-4(x/l)^3]$	0.37 ml	$192EI/l^3$
Pinned — Pinned	Blast	$\Psi(x)=16/4[(x/l)^4-2(x/l)^2+(x/l)]$	0.5 ml	$48.7EI/l^3$
	Impact	$\Psi(x)=3[(x/l)-2(x/l)^3]$	0.49 ml	$48EI/l^3$
Fixed-Free	Blast	$\Psi(x)=1/3[(x/l)^4-4(x/l)^3+6(x/l)^2]$	0.26 ml	$3.2EI/l^3$
	Impact	$\Psi(x)=1/2[3(x/l)^2-(x/l)^3]$	0.24 ml	$3EI/l^3$

The equilibrium equation of the SDOF system model is given by

$$m^* \ddot{x} + k^* x = F(t) \tag{5}$$

where $F(t)$ is the pre-determined pulse loads.

Application Example

The application of the unified approach based on SDOF is demonstrated herein by an example of a simply supported steel beam with length of 4 m. The cross section of the beam is 0.05 m × 0.05 m. The values of ρ, E and I are 7 850 kg/m^3, 200 GPa and 5.2×10^{-7} m^{-7} respectively. The generalized properties m^* and k^* according to Table 1 are 39.5kg and 7.81×10^4 N/m respectively. The yielding of the beam element is adopted as the iso-damage criteria. By assuming a yield strength of 250 MPa, the maximum allowable deflection of the target beam based on static deflection equation is about 0.083 m for a blast load and 0.067 m for an impact load. As an illustrative example, the pulse load function is idealized as triangular pulse for blast action, whereas a half-cycle sine pulse is assumed to be the pulse load function for impact action. Varying force (pressure) — impulse combinations were used to establish performance indicators for both blast and impact actions. The force (pressure) — impulse curves associated with the iso-damage criteria of the beam element are plotted in Figure 3. It should be noted that the abnormal behavior of the performance indicator for impact action at dynamic regime is presumably due to the magnification of response caused by resonance when the T_d is approximately equal to T_n.

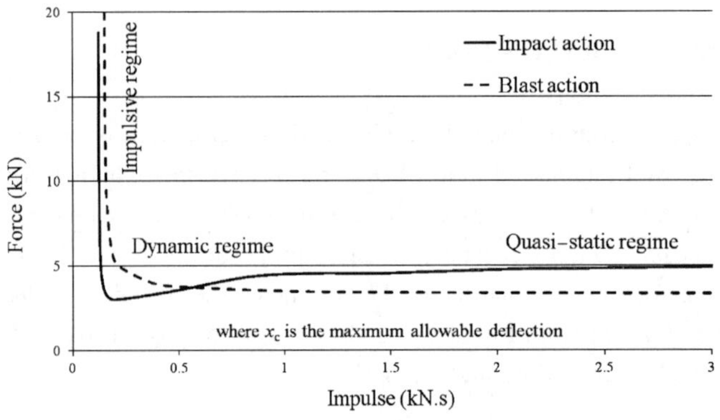

Figure 3 Force (Pressure) — Impulse curves of application example obtained from unified approach based on SDOF system for both blast and impact actions

In general, the damage of the structures by various combinations of mass and impact velocity of an impactor is of particular interest to engineers and designers. However, the force (pressure) — impulse curve for impact action does not provide direct indication of the velocity and mass combinations required to induce the iso-damage criteria of the target beam. Thus, it is important to convert the force (pressure) — impulse curve to a velocity — mass curve which will intuitively assist engineers in structure design. This can be achieved by assuming the T_d is one half of the natural period of the applied excitation caused by contact, T_M. The impactor mass M can then be expressed as a function of contact stiffness K and T_d, i.e.

$$T_m/2 = T_d = \pi\sqrt{\frac{M}{K}} \Rightarrow M = K\left(\frac{t_d}{\pi}\right)^2 \tag{6}$$

Once the impactor mass is determined, the impacting velocity V can be calculated in accordance with the impulse and momentum relationship, i.e.

$$I = (1+e)MV \Rightarrow V = \frac{I}{(1+e)M} \tag{7}$$

where e is the coefficient of restitution ($e = 1$ for elastic impact and $e < 1$ for inelastic impact).

The contact stiffness K is assumed to have a value of 5 000 N/m which is much smaller than k^* in order to diminish higher mode contribution caused by hard object impact. In other words, only the fundamental response mode governs the dynamic behavior of the beam. The coefficient of restitution e is given a value of 1 by assuming a perfectly elastic impact.

The velocity and mass curve required to trigger the iso-damage condition in this example is shown in Figure 4. This type of structural performance indicator, which involves impactor mass and impacting velocity combination to distinguish damage and non-damage areas, can be easily used by end-users to perform damage assessment due to projectile impact actions.

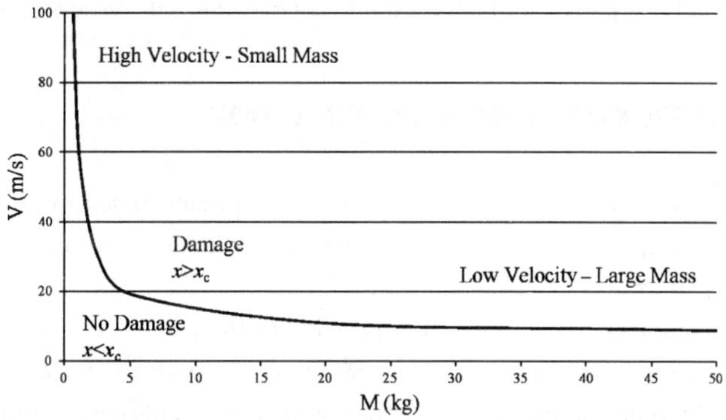

Figure 4 Iso-damage curve expressed in terms of impacting velocity — impactor mass

SIMPLE HAND CALCULATION PROCEDURE

Contemporary method of modeling impact action is based on the assumption that all kinetic energy carried by the impactor is fully converted to the internal strain energy of the target structure. However, this calculation method tends to overestimate on the displacement demand in instances where there is some energy loss due to rebound of the impactor for elastic impact, or irrecoverable deformation of the projectile for inelastic impact. A simple hand calculation procedure based on equal energy and momentum principles taking mass effect into consideration can be used to address the energy lost in the above-mentioned forms. The maximum displacement estimation is given by

$$x_{\max} = \beta \frac{MV}{\sqrt{kM}} \tag{8}$$

where β is Mass Reduction Factor ($\beta = \sqrt{\frac{4\alpha}{(1+\alpha)^2}}$ for elastic impact and $\beta = \sqrt{\frac{1}{1+\alpha}}$ for inelastic impact);

α is mass ratio, $\alpha = \frac{m}{M}$

Figure 5 shows that the β factor decreases with increasing values of α, which implies the amount of energy that is transferred into the target structure is highly sensitive to changes in the mass. Furthermore, the nature of the impact condition (i. e. elastic and inelastic) has a great influence on the trends of the values of β factors as indicated in Figure 5. Significant discrepancies between elastic and inelastic impacts are observed when α value is close to unity. However, the distinction between the two types of impact becomes negligible with high α values.

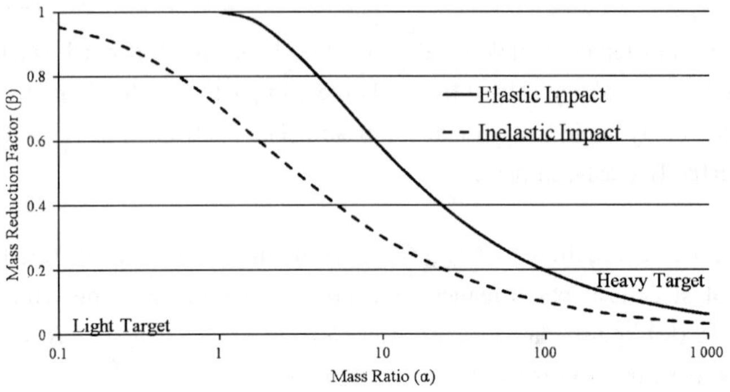

Figure 5 Mass reduction factor β changes with increasing mass ratio α for both elastic impact and inelastic impact

TWO-DEGREE-OF-FREEDOM-SYSTEM MODEL (2DOF)

The two aforementioned modeling techniques (i. e. unified approach based on SDOF system and equal energy and momentum principles) have their own shortcomings in modeling impact action. The unified approach requires a pre-determined forcing function which cannot be readily identified to represent the impact action. With equal energy and momentum principles, the rigidity of impactor in relation to the surface of the target structure is neglected. A more accurate approach to model impact action is to add additional lumped mass M representing the impacting object, and additional spring K representing the deformation of the projectile and indentation into the surface of the target structure. The displacement calculation of the 2DOF system model is based on elastic modal analysis and modal superposition principles. The displacement-time histories of both the impactor and the target are given by

$$\begin{Bmatrix} x_1(t) \\ x_2(t) \end{Bmatrix} = \left[\sum_{j=1}^{2} \dot{Y}_{oj} h_j \{t-\tau\} \begin{Bmatrix} \phi_{1,j} \\ \phi_{2,j} \end{Bmatrix} \right] \tag{9}$$

where $h_j(t-\tau) \approx \frac{I}{\omega_j} e^{-\xi\omega_j(t-\tau)} \sin(\omega_j(t-\tau))$ is the impulse response function and

$\dot{Y}_{oj} = \frac{\{\phi_j\}^T[M]\{r\}}{\{\phi_j\}^T[M]\{\phi_j\}} V_0$ is the modal velocity during the time of contact;

where $\phi_{i,j}$ is the mode shape vector, τ is the time delay, $[M]$ is the mass matrix, and $\{r\} = \{1, 0\}^T$, $x_1(t)$ is the impact displacement and $x_2(t)$ is the target structure displacement over time, respectively.

The contact force time history between the impactor and target structure is, therefore, given by

$$F(t) = K[x_1(t) - x_2(t)] \tag{10}$$

The 2DOF system model has been applied to estimate the dynamic response of beam and plate structures subject to low velocity impact by a solid object[12-14].

FUTURE RESEARCH

Currently, the unified approach capability is limited to the prediction of protective element response within an elastic region, whereas in its application, the failure criteria of a protective structure subject to either blast pressures or impact loads are often defined as a protective element rupture. For example, the failure criteria of a protective steel panel subjected to blast or impact would be the panel rupture or penetration. Similarly, the failure criteria of the reinforced concrete blast wall would be the breach of the wall, whereas the performance criteria of laminated glass panel is defined based on the fragments projectile and rupture of the Polyvinyl Butyral (PVB) membrane. As a performance assessment tool, the unified approach would need to be improved to include post-damage performance prediction of the structural component.

The iso-damage performance curve can be determined for post damage failure criteria. However, the concern with simplifying the post damaged behavior of a protective element to a SDOF or 2DOF model is that the deflection of the system under dynamic impact or blast pressure does not follow the assumed idealized deflected shape. In order to derive the iso-damage curves, key understandings on material parameters, which influence the load displacement behavior and the deflected shape of the protective component within the post damage region, are required. These are often achieved through a combination of experiments and numerical analysis.

As an illustrative example, laminated glass is one of such materials where its complex composite behavior leads to varying critical material parameters. Figure 6 shows a typical behavior of laminated glass observed from a series of test carried out as described in[15].

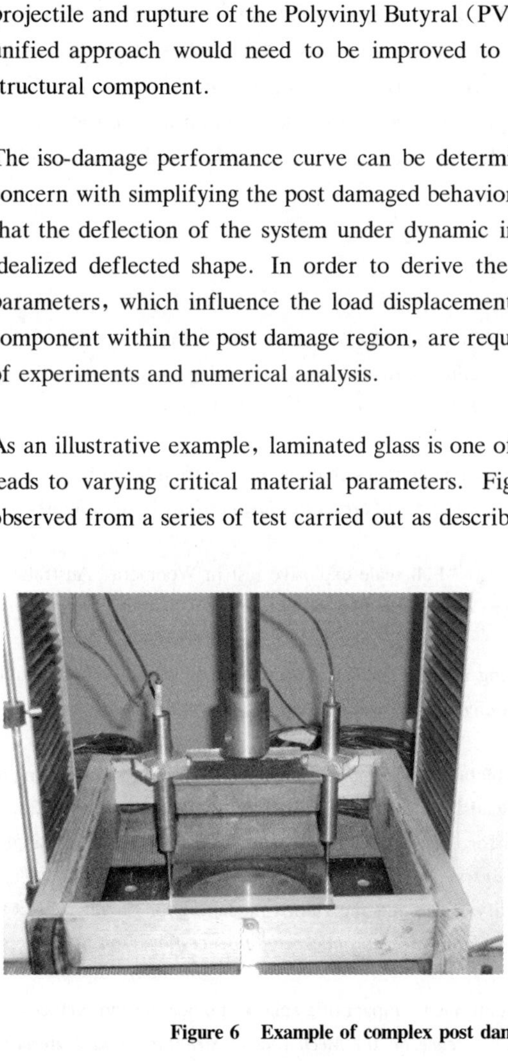

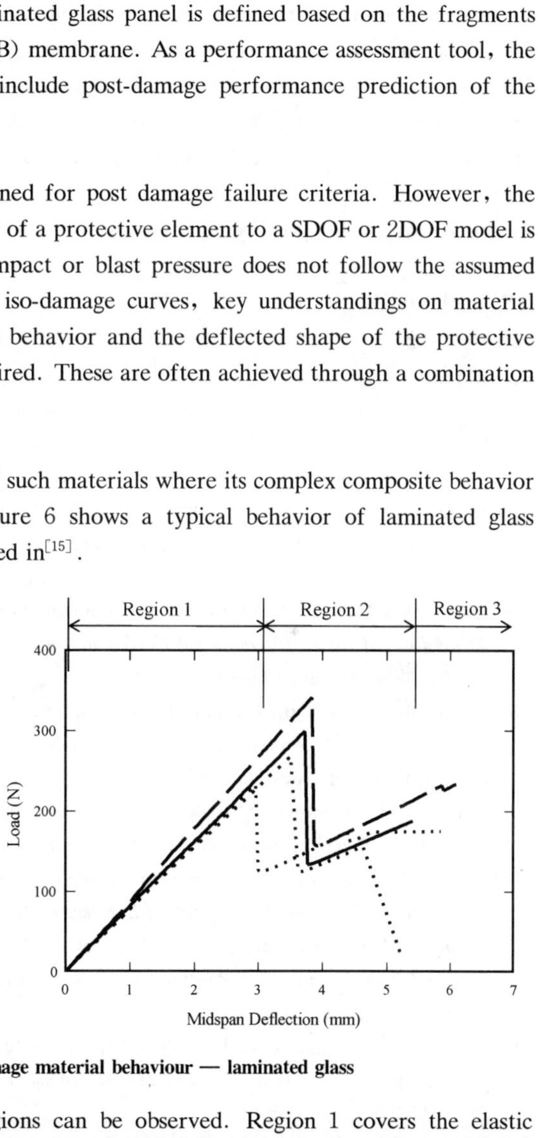

Figure 6 Example of complex post damage material behaviour — laminated glass

In the load-displacement curve, three distinct regions can be observed. Region 1 covers the elastic material behavior. In this region, the flexural stiffness of the system is governed by the Young's Modulus of the glass panels and the shear modulus of the PVB laminates, which varies with respect to the loading rate. Region 2 is observed when the tension pane of the glass panel has fractured, while the compression pane still behaves elastically. Although, region 3 is not captured in the results, it was observed in the model that the PVB layer behaved similar to a membrane before rupture.

In order for the unified approach to be applicable in protective structure design, it would need to be able to take into account the post damage behavior of a protective component. The general intention is to improve the unified approach so that with basic knowledge of the post-damage load displacement curve and the generic dynamic deflected shape of the structural component, engineers can determine the performance of the structural component when subject to extreme loadings.

CLOSING REMARKS

This paper presents a series of techniques for modeling explosive and projectile-induced impact actions. The application in damage assessment of protective elements using a simple unified approach based on SDOF system was demonstrated herein by a case study. Two alternative approaches, namely, equal energy and moment principle and 2DOF system model were described herein for impact action modeling. The limitations of the simplified modeling methodologies were also discussed. Extension of the unified approach on modeling inelastic behavior of structure elements requires better understanding of material parameters within the post damage region.

REFERENCES

[1] Abrate, S., "Modeling of impacts on composite structures", *Composite Structures*, 2001, 51, 129-138

[2] Lumantarna, R., Ngo, T., and Mendis, P., "An investigation on the behaviour of glazing components subjected to full scale blast test", *Proceedings of the 1st International Conference on Modern Design, Construction and Maintenance of Structures*, 2007, Hanoi, Vietnam.

[3] Li, B., Huang, Z. W., and Lim, C. L., "Verification of nondimensional energy spectrum-based blast design for reinforced concrete members through actual blast tests", *Journal of Structural Engineering*, 2010, 136(6), 627-636.

[4] Gupta, A., Mendis, P., Ngo, T., and Lumantarna, R., "Full scale explosive test in Woomera, Australia", *Proceedings of the 1st International Conference on Analysis and Design of Structures against Explosive and Impact Loads*, 2006, Tianjin, China.

[5] Lumantarna, R., Ngo, T., and Mendis, P., "Modelling and performance assessment of window glazing units under blast loads", *Proceedings of 20th Australasian Conference on the Mechanics of Structures and Materials*, 2008, Queensland, Australia.

[6] Jama, H. H., Bambach, M. R., Nurick, G. N., Grzebieta, R. H. and Zhao, X. L., "Numerical modelling of square tubular steel beams subjected to transverse blast loads", *Thin-Walled Structures*, 2009, 47, 1523-1534.

[7] Li, Q. M., and Meng, H., "Pressure-impulse diagram for blast loads based on dimensional analysis and single-degree-of-freedom model", *Journal of Engineering Mechanics*, 2002, 128(1), 87-92.

[8] Steward, M. G., and Netherton, M. D., "Blast reliability curves and uncertainty modelling for glazing subject to explosive blast loading", *Proceedings of 6th Asia-Pacific Conference on Shock and Impact Loads on Structures*. 2005, Perth, Australia.

[9] Yang, Y., Zhang, L., and Lam, N., "Approximate solutions to impact of a spherical object on the surface of a flat steel plate", *Proceedings of the 21st Australasian Conference on the Mechanics of Structures and Materials*, 2010, Melbourne, Australia.

[10] Sojblom, P., Hartness, J., and Cordell, T., "On low-velocity impact testing of composite materials", *Journal of Composite Materials*, 1988, 22, 30-52.

[11] Timmel, M., Kolling, S., Osterrieder, P., and Dubois, P. A., "A finite element model for impact simulation with laminated glass", *International Journal of Impact Engineering*, 2007, 34, 1465-1478.

[12] Yang, Y., N, Lam., and Zhang, L., "Evaluation of simplified methods of estimating beam response to impact", *International Journal of Structural Stability and Dynamics*, 2012, 12(3).

[13] Yang, Y., N, Lam., and Zhang, L., "Estimation of response of plate structure subject to low velocity impact by a solid object", *International Journal of Structural Stability and Dynamics*, in press.

[14] Lam, N., H. H. Tsang and E. Gad., "Simulations of response to low velocity impact by spreadsheet", *International Journal of Structural Stability and Dynamics*, 2010, 10(3), pp. 483-499.

[15] Lumantarna, R., Lam, N., Mendis, P. and Gad, E., "Analytical model of glazing panel subject to impact loading", *Proceedings of the* 19*th* *Australasian Conference on the Mechanics of Structures and Materials*, 2006, Christchurch, NewZealand.

INDEX OF CONTRIBUTORS

图书在版编目(CIP)数据

2012 国际结构稳定与动力学进展会议论文集 = Proceedings of the IJSSD Symposium 2012 on Progress in Structural Stability and Dynamics：英文/(新加坡)王建明，杨永斌，(美)雷迪(Reddy，J.N.)主编.—南京：东南大学出版社，2012.4

ISBN 978-7-5641-3399-3

Ⅰ.①2… Ⅱ.①王… ②杨… ③雷… Ⅲ.①结构稳定性—国际学术会议—文集—英文 ②结构动力学—国际学术会议—文集—英文 Ⅳ.①TU311.2-53 ②O342-53

中国版本图书馆 CIP 数据核字(2012)第 046116 号

Proceedings of the IJSSD Symposium 2012 on Progress in Structural Stability and Dynamics

Edited by C.M. Wang, Y.B. Yang, J.N. Reddy

Published by Southeast University Press, Nanjing, China
ISBN 978-7-5641-3399-3

Printed in Nanjing, China
Printing History April 2012, First Edition